Sludge Management

Sludge Management

Bhola R. Gurjar

Indian Institute of Technology, Roorkee, India

Vinay Kumar Tyagi

Nanyang Technological University, Singapore

CRC Press
Taylor & Francis Group
Boca Raton London New York

CRC Press is an imprint of the
Taylor & Francis Group, an **informa** business

A BALKEMA BOOK

CRC Press
Taylor & Francis Group
6000 Broken Sound Parkway NW, Suite 300
Boca Raton, FL 33487-2742

First issued in paperback 2020

ISBN 13: 978-0-367-57384-3 (pbk)
ISBN 13: 978-1-138-02954-5 (hbk)

Typeset by MPS Limited, Chennai, India

Library of Congress Cataloging-in-Publication Data

Names: Gurjar, B. R., author. | Tyagi, Vinay Kumar, author.
Title: Sludge management / Bhola R. Gurjar, Indian Institute of Technology,
   Roorkee, India, Vinay Kumar Tyagi, Nanyang Technological University, Singapore.
Description: Leiden, The Netherlands : CRC Press/Balkema, [2017] | Includes
   bibliographical references and index.
Identifiers: LCCN 2017003649 (print) | LCCN 2017003786 (ebook) |
   ISBN 9781138029545 (hardcover : alk. paper) | ISBN 9781315375137 (ebook)
Subjects: LCSH: Sewage sludge.
Classification: LCC TD767 .G865 2017 (print) | LCC TD767 (ebook) | DDC 628.3—dc23
LC record available at https://lccn.loc.gov/2017003649

Visit the Taylor & Francis Web site at
http://www.taylorandfrancis.com

and the CRC Press Web site at
http://www.crcpress.com

Table of contents

Foreword

Sludge is the main residue of water and wastewater treatment. Varying according to the treatment process, sludges are composed of solid particles separated from water, microorganisms grown on organics and their metabolic products, in particular exopolymeric substances as well as non-degraded organic matter and mineral matter originally present in wastewaters. Sludge generation implies various mechanisms such as separation technologies, microbial conversion and adsorption phenomena. As sludge contains large amount of water and putrescible organic matter, its management includes various operations of thickening, conditioning, dewatering, stabilisation, disinfection, drying and final disposal.

Over the years, the global sludge production has increased significantly due to increase in the capacity and efficiency of wastewater treatment facilities and more stringent regulations for effluent quality. Thus management and disposal of sludge is both societal and environmental issue and a challenge for legislators, researchers and engineers. However, earlier research efforts have made it possible to recycle the waste sludge to beneficial resources, thus turning the sludge image from "wastewater pollution concentrate" to "wastewater resource concentrate". Anaerobic digestion processes play a prominent role in recovering resources from sludge, providing energy as biogas and fertilizer or soil conditioner. Some other promising processes will allow recovering other valuable products from sludge such as biofuel and syngas, bio-oil, biodiesel, electricity, heavy metals, proteins and enzymes, and also the construction materials.

With the context and background illustrated as above, this book is a well informative and timely contribution from Prof. Dr. Bhola Gurjar and Dr. Vinay Kumar Tyagi in the form of a comprehensive guide on overall sludge management. The book would fulfill the needs of undergraduate and graduate students, academics, researchers, policy makers and executives of chemical and environmental engineering disciplines, especially those dealing with wastewater and sludge. A detailed and up to date information on all the aspects of sludge management i.e. sludge generation and characterization, sludge treatment and disposal and associated risks as well as sludge minimisation at source and resource recovery from waste sludge are the hot topics at present which are included in the book.

Prof. Gurjar and Dr. Tyagi have extensive national and international industrial, teaching, training and research experience in the area of environmental risk analysis, sludge management and biomass to bioenergy recovery, integrated cross-disciplinary study of science and technology issues of the environment, health, energy, infrastructure and resources – particularly from the perspectives of global change, sustainable development and risk assessment. I congratulate the authors of *Sludge Management-A comprehensive guide on sludge treatment, reuse and disposal* for publishing such an important and useful piece of work.

Hélène Carrère
INRA, UR0050 Environmental Laboratory Technology
Narbonne, FRANCE

Preface

There is extensive literature of scientific and technical nature on the subject of water purification and wastewater treatment. However, till recently the topic of sludge treatment and disposal has been treated merely as a part of water and wastewater treatment rather than as a separate subject. Over the years new body of knowledge has evolved in this field, which has subsequently made sludge treatment and disposal as an independent area of study, research and development. Thus it needs to be compiled and documented in the form of a separate text and reference book. To fill this gap, the present book has been written to cover the basic principles, conventional methods, and advanced practices of sludge treatment and its safe disposal.

The general processes used for sludge treatment and disposal can be classified according to their generic phenomenon viz. physical (e.g. thickening and dewatering), biological (e.g., anaerobic digestion, aerobic digestion, and composting), thermal (oxidation, pyrolysis and hydrogenation), and chemical (acid or alkaline hydrolysis) processes. In addition, the unit operation and/or processes of sludge treatment and disposal involve transportation, concentration/thickening/digestion/dewatering/drying/incineration and finally its safe disposal. All such operations and processes have been described in detail in this book. Furthermore, some common properties of sludge deserve special mention in relation to sludge treatment processes. Among these are (1) moisture/weight/volume relationships, (2) density, viscosity, rigidity and other flow characteristics, (3) response to concentration or thickening and to filtration (i.e. drainability), (4) fuel value, (5) digestibility, and (6) fertilizer value. These properties of sludge have also been covered in the present text. Two most important and emerging areas in sludge management are (1) advanced anaerobic digestion of sludge by pretreating the sludge using physical, chemical and mechanical methods, and (2) energy and resource recovery from sludge, rather than considering the sludge as a waste to get rid of. Both topics are covered in details with up-to-date technological advancements.

Due to its offensive nature, the treatment and disposal of sludge is a worldwide concern. Nevertheless, depending upon the peculiarity in geography and politics, different approaches have been emphasised in different parts of the world. For instance, the relevant literature documents that Germany and UK incinerates 51% and 65% of sludge produced, respectively. Nonetheless, a major quantity of sludge is being recycled (as soil conditioner) to agriculture land in England and Wales. Further, largely because of its high population density and its refraining the ocean disposal option, Japan has to incinerate more than half of its total sludge production. Usually incineration is a high cost option, but the Japanese pursued an effective technique by using the smelting sludge and incinerator ash to produce a non-leaching hard aggregate, which has commercial value. All the conventional methods for sludge disposal i.e. landfilling, incineration, oceans disposal and land application, are under scrutiny mainly due to social, economical and environmental issues. Therefore, the current efforts are toward to use the beneficial properties of sludge for resources and energy recovery. In United States, several well-developed technologies are under practice to generate electricity through biogas produced from anaerobic digestion of sludge. In China, yearly biogas generation (methane) from feedstock including sludge is 720 million cubic meters. The Swedish government stressed out on nutrient recycling with a target of recycling 75% of phosphorus supply from sludge and waste. Moreover, sludge derived biogas is used as biofuel in transportation sector. Japan is the leading country to using the sludge in the production of construction material. All such issues and efforts have been given due importance in this book at appropriate places.

To present a clear picture of general principles and methods of sludge treatment and disposal to the concerned students, teachers and practicing engineers of civil, chemical, public health and environmental disciplines, this book has been divided into 14 chapters. These chapters cover the topics in detail about sludge classification and its characteristic parameters, preliminary operations of sludge treatment, sludge thickening operations, sludge stabilisation processes (biological, physical and chemical), sludge conditioning and dewatering, sludge minimisation technologies, sludge disinfection and heat drying processes, thermal reduction and disposal of sludge, sludge disposal options, problems and solutions, and lastly energy and resource recovery from sludge. Each chapter starts with a brief introductory note incorporating its necessity and scope. New developments having potential for their application and viable use in treatment and disposal of sludges have been given due consideration in their respective sections in the text. However, those methods which are still in the state of research and development, and/or are less popular because of their high costs, have not been discussed in detail. It is hoped that this book will prove as a valuable resource to the students, teachers and practicing engineers who deal with the subject of water and wastewater treatment in general and sludge treatment and disposal in particular. Creative comments and suggestions are solicited from the readers and users of this book to make improvements in and add value to its future additions.

January 2017

Bhola R. Gurjar
Vinay Kumar Tyagi

Acknowledgement

We are grateful to a number of academics, colleagues, students, practitioners and friends who have contributed in various ways during the preparation of the draft manuscript of this book. We particularly wish to express our thanks to Prof. Hélène Carrère for writing the foreword for the book.

We wish to record our appreciation for the patience and support received from our family members to help us complete the project of this book. We sincerely admire Mr. Rajeev Grover and his team for word processing the text and to produce the diagrams.

We are grateful to our respective institutes we have been affiliated to viz., Indian Institute of Technology (IIT) Roorkee, National Taiwan University (NTU), University of Cadiz (UCA), The Energy and Resources Institute (TERI) University and Nanyang Technological University (NTU) for providing the comfortable ambience that helped us to extend our work hours in order to complete this book.

We are indebted to the researchers, authors and editors whose works we have accessed, reviewed, referred and cited to produce this book. As William Turner states: *If the honey that the bees gather out of so manye floure of herbes......that are growing in other mennis medowes..........may just be called the bees' honeye............so may we call it that we have........gathered of manye good authores..........our booke* (quoted by A. Scott-James in The Language of the Gardens: A Personal Anthology).

January 2017   Bhola R. Gurjar
                                                                                                      Vinay Kumar Tyagi

Introduction

With the growth of civilization, increasing population and flourishing industrialization, there arose a problem of water scarcity. To circumvent this problem, man devised methods to treat the naturally available water (i.e. raw water) to make it wholesome and palatable, and started to clean the sewage and industrial wastewater before returning it to the earth. The principal end products of raw water and wastewater treatment consist of:

- product water,
- treatment-plant effluent, and
- by-product slurry or sludge.

Here, the product water is sent to the cities or industries, treatment-plant effluent is discharged into receiving waters or on to receiving soils/lands, and sludge is treated before its final disposal. It is because the product water and treatment-plant effluent are finished products, but sludge is not. In fact the sludges, in the form of slurry or semisolid-liquids, typically contain from 0.25 to 12% by volume solids having obnoxious characteristics and remaining portion is water depending on the type of operations and processes used for the treatment of raw water/wastewater (Fleming, 1986; Moseley *et al.*, 2003; Parker *et al.*, 1992; MWST, 1991).

In conventional methods of wastewater treatment such as activated sludge and trickling filter processes, large volumes of primary sludge is produced in addition to the excess settled secondary sludge (activated sludge). In case of the activated sludge process, the secondary sludge is mainly the microbial biomass produced by the metabolism of the organic material. The microbial yield on settled sewage is about 50 per cent. About 20 per cent proportion of this biomass is recycled and the remainder is combined with the primary sludge for disposal. In the trickling filters with a lower loading rate less sludge is produced and there is no recycle of sludge; In general, large volumes of sludge having solids content of about 1–4 per cent are formed in wastewater treatment processes and represent one of the main problems of disposal. This is because the excess waste sludge is a mixture of organic material and microbial cells, which can be degraded by other microorganisms.

More recently, anaerobic processes have been used to treat industrial wastes and effluents containing a high content of insoluble or organic compounds. The advantages of anaerobic digestion are that the process:

- produces much less biomass or sludge,
- requires no aeration,

- forms methane (biogas), and
- the associated smell is less as the process is enclosed.

Here, the methane can be utilised as a fuel to run boilers or to be used as vehicle biofuels or to generate electricity at the rate of about 1.16×10^7 kJ produced per · 1000 tonnes of COD removed (Scragg, 1999). However, the disadvantages of anaerobic digestion are that the process requires:

- good mixing,
- a temperature in the range of 35°C–55°C,
- a substrate with a high BOD (1.2–2 g/l), and
- long retention times of 30–60 days (for sludge digestion).

The requirement for a high BOD (1.2–2 g/l) waste means that the anaerobic digestion process is suitable for some agricultural and industrial wastes (Scragg, 1999).

Nevertheless, whatsoever be the operations or processes used for the treatment of wastewater, sludge is the largest in volume than the other constituents (by-products) removed during the treatment processes. The treatment and disposal of sludge is per-haps the most complex problem faced by an environmental engineer in the field. The problem of dealing with sludge is complex because:

- It is composed largely of the substances responsible for the offensive character of the wastewater. The portion of sludge produced from biological treatment of wastewater is composed of the organic matter, mineral contents and microorgan-isms originally present in the wastewater and it, too, will decompose and become offensive if untreated sludge is disposed off.
- Only a small part of the sludge is solid matter and the rest is a liquid, which can contaminate surface or underground water sources if it is dumped in water bodies or on land without any treatment.

Further, it is to be mentioned that the dry weight of waste solids present in sludge is the weight of solids settleable at the time of separation of solids or phase transfer from the suspending water. These may include (Bowen et al., 1990; Metcalf & Eddy, 2003):

- settleable solids naturally present in water and wastewater,
- additives/chemical coagulants and precipitants, produced by converting unwanted non-settle able or colloidal solids into settleable solids,
- sloughed biological films and wasted biological floes or other biomasses generated by living organisms from dispersed and dissolved nutrient organic matter during wastewater treatment.

Thus, the solids from the sludge are the waste solids derived from the treat-ment process in which they originate. Examples are, plain-sedimentation, chemical-precipitation, chemical coagulation, trickling-filter humus and waste activated sludges in wastewater treatment plants, and chemical softening/deferrization of sludges or slur-ries in water-purification plants. On reaching the bottom of settling units, most organic and mineral solids form loose, honey-combed structures of particulate and flocculent matter united with relatively large volumes of water. As deposits build up, they consolidate under their own weight. However, water is not displaced from them with ease, and the moisture content of most of them remains high. As a result sludges contain considerable volume of water as well as organic content. Due to this, satisfactory treatment and disposal of slurries and sludges create economic problems of considerable magnitude. Chief among them are transportation and final storage/disposal (Girovich, 1990; Martin & Bhattarai, 1991; Topping, 1986).

From the above discussion it is clear that because of their origin, bulk and watery consistency, and putrescibility, most of the sludges need to be treated prior to disposal. Treatment ensures the hygienic safety and sensory acceptability of the sludges. It also reduces the volume and weight of the materials to be handled, transported and disposed off. Moreover, the processing or treatment of sludge not only ensures its safe disposal, but it may also be designed to produce useful by-products. This way sludge can be utilised as a valuable resource rather than to be considered merely as a waste to be got rid off. Thus sludge treatment as a part of management of solids is indeed one of the prime determinants in choosing between specific water-purification and wastewater treatment processes. As an instance, the nature of sewage sludge depends on the wastewater treatment process and on the source of the sewage. It can contain not only organic and inorganic matter, but also bacteria and virus, oil and grease, nutrients such as nitrogen and phosphorus, heavy metals, organo-chlorines and other persistent organic pollutants. Hence, each component of the sludge has its own environmental impact, which needs to be taken into account when choosing its treatment technique and the disposal route.

There are a number of methods employed to dispose off the sludge such as landfill, dumping at sea, incineration, drying, spray irrigation (i.e. agricultural disposal), composting, and anaerobic digestion. In the last method anaerobic bacteria break down in the range of 30 to 40 per cent of the organic matter of sludge to form a gaseous mixture of methane and CO_2 in the ratio of about 3:1. Now a day, this process is widely practiced in order to stabilise the sludge before its final disposal. This is a relatively simple process to engineer and it produces a valuable energy source in the methane containing biogas. As conventionally practices, anaerobic digestion of sewage sludge converts only about 30 per cent of the organic matter to methane and carbon dioxide. Although this reduction is sufficient to stabilise the sludge, a large volume remains for disposal, which result into transportation and handling problems. To mitigate such problems, there are a number of ways in the stage of research and development as well as numerous full scale processes i.e. thermal process (Cambi), chemical (acid, alkaline, ozonation), mechanical (sonication, high pressure homogenizer: Micro-Sludge™), which can significantly increase the conversion of organic matter to methane in anaerobic digesters. Those methods need to be further researched and refined as they promise a great reduction in subsequent sludge handling problems (Priestley, 1992; Carrere et al., 2010; Tyagi & Lo, 2011).

Further, the large-scale incineration of sludges is an expensive alternative due to high capital costs, and is only a partial disposal option as the ash formed creates further disposal problem. However, the development of autothermic incineration process has made the incineration technique more attractive. In this process primary and secondary sludges are mixed together and pressed to remove the water content. This produces a cake of 30 per cent solids, which can support autothermic combustion. Advanced combustion systems viz., fluidised beds working at high temperatures of 750–850°C create more heat than it is required to heat the inlet air and remove the water from the sludge. This means that once the process has started, no more fuel is needed to be added as the sludge itself generates sufficient heat. The ash formed in this process can be removed by an electrostatic precipitator and the wet scruber can remove sulphur dioxide, hydrogen fluoride and hydrogen chloride. Here, the ash contains the heavy metals, which are present in the sludge and represents 30 per cent of the original dry mass and 1–2 per cent of the volume and is normally disposed off in landfill sites (Scragg, 1999).

Furthermore, some sludges can be disposed off by making them useful for some agricultural purposes. Almost all sludges contain more or less concentration of heavy metals, as microorganisms have the ability to segregate metals. Hence the application of sludge to soils carries the risk of producing high levels of heavy metals in the soil. Any sludge, which is to be applied to agricultural land, is required to have some sort of chemical or biological treatment to reduce the levels of pathogens unless it is injected below the surface. Such treatment can be any of the following ones:

- alkali stabilisation, pH > 12 for 12 hours,
- anaerobic digestion at 35°C for 12 days,
- composting, 4–5 weeks,
- drying and storage, 3 months,
- liquid storage, 3 months,
- pasteurization, heating at 70°C for 30 min. or more,
- thermophilic aerobic and anaerobic digestion, at 55°C.

Any of these methods can be used to reduce the pathogens content of the sludge before it is applied to the land. The sludge is normally conditioned prior to drying so that its ability to settle before dewatering is improved. Often dewatering is carried out in drying beds, but the filtration and centrifugation have also been used to produce a compact cake. Nevertheless, in addition to the above-mentioned methods, the basic principles and conventional as well as advanced practices of treating and disposing off the sludges have been discussed and described in detail in the subsequent chapters.

Building the sludge derived resources recovery system can help to produce environmentally benign products, reduce the dependency on non-renewable resources thus facilitate the conservation of natural resources, decrease the human health risk and environmental pollution as well as offers the routes for sustainable management of waste sludge. Several value added products can be recovered from sludge, viz., energy rich biogas (methane, hydrogen, syngas), liquid bio-fuels (bio-diesel, bio-oils), construction material (bricks, cement, pumice, slag, artificial lightweight aggregates), bio-plastic, proteins and hydrolytic enzymes, bio-fertilizers, bio-sorbent, bio-pesticides, electricity generation using microbial fuel cells, nutrients (nitrogen and phosphorus) and heavy metals. An up to date information about resource recovery from sludge is presented in the last chapter.

Chapter 1

Sludge: An overview

1.1 INTRODUCTION

Sludge is universally considered as waste, something to be disposed of out of sight and out of mind. As a result, the word "sludge" is usually associated with pollution, contamination and disease. Nobody wants it in their backyard, and everybody thinks it should be somewhere else.

Sludge can be defined as a soft mud or mire, a slimy precipitate produced during the treatment of waste water. As far as an environmental engineer is concerned, sludge is a byproduct of several processes (Bowen *et al.*, 1990; Fleming, 1986):

- Water treatment plants
- Sewage treatment plants
- Dredging of rivers and harbours
- Coal and sand washeries
- Industrial manufacturing
- Agriculture.

Sludge can be regarded as consisting of particles aggregated into flocs that act hydrodynamically as single particles. These flocs can be in suspension, separated from other flocs (e.g. alum floc in water treatment prior to settling), or in a solid matrix where individual flocs cannot be identified and the sludge mass forms a continuum (e.g. waste-activated sludge).

In terms of the quantity of sludge, overall sludge production is dependent on influent loading, type of treatment, treatment performance, type of sludge-handling facilities, and effluent treatment requirements. In general, sludge production is expected to be directly related to the average dry-weather flow rate through the plant, assuming no changes in influent unit loading or the treatment process.

1.2 SOURCES OF SLUDGE
(Balmer & Frost, 1990; Girovich, 1990; Martin & Bhattarai, 1991; Supernant *et al.*, 1990; Topping, 1986)

Sludge has several sources, namely, water treatment plants, municipal waste water treatment plants and industrial effluent treatment plants. Is sludge purely a waste?

Can it serve as a resource? These questions require closely considered definitions of the terms "waste" and "resource". Obviously, waste can be defined as something which is lying unused, unproductive, uncultivated, superfluous, or which is in disadvantageous (useless) condition. On the other hand, a resource is a means of generating wealth or money, that is, something that can be used in one or more ways, a thing from which some useful byproduct can be extracted, or something that helps in the manufacture of some advantageous product/item. Clearly, from such general definitions, any form of sludge that is put to use is *not* waste, but those sludges that are just dumped most certainly are.

1.2.1 Water treatment plants

In the case of water treatment plants, sludge comprises either settled particulate matter in sedimentation tanks, flocculated and precipitated material resulting from chemical coagulation, or the residue of excess chemical dosage, plankton, etc. In such cases, for continuous sludge removal, the feasibility of discharging sludge to existing sewers nearby is normally considered. For lime-softening plant sludge, reclamation by calcination and reuse can be explored.

Sludge from clarification units using iron and aluminium coagulants can be dewatered to a cake by vacuum filtration, using lime as a conditioner, and can conveniently be trucked to landfill. Recovery of alum from sludge by treatment with sulphuric acid offers possibilities for reduction in the quantities of sludge to be handled. Sand drying beds are an acceptable method for dewatering certain types of sludge from settling tanks or clarifiers for subsequent disposal to landfill. Simple lagooning of sludge can also bring about a reduction in the bulk of sludge to be handled, followed by disposal to landfill. However, acceptable rates of application of water treatment sludges to various soil types are related to the phosphorus-fixing capacity of the sludge.

1.2.2 Sewage treatment plants

Sewage sludge is slurry with a water content usually in excess of 95%. The solid phase consists principally of organic matter, derived from human, animal and food wastes. Other constituents are trace contaminants (metals and persistent organic compounds), mainly from industrial effluents and bacteria, some of which may be pathogenic.

Girovich (1990) describes the sewage sludge in its initial form as a liquid with 2–6% total solids (TS). On a dry basis, sludge contains 35–65% of organic matter, with the remainder being non-combustible mineral ash. Nutrients like nitrogen, phosphorus, potassium and some trace metals are present in sludge and it can be used as an effective fertiliser. However, the sludge also contains pathogens and, in some cases, constituents such as heavy metals and hazardous organics.

1.2.3 Industrial effluent treatment plants

Industrial waste water treatment facilities generate different types of sludge, which can be generally classified as either organic or inorganic, and both may contain toxic materials, mainly heavy metals. The presence of toxic compounds (e.g. from pharmaceutical and metal industries) contaminates the sludge and thus restricts the safe

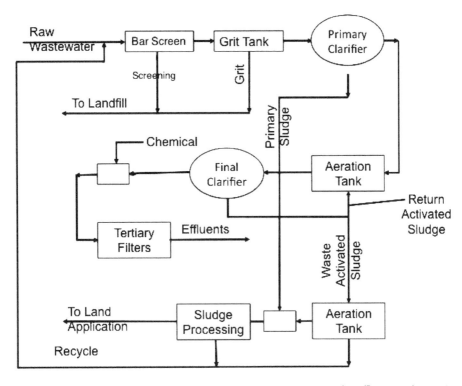

Figure 1.1 Various stages of sludge generation in wastewater treatment plant (Reuse with permission of John Wiley and Sons; Source: *Turobskiy & Mathai, 2006*).

disposal of industrial sludge. Industrial sludges classified as hazardous contain toxic substances (heavy metals, endocrine disruptors, other dangerous toxins), making them unfit for agricultural use.

1.3 SLUDGE CATEGORISATION
(Coker *et al.*, 1991; MWST, 1991; Turovskiy & Mathai, 2006; Bahadori, 2013)

Sludge can be categorised on the basis of several criteria, such as chemical composition, waste water source and the stage of sludge generation in a waste water treatment plant. Thus, in terms of chemical composition, there is mineral sludge, in which the amount of suspended mineral solids exceeds 50%, and organic sludge, in which the amount of suspended organic solids exceeds 50%. In terms of waste water origin, there are domestic, municipal and industrial sludges. The sources and types of sludge generated in a waste water treatment plant with primary, biological and chemical treatment facilities are shown in Figure 1.1. Sludge can be categorised as: primary sludge (sludge from primary settling tanks), chemically precipitated sludge (from settling tanks after chemical precipitation), secondary or biological sludge (from secondary settling tanks after biological filtering, i.e. trickling-filter sludge), activated sludge, sludge from septic tanks (i.e. septage), and digested sludge.

1.3.1 Primary sludge

In general, waste water treatment plants apply physical treatment in the form of settling to remove settleable solids from raw waste water. Sludge produced from primary settling tanks is a grey suspension with solids of different sizes and composition. The concentration of total solids in primary sludge can range between 2 and 7%. Due to the presence of high levels of organic content, primary sludge decomposes rapidly and becomes septic, which can be identified by its change to a dark grey or black colour and an objectionable odour. If compared with biological and chemical sludge, primary sludge is easy to dewater due to the presence of discrete particles and debris, and will produce a drier cake and give better solids capture with low conditioning requirements. The quantity of raw primary sludge can be approximately 0.4–0.5% by volume of the plant influent flow, or approximately $1.1\,m^3$ ($39\,ft^3$) per 1000 people.

1.3.2 Chemical sludge

Chemicals are generally used in waste water treatment, mainly in industrial waste water treatment, in order to precipitate and eliminate hard-to-remove substances, and in some instances, to improve removal of suspended solids. The chemicals commonly used are lime, alum, ferrous chloride, ferric chloride, ferrous sulfate and ferric sulfate. Sludge from chemical precipitation tanks is usually dark in colour, though its surface may be red if it contains a high amount of iron. Its odour may be objectionable, but not as strong as from primary sludge. While it is somewhat slimy, hydrates of iron or aluminium will make it gelatinous. If left in the tank, it undergoes decomposition like the sludge from primary sedimentation but at a slower rate. It gives off gas in substantial quantities and its density increases if it is allowed to stand.

1.3.3 Biofilter sludge

Biofilter sludge is brownish and flocculent. The sludge coming from low-rate biological filters contains many dead worms and so its odour can be quite offensive. It does not drain readily and it must be very well-digested before dewatering. It is good practice to recycle and mix this sludge together with sludge from primary settling tanks and then digest it, and this is done in all treatment plants.

1.3.4 Activated sludge

Activated sludge is flocculent and air-dries slowly, even when it is spread in thin layers. That is, it has poor dewaterability/drainability. If its biological treatment is efficient, the sludge is golden-brown and has an earthy odour. If the colour is lighter than usual, there may have been under-aeration, with a tendency for the solids to settle slowly. If the colour is quite dark, it may be approaching a septic condition. It digests easily at temperatures of 30–35°C. Before being conveyed to sludge digesters, it is mixed with primary sludge.

1.3.5 Aerobically digested sludge

Aerobically digested sludge is brown to dark brown in colour and has a flocculent appearance. The odour of aerobically digested sludge is unobjectionable, having a musty character. Aerobically digested sludge can be dewatered rapidly and offers good nutrient value (Bahadori, 2013).

1.3.6 Anaerobically digested sludge

Anaerobically digested sludge is dark brown to black in appearance and comprises a sizeable amount of gas. Anaerobically digested sludge is non-offensive, less odorous and looks like hot tar, burnt rubber or sealing wax. When drawn off on porous beds in thin layers, the solids are first carried to the surface by the entrapped gases, leaving a sheet of comparatively clear water below them. This drains off rapidly and allows the solids to sink down slowly on to the bed. As the sludge dries, the gases escape, leaving a well-cracked surface with an odour resembling that of garden loam (Bahadori, 2013).

1.3.7 Septage

Sludge from septic tanks is black. Unless well-digested by long storage, it is offensive because of the hydrogen sulfide and other gases it gives off. The sludge can be dried on porous beds if spread out in thin layers, but objectionable odours are to be expected while it is draining, unless it has been well-digested (Bahadori, 2013).

1.3.8 Industrial sludge

An overview of the characteristics of different types of industrial sludge, which are hazardous in nature when compared with sewage sludge, is provided here:

(a) *Petroleum refining*
   The sludge produced from petroleum refineries is large in volume and contains oil, wax, sulfides, chlorides, mercaptans, phenolic compounds, cresylates and, sometimes, large amounts of iron.
(b) *Pesticides industry*
   Sludge generated from the pesticide industry contains dangerous chemicals such as dichlorophenol (DCP), ethyl hydrogen sulfate and chlorobenzene sulfonic acid.
(c) *Paint industry*
   Oils, resins, dyes, solvents, plasticisers and extenders all contribute to the chemical characteristics of sludge from the paint industry.
(d) *Electroplating industry*
   The most important toxic contaminants in electroplating industry sludge are acids, metals such as chromium, zinc, copper, nickel and tin, and cyanides. Alkaline cleaners, grease and oils are also found in the sludge.
(e) *Pulp and paper industry*
   Sludge from the paper industry contains pulp, bleaching chemicals, mercaptans, sodium sulfides, carbonates and hydroxides, casein, clay, dyes, waxes, grease, oil and fibre.

(f) Tanneries/leather industry
The sludge produced from the leather industry, such as from tanneries, contains high concentrations of sulfides and chromium.

1.4 SLUDGE CHARACTERISTICS
(Coker *et al.*, 1991; Eriksson *et al.*, 1992; MWST, 1991; Turovskiy & Mathai, 2006)

Sludge may be distinguished by its physical, chemical, bacteriological and biological properties. These properties or characteristics depend on the nature of the origin or source from which the sludge is obtained, and the unit operations and processes by which it is produced. The major component of sludge is water, up to 95% by weight. The remaining dry solids contain variable proportions of nitrogen, phosphorus, potassium, heavy metals, pathogens, polychlorinated biphenyls (PCBs) and other constituents, depending on the source of the sludge.

Table 1.1 summarises the principal characteristics of waste water sludge. However, each physico-chemical and biological characteristic of sludge is discussed in detail in the following subsections.

1.4.1 Physical characteristics

The physical properties of a sludge determine, to a great extent, the possibilities and conditions for digestion and disposal of that sludge. The most important physical properties are described below.

Colour and odour

Fresh sludge from municipal waste water is light greyish or yellowish in colour. Fully digested sludge is black (due to iron sulfides) and has a tarry odour. Sludge produced by aerobic digestion is brown and has a humus-like odour. It can be air-dried without odour release in about two weeks in normal dry weather.

Water content

The volume of sludge depends mainly on its water content and only slightly on the character of the solid matter. The 10% of solids in sludge, for example, comprise 90% water by mass. If the solid matter is composed of fixed (mineral) solids and volatile (organic) solids, the specific gravity of all of the solid matter can be computed as follows:

$$\frac{M_s}{S_s P_w} = \frac{M_f}{S_f P_w} + \frac{M_u}{S_v P_w} \tag{1.1}$$

where,
M_s = mass of solids
S_s = specific gravity of solids
P_w = density of water
M_f = mass of fixed solids (mineral matter)

Table 1.1 Wastewater sludge characteristics (Source: USEPA, 1979).

Parameters	Primary sludge Range	Typical	Activated sludge Range	Typical
pH	5–8	6	6.5–8.0	7
Alkalinity (mg/L as CaCO$_3$)	500–1500	600	580–1100	–
Total dry solids (TS) (%)	2–7	5	0.4–1.5	1
Volatile solids (% of TS)	60–80	65	60–80	75
Specific gravity	–	1.02	–	1.01
Grease and fats				
Ether soluble (% of TS)	6–30	–		–
Ether extract (% of TS)	7–35	–	5–12	–
Protein (% of TS)	20–30	25	32–41	–
Nitrogen (N, % of TS)	1.5–4.0	2.5	2.5–5.0	–
Phosphorus (P$_2$O$_5$, % of TS)	0.8–2.8	1.6	2.8–11.0	–
Potash (K$_2$O, % of TS)	0–1	0.4	0.5–0.7	–
Cellulose (% of TS)	9–13	10	–	7
Iron (% of TS)	2–4	2.5	–	
Silica (SiO$_2$, % of TS)	15–20	–	–	8
Organic acids (mg/L as HAc)	200–2000	500	1100–1700	–
Energy content				
kJ/kg	23300	18600	23300	–
Btu/lb	10000	8000	10000	–
Major mineral components of sludge[a]				
SiO$_2$	21.5–55.9		17.6–33.8	
Al$_2$O$_3$	0.3–18.9		7.3–26.9	
Fe$_3$O$_4$	4.9–13.9		7.2–18.7	
CaO	11.8–35.9		8.9–16.7	
MgO	2.1–4.3		1.4–11.4	
K$_2$O	0.7–3.4		0.8–3.9	
Na$_2$O	0.8–4.2		1.9–8.3	
SO$_3$	20–7.5		1.5–6.8	
ZnO	0.1–0.2		0.2–0.3	
CuO	0.1–0.8		0.1–0.2	
NiO	0.2–2.9		0.2–3.4	
Cr$_2$O$_3$	0.1–3.1		0.0–2.4	

[a]Values in percentage of total mineral constituents.

S_f = specific gravity of fixed solids
M_u = mass of volatile solids (organic matter)
S_v = specific gravity of volatile solids

The volume of sludge (V_{sl}) may be computed with the following formula:

$$V_{sl} = (M_s/P_w S_{s1} P_s) \qquad (1.2)$$

where,
S_{s1} = specific gravity of sludge
P_s = % of total solid content (as decimal fraction)
M_s and P_w are defined as above

Thus, the sludge moisture (or percentage water content) is of great importance in establishing the appropriate processing/treatment method (digestion, dewatering, etc.) and transport of the sludge. The moisture content is determined by estimating the sludge's weight loss as a result of evaporation to complete dryness in an oven at 105°C. In this case:

$$\text{Moisture } (P_w) = (\text{Loss in weight/Initial weight}) \times 100 \text{ (\%)} \tag{1.3}$$

Often, the moisture content is replaced by the amount of total solids in the sludge (P_s).

Specific gravity

The specific gravity of sludge depends on the nature and proportion of fixed and volatile solids, and the water content of the sludge. Generally, the specific gravity of sludge varies between 0.95 and 1.03 (maximum 1.25) as a function of the origin of the sludge. Variation in volume, as a result of changing the sludge moisture (P_w) or total solids content (P_s), is determined with the equation:

$$V_{s1(1)}/V_{s1(2)} = [100 \cdot P_w + P_{w2}(P_s - P_w)] \cdot (100 - P_{w2})/$$
$$[100 \cdot P_w + P_{w2}(P_s - P_w)] \cdot (100 - P_{w1}) \tag{1.4}$$

For approximate calculations for a given moisture or solids content, it is simple to remember that the volume varies inversely with the percentage of solid matter contained in the sludge, given by:

$$V_{s1(1)}/V_{s1(2)} = (100 - P_{w2})/(100 - P_{w1}) = P_{s2}/P_{s1} \text{ (approx.)} \tag{1.5}$$

where,
$V_{s1(1)}$, $V_{s1(2)}$ = sludge volumes in two stages
P_{s1}, P_{s2} = % of total solid content in two stages

Drainability

Drainability, or the drying properties of sludge, is determined, briefly, by observing the time required for the surface of the sludge to become uniformly cracked after it has been placed on a layer of sand or filter paper.

In the laboratory, the drainability of sludge can be determined by placing a measured weight of sludge on a filter paper in a Buchner funnel in the apparatus shown in Figure 1.2 and noting the time required for the collection of increasing volumes of filtrate (water-supernatant), with or without suction. This test is particularly useful in determining the influence of various chemicals upon the drainability or filterability of the sludge. The drainability of digested sludge is greater than that of the fresh sludge.

Fuel value or thermal content

Fuel value or thermal content of sludge is a function of the volatile or total organic dry solids contained in the sludge. It is normally determined in a bomb calorimeter.

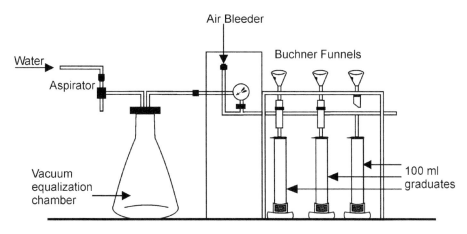

Figure 1.2 Apparatus for the determination of the drainability of sludge.

For municipal waste solids/sludges, statistical correlation between pairs of observed fuel values and volatile-solids values is high and can be expressed as follows:

$$Q = a[100P_v - b][100 - P_c]/[(100 - P_c) \times 100] \tag{1.6}$$

where,
Q = the fuel value of the solids/sludge in British Thermal Units (BTUs) per pound dry weight
P_v = proportion of volatile matters (%)
P_c = proportion of chemical, precipitating or conditioning reagent (%)
a, b = coefficients for different classes of waste solids/sludges
    Note: For plain-sedimentation municipal waste water sludges (fresh and digested), $a = 131$ and $b = 10$, while for fresh activated sludge, $a = 107$ and $b = 5$.

The thermal content of sludge is important where incineration or some other combustion process is considered, and accurate bomb-calorimeter tests should be performed so that a heat balance can be established for the combustion system. The thermo-physical characteristics of sludge are summarised in Table 1.2. The thermal content of untreated primary sludge is the highest, especially if it contains appreciable amounts of grease and skimmings. The fuel value of untreated sludge ranges from 11 to 23 MJ/kg of dry solids, depending on the type of sludge and the volatile content. This is equivalent to some of the lower grades of coal. Digested sludge has a heat content ranging from 6 to 13 MJ/kg.

Other physical properties

Sometimes, it is necessary to determine the granularity (drawing the respective granulometric curves), compressibility and viscosity of sludge, which are especially important parameters for sludge pumping. The formation of granules from anaerobic sludge and their maintenance are influenced by a number of parameters. The composition

Table 1.2 Thermo-physical characteristics of sludge (Turovskiy & Mathai, 2006).

Sludge Type	Temperature Conductivity ($10^8 m^2/s$)	Thermal Conductivity ($W/m \cdot K$)	Specific Heat ($kJ/kg \cdot K$)
Raw primary and waste activated sludge	–	0.4–0.6	3.5–4.7
Vaccum filter dewatered	10.9–14.3	0.2–0.5	2.1–3.0
Centrifuge dewatered	8.5–12.1	0.1–0.3	2.0–2.4
Thermally dried	14.0–21.6	0.1–0.3	1.7–2.2

and temperature of the waste water, reactor configuration, loading rate and hydrodynamic conditions seem to be the most important parameters in this regard. The type of moisture and type of solids are also important physical properties.

1.4.2 Chemical characteristics

Chemical properties are significant parameters in sludge digestion and, less importantly, for the dewatering process. Many chemical constituents, including nutrients, are important when considering the ultimate disposal of the processed sludge and the liquid removed from the sludge during processing. The most important chemical characteristics are enumerated below.

pH value

The measurement of pH, alkalinity and organic acid content is important in process control, particularly for anaerobic digestion. The value of pH during the digestion process should be maintained around 7; values above 8.5 or below 6.0 indicate a deficient process.

Total solids (dry)

Total solids can be mineral (fixed) and organic (volatile). Their proportion and total amount depend on the nature of the waste water. Their determination is important, because they influence the volume–mass relationship of sludge and thus the dewatering efforts too.

Total solids (dry) can be determined by evaporating the sludge in an oven at 105°C and weighing the remainder (residue). After weighing, this residue is introduced into an electric muffle furnace for ignition. The weight of ash remaining represents the total mineral solids (dry weight), and the difference from the weight before ignition represents the total organic solids (dry weight). In fresh raw sludge in primary settling tanks, as for raw water, the total organic dry solids vary between 60 and 70% of the total solids, the balance being mineral solids. After digestion, organic dry solids vary between 30 and 40%, while the mineral dry solids vary between 70 and 60%.

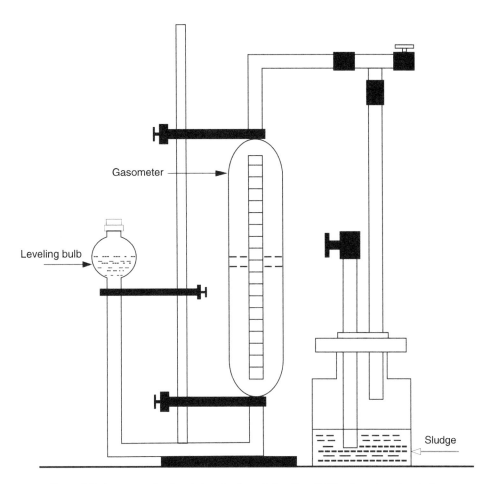

Gasometer

Leveling bulb

Sludge

Figure 1.3 Apparatus for the determination of the digestibility of wastewater sludge.

Agricultural or fertiliser value

The fertiliser value of sludge, which should be evaluated where the sludge is to be used as a soil conditioner, is based primarily on its content of nitrogen (N), phosphorus (P) and potassium (K). Generally, phosphorus and nitrogen are found in sludge in sufficient quantities to meet the requirements of agriculture. Potassium, however, is not found in sufficient amounts. Tests carried out regarding the utilisation of sludge in agriculture showed that one ton of digested and dewatered sludge contained nitrogen equivalent to 60 kg of ammonium sulfate, and calcium equivalent to 150 kg of calcium carbonate.

Digestibility

The digestibility of sludge is conveniently measured by placing a 2:1 mixture (dry solids) of fresh and well-digested sludge in the apparatus illustrated in Figure 1.3.

Table 1.3 Values of indicator bacteria and pathogens in sludge (Lue-Hing *et al.*, 1998).[a]

Microbes	Range (number/g)[b]	Average (number/g)
Total coliform	1.1×10^1–3.4×10^9	6.4×10^8
Fecal coliform	ND–6.8×10^8	9.5×10^6
Fecal streptococci	1.4×10^4–4.8×10^8	2.1×10^6
Salmonella sp.	ND–1.7×10^7	7.9×10^2
Shigella sp.	ND	ND
Pseudomonas aeruginosa	1.5×10^1–9.4×10^4	5.7×10^3
Enteric virus	5.9–9.0×10^3	3.6×10^2
Parasite ova/cysts	ND–1.4×10^3	1.3×10^2

[a]Values for primary, secondary and mixed sludge.
[b]Dry weight basis.
ND, none detected.

At the end of fermentation, if the amount of gas measured is 500–700 l/kg of total organic dry solids, the ratio between total organic and total mineral dry solids is 0.40–0.50, and the amount of volatile acid is 300–2000 mg/l (as acetic acid), then the sludge is termed digestible, otherwise it is not. (Digestibility can be measured in terms other than rate of gas production: destruction of volatile matter and loss of calorific power are examples).

Other chemical properties

Toxic substances, such as copper salts, cyanides, arsenic or metal salts, can supply useful information about the digestion process. Significant amounts of toxic substances lead to the inhibition of the process. Fats and detergents in large amounts impede the digestion process. An amount of 7–35% fat as a fraction of the total solids is normal in primary sludge.

1.4.3   Biological characteristics

Fresh sludge from primary settling tanks usually has the same biological and bacteriological properties as waste water. Pathogenic bacteria, viruses, protozoa (cysts) and worms (eggs) can survive during waste water treatment and are included in the waste solids. They are not fully destroyed during the normal course of digestion and air-drying. Table 1.3 shows the value of indicator bacteria and pathogens in primary, secondary and mixed sludges. Primary sedimentation and secondary biological treatments of waste water are efficient in removing microorganisms from waste water and transferring them to the sludge. Primary and secondary treatments are able to reduce the microorganisms in sewage by 30–70% and 90–99%, respectively.

The majority of bacteria in activated sludge are floc-forming. However, a few filamentous microorganisms are also present. An abundance of filamentous microbes in an activated sludge reactor can lead to sludge bulking, which will result in poor sludge settlement.

1.5 SLUDGE CHARACTERISATION PARAMETERS

(Baier & Zwiefelhofer, 1991; Coker, 1991; Fleming, 1986; Hao & Kim, 1990; Korsaric et al., 1990; Nishioka et al., 1990; Oku et al., 1990; Webb, 1964)

Several parameters have been identified by researchers that influence sludge processing. The main parameters used in characterising sludge for this purpose are: specific resistance to filtration (R_c), coefficient of compressibility (k), initial solid content (F), capillary suction time (t_{cs}), sludge viability (S_v), sludge stability (S_s), sludge volume ratio (SVR) and sludge volume index (SVI). The factors R_c, F, t_{cs}, SVR and SVI mainly govern the dewatering/thickening/settling characteristics of sludge, while S_v and S_s chiefly concern the stabilisation/digestion of sludge.

1.5.1 Specific Resistance to Filtration (SRF)

SRF provides an empirical measure of the resistance offered by sludge to the withdrawal of water. SRF can be derived as follows:

$$R_c = 2 \cdot b \cdot H_L \cdot A^2 / \mu \cdot w \tag{1.7}$$

where R_c = specific resistance to filtration (SRF), b = constant, H_L = head loss, A = filtering area, μ = dynamic viscosity of filtrate (displaced water), and w = weight of the suspended solids yielded from a unit volume of filtered sludge water.

Figure 1.4 shows the Buchner funnel test apparatus used for the determination of the specific resistance of sludge to filtration. Table 1.4 shows the typical SRF values for various biological sludges.

1.5.2 Compressibility coefficient

The influence of cake compressibility upon the specific resistance of the cake is given by the relationship:

$$R'_c = R_o \cdot H_L^k \tag{1.8}$$

Taking the log of Eqn. 1.8, we get:

$$\log\left(\frac{R'_c}{R_o}\right) = l \cdot \log H_L \tag{1.9}$$

where,

R'_c = constant, representing the specific resistance of the cake for $H_L = 1$, i.e. specific resistance to filtration of 1 cm^3 of cake

k = coefficient of compressibility; a characteristic for each cake

R_o = specific resistance to filtration when $k = 0$, i.e. SRF of incompressible cake (specific resistance is independent of pressure)

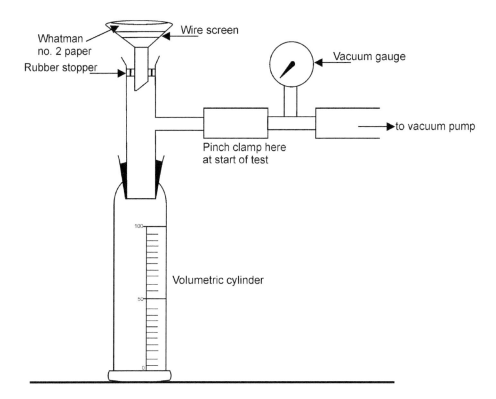

Figure 1.4 Buchner funnel test apparatus for the determination of the specific resistance of sludge.

Table 1.4 Typical specific resistance to filtration for various sludges.

Sludge	Specific Resistance to Filtration (R_c) (m/kg)
Primary	$1.5–5.0 \times 10^{14}$
Activated	$1–10 \times 10^{13}$
Digested	$1–6 \times 10^{14}$
Digested and Coagulant	$3–40 \times 0^{11}$

The specific resistance to filtration (R_c) and the coefficient of compressibility (k) are determined in the laboratory with various devices, incorporating Buchner funnels, vacuum pumps, manometers, etc. (see Figure 1.4).

1.5.3 Viability of sludge

The ratio of microbial activity to the volatile suspended solids (VSS) is called the sludge viability. Viability, therefore, represents the active fraction of the VSS (i.e. that fraction of the VSS which constitutes active bacteria). The adenosine triphosphate (ATP, an organic compound of microbial cells which stores energy) content per viable cell remains constant over a wide range of growth rates and, thus, ATP can be used as a rapid and convenient indicator of the number of viable organisms in the activated sludge.

Chung and Neethling (1990) showed that both ATP and dehydrogenase activity (DHA) responses can be correlated to other indirect activity measurements, such as gas production rate and pH changes. However, because VSS measurements in anaerobic sludge digesters also include VSS present in the primary sludge feed, viability expressions based on total VSS are rendered meaningless for anaerobic sludge digesters. However, this parameter is significant to control of the anaerobic sludge digestion process.

1.5.4 Stability of sludge

The stability of an aerobically digested sludge can be defined and calculated by the following equation, suggested by Paulsrud and Eikum (1975):

$$S_s = 100 \cdot a[1 - OUR_{meas}/OUR_{max}] \tag{1.10}$$

where S_s = sludge stability (%), a = constant (1.035), OUR_{meas} = oxygen uptake rate measured in digested sludge, and OUR_{max} = maximum oxygen uptake rate in the activated sludge plant producing the sludge to be digested.

For calculation purposes, both OUR_{meas} and OUR_{max} are adjusted to the same temperature using the Streeter-Phelps temperature sensitivity coefficients as described by Paulsrud and Eikum (1975).

The higher the value of S_s, the greater is the stabilisation or better is the digestion of sludge.

1.5.5 Sludge Volume Index (SVI)

Sludge volume index (SVI) indicates the settling characteristics of the activated sludge. It is defined as the volume, in ml, occupied by one gram of activated-sludge mixed-liquor solids (dry weight) after settling for 30 minutes in a graduated 1000 ml cylinder. In practice, it is taken to be the percentage volume occupied by the sludge in a mixed-liquor sample (taken at the outlet of the aeration tank) after 30 minutes of settling (P_s), divided by the percentage of suspended solids concentration of the mixed liquor (P_x). Thus,

$$SVI = P_s/P_x \tag{1.11}$$

SVI can also be expressed as follows:

$$SVI = (V_{oc} \times 1000)X_{ss} \tag{1.12}$$

where V_{oc} = volume occupied by settled sludge (ml), and X_{ss} = concentration of mixed-liquor suspended solids (MLSS) (mg/l).

Generally, activated sludge with an SVI of less than 50 is considered as excellent in settling. Sludge with an SVI between 50 and 100 is considered very good, between 100 and 150 is good to moderate, and above 150 is poor, as in this case sludge bulking occurs. However, the index value varies with the characteristics and concentration of the mixed-liquor solids, so values observed at any given plant should not be compared with those reported for other plants or in the literature. For example, if the solids did not settle at all but occupied the entire 1000 ml at the end of 30 min, the maximum index value would be obtained and would vary from 1000 for a mixed-liquor solids concentration of 1000 mg/l to 100 for a mixed-liquor solids concentration of 10,000 mg/l. For such conditions, the computation has no significance other than in the determination of limiting values.

1.5.6   Sludge Volume Ratio (SVR)

Sludge volume ratio (SVR) is a controlling parameter for gravity thickening. In operation, a sludge blanket is maintained at the bottom of a gravity thickener to aid in concentrating the sludge. To operate the thickener effectively, SVR is used as an operating variable to control the length of time sludge is retained in the thickener, and is the volume of the sludge blanket held in the thickener divided by the volume of the thickened sludge removed daily. Values of SVR normally range between 0.5 and 2 days, with the lower values (i.e. higher sludge withdrawal rate) required during warm weather as the sludge settles and turns septic more quickly.

1.5.7   Other parameters

Some other characteristic parameters are also important in control and operation of sludge processing operations. These include pH, specific gravity, alkalinity, type of moisture, type of solids (e.g. flocculent, granules) and degree-day coefficient.

The types and nature of sludge moisture (described in section 1.6) are defined in terms of moisture-to-solid bond strengths. The relative proportions of the moisture types influence the specific energy requirements for separation of solids. Knowledge of these is a prerequisite for rational selection of the most cost-effective sludge volume reduction process. Consideration of the degree-day coefficient is taken into account in the process of sludge freezing–thawing.

1.6   MOISTURE CONTENT OF SLUDGE

Given that we have seen that moisture/water occupies a large fraction of sludge, one of the main objectives of sludge treatment is to reduce its moisture content in order to reduce the total volume of the sludge, so that it can be handled, transported, treated and disposed of in an economical way. Sludge moisture may be present in different forms (e.g. free moisture, bound moisture), the distribution and proportion of which may influence the sludge treatment method, particularly the sludge-dewatering process. It is very important, therefore, to understand the characteristics and behaviour of sludge moisture to assess the most suitable dewatering mechanism and the associated operations or processes.

1.6.1 Sludge moisture hypothesis
(Girovich, 1990; Marklund, 1990; Martin & Bhattarai, 1991; Topping, 1986; Vesilind & Martel, 1990)

Vesilind and Martel (1990) state that water exists in sludge in several readily identifiable forms, such as free water, interstitial water, surface water and bound water, as shown in Figure 1.5 and described below.

(a) Free water

Water that is not associated with the sludge solids is defined as "free water". Free water surrounds the sludge flocs and does not move with the solids. Free water includes void water not affected by capillary force.

(b) Interstitial water

Water that is trapped within the floc structure and travels with the floc or is held between the particles by capillary forces is defined as "interstitial water". Thus, this is the floc moisture when sludge is in suspension and is present in the capillaries when sludge-cake is formed. It is assumed that some of the interstitial water can be removed by mechanical dewatering devices such as belt filters and centrifuges that compress the flocs and expel the water.

(c) Surface water

The third type of water is associated with individual sludge particles. It is held on to the particles by surface forces (i.e. by adsorption and adhesion) and is, therefore, called "surface water". It cannot be removed by purely mechanical means, but its removal can be assisted by pre-chemical conditioning.

(d) Bound water

Lastly, some water is chemically bound to the particles and is referred to as "bound water".

To explain the theory of natural dewatering, Marklund (1990) argued that the moisture content in sludge could be divided into three types:

1   Gravitational water: Water removable by the force of gravity, when conditions for free drainage exist.
2   Capillary water: Water held by cohesion as a film around the soil particles and in the capillary spaces.
3   Hygroscopic water: Water tightly adhered to solids; can only be removed by evaporation.

Further, Girovich (1990) described sludge moisture as having four major phases, which are more or less similar to those outlined by Vesilind and Martel (1990) and described above. Girovich's phases are:

- Free water
- Capillary water
- Colloidal water
- Intercellular water.

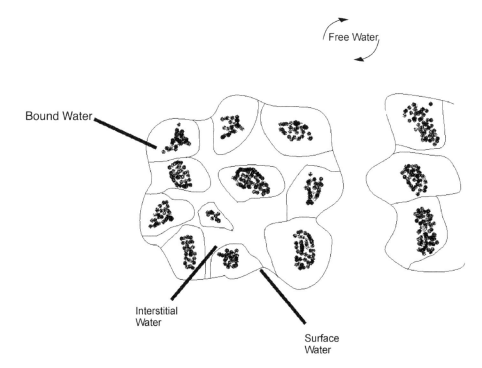

Figure 1.5 A conceptual visualisation of the moisture distribution in sludge.

Free water can easily be separated from sludge by gravity. Capillary and colloidal water can be removed, usually after chemical conditioning, by mechanical forces, including centrifuges, belt presses and vacuum filters. Intercellular water can only be removed from the sludge particle by disrupting the cell structure with thermal treatment.

1.6.2 Moisture distribution

Water and waste water sludge disposal often require the removal of substantial quantities of water, usually by such mechanical means as belt filters, filter presses or centrifuges. The performance of these mechanical devices varies widely according to the characteristics of the moisture present in the sludges to be dewatered. Tsang and Vesilind (1990) have attempted to measure the different types of moisture in sludge and studied the relationship between different dewatering procedures and moisture distribution.

Measurement of moistures

Using a thin layer of sludge in a thermally controlled tube dryer (Figure 1.6) and measuring the weight of the sludge over time using a continuous readout, three clearly defined parts of the drying curve can be identified, as shown in Figure 1.7.

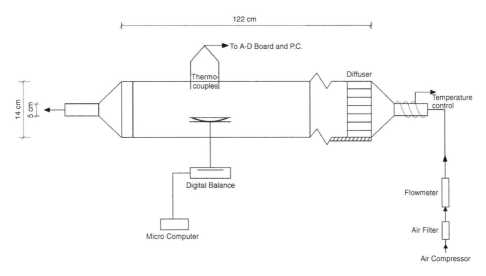

Figure 1.6 The drying apparatus.

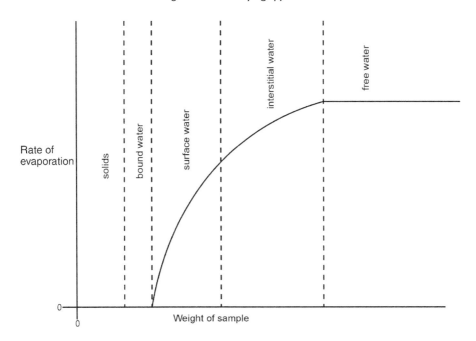

Figure 1.7 Dying curve for identifying four different types of water in sludge.

While free water is being evaporated the drying rate is constant and the slope (differential) does not vary. Once the free water has been eliminated, the interstitial water begins to be removed, and this results in a progressively decreasing rate of drying. When this interstitial water has also fully evaporated, the surface water begins to be

Table 1.5 Moisture distribution of sludge after treatment by different dewatering processes.

		Dewatering Procedure		
S. No.	Type of Moisture	Drained	Filtered	Centrifuges
1	Moisture removed by dewatering (%)	91.2	92.8	88.4
2	Moisture remaining (%)			
	(i) Free	5.2	3.4	4.5
	(ii) Interstitial	2.9	3.1	6.4
	(iii) Surface	0.5	0.5	0.5
	(iv) Bound	0.2	0.2	0.2
	Total	100.0	100.0	100.0

Note: All moisture content expressed in % of initial moisture.

eliminated, and the rate of water removal is again reduced. By measuring the mass of the sludge at the time the transitions occur, the three different types of water can be identified. Bound water is measured as the remaining water that is thermally driven from the sludge sample when it is placed in a 105°C oven and the weight difference recorded.

Dependencies between moisture distribution and dewatering processes

A variety of sludge-dewatering processes usually result in cakes of differing solid content. It would, therefore, be logical to assume that the moisture distribution in sludges dewatered by different processes will also be different, if moisture distribution does indeed affect dewaterability. The moisture distribution in sludges dewatered by three different processes, viz. vacuum filtration, centrifugation and sand-bed drying, were investigated by Tsang and Vesilind (1990). These are the most commonly used dewatering processes and can easily be simulated in the laboratory. The moisture distributions observed in sludges dewatered by these processes are summarised in Table 1.5.

From Table 1.5, three observations can be made:

1 All four types of moisture exist in the dewatered sludge cakes obtained from the three processes.
2 The quantities of surface moisture and bound moisture remain constant regardless of the process used to dewater the sludge.
3 The poor performance of the centrifugation process is caused by an exceptionally high content of interstitial moisture in the sludge.

1.6.3   Moisture effects on sludge drying
(APHA, 2005; Karr & Keinath, 1978; Katz & Mason, 1970; Logsdon & Edgerley, 1971)

In the classical approach to drying for the purpose of obtaining drying characteristics, a small sample of the material is dried under constant conditions, such as temperature, air velocity and humidity. The results of such an investigation are usually presented as a drying rate curve and the nature of the curves is used to interpret the drying behaviour. The classical instantaneous rate of drying is shown in Figure 1.8.

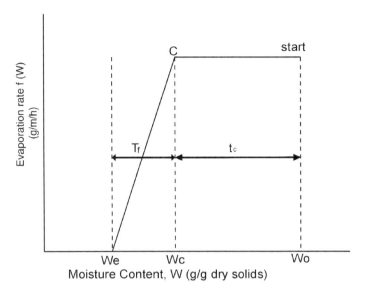

Figure 1.8 Typical drying rate curve.

Three values of moisture content are marked on this diagram:

1   W_o = original moisture content (g water/g dry solids).
2   W_c = critical moisture content (g water/g dry solids), which refers to a sharp inflex-
    ion in the drying rate. Since the critical moisture is the average moisture through
    the sample, its value depends on the rate of drying, the thickness of the material,
    and other factors influencing moisture movement and resulting gradients within
    the solid.
3   W_e = equilibrium moisture content (g water/g dry solids), which represents the
    limiting moisture content for given conditions of humidity and temperature.
    It depends greatly on the nature of the solids.

Similarly, two periods of time are shown in the diagram:

1   t_c = Constant-rate period, during which the rate of evaporation remains at its
    initial level until the critical moisture content (point C) is reached. The rate of
    drying is controlled by the rate of heat transferred to the evaporating surface and
    is essentially independent of the nature of the solids. The internal mechanism of
    liquid flow does not affect the constant rate.
2   t_f = Falling-rate period (as shown by line CE), which is typified by a continuously
    changing rate throughout the remainder of the drying cycle. The rate of internal
    moisture movement controls the drying rate. It reduces progressively, until the
    equilibrium moisture content (point E) is reached, and no further evaporation
    occurs.

Table 1.6 Moisture Bonds (Katz & Mason, 1970).

Type of moisture attachment	Chemical		Physical		Mechanical		
Order of bound energy	Ionic Molecular	Adsorptive	Osmotic	Structural	Micro-capillary	Macro-capillary	Unbound
kJ/k mol	5000	3000			>100	<100	0

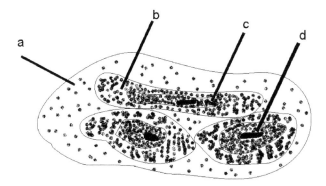

Figure 1.9 Distribution of (a) free; (b) immobilised; (c) bound; (d) chemically bound moistured in sludge sample.

At any instant during the drying phases, the moisture is present as bound and free water. The free moisture content is equal to the difference between the total moisture content and the equilibrium moisture content.

1.6.4 Additional remarks

The classical theory described above may be applicable to the drying of a clearly defined crystalline material. However, the classification of moisture content as "free" and "bound" can be misleading when applied to biological sludges. Moisture bonds in sludges are of more than two different types. Thus, three different types of moisture bonding and the energy associated with each are included in Table 1.6. These can be linked to the moisture categories, viz. chemically bound, physico-chemically bound, physico-mechanically bound, immobilised and free unbound moisture. A diagrammatic illustration of these moisture types is provided in Figure 1.9.

The different types of moisture present in sludge increase the complexity of the internal flow mechanism during its drying. However, Keey (1972) states that the quantification of different types of sludge moisture content is possible using vacuum drying, and these moisture types allow elucidation of the effect of the addition of polyelectrolyte on the process of dewatering. The addition of polyelectrolyte increases the rate of filtration by releasing some of the immobilised water, but simultaneously increases the physically bound moisture content (Logsdon & Edgerley, 1971). This behaviour appears to provide a reason for the problems experienced in the practical use of mechanical dewatering equipment, producing sludge-cake with an unsatisfactorily low solids concentration.

Existing dewatering parameters such as specific resistance to filtration (SRF) and capillary suction time (CST) do not correlate with chemically bound, physically bound and immobilised moisture fractions. These tests give information on the mobilised water fraction, but give no information on the ease of removal of the bound moisture fraction. In centrifugal dewatering, it is the bound moisture content that gives the highest correlation with the moisture content of the final cake.

1.7 SLUDGE GENERATION

The amounts of sludge generated in different treatment processes are measured routinely in water-purification and waste water treatment plants. The estimation of the sludge volumes depends on expected water or solid content, and on the proportion and specific gravity of volatile components and fixed solids. The best method of estimating the level of sludge production is to base the estimate on past data from similar facilities and the projected strength of the influent. In general, sludge production rates range between 0.2 and 0.3 kg/m^3 (0.8 to 1.2 dry tons/million gallons) of waste water treated. In the absence of previous data, a rule-of-thumb approximation for sludge produced in a waste water treatment plant is 0.24 kg/m^3 (1 dry ton/MG) of waste water treated (WEF, 1998).

The nature and characteristics of the solid fractions in municipal waste water can be described as follows (Parker *et al.*, 1992).

1.7.1 Sedimentation sludge

Typically, primary sludge generation can vary from 0.1 to 0.3 kg/m^3 (800 to 2500 lb/MG) of waste water. A rule-of-thumb approximation is 0.05 kg/capita (0.12 lb/capita) per day of primary sludge generation. The most common method of estimating primary sludge generation is to calculate the amount of suspended solids entering a treatment plant and assume a removal rate, which is typically in the range 50–65%. The fresh solids contain most of the settleable solids and about 60% of the suspended solids of the raw waste water. A removal rate of 60% is commonly used for estimating purposes, provided that the effects of industrial contribution are minimal and no major side streams from the sludge processing units are discharged to the primary influent of the clarifier.

Daily variations in primary sludge generation can occur, usually in proportion to the quantity of solids entering a waste water treatment plant. Peak rates of sludge production can be several times the average (Turovskiy & Mathai, 2006).

1.7.2 Chemical coagulation sludge

Chemicals are widely used in waste water treatment to precipitate and remove phosphorus and, in some cases, to improve the efficiency of suspended solids removal. Although theoretical rates of chemical sludge production can be estimated from the anticipated chemical reactions, competing reactions can make the estimation difficult

(Turovskiy & Mathai, 2006). The fresh solids incorporate between 70 and 90% of the solids suspended in the raw waste water, depending on the effectiveness of chemical dosage. For example, ferric chloride, of molecular weight 162.2, is precipitated as ferric hydrate, having molecular weight 106.9. Thus, each milligram per litre of $FeCl_3$ will produce 0.66 mg/l of $Fe(OH)_3$. Conventional jar tests are helpful in estimating the quantities of chemical sludge. However, quantities of precipitates in chemical sludge are influenced by several factors, for example, pH and mixing and reaction times.

The following are some of the types of precipitates that must be considered in the measurement of total sludge production (U.S. EPA, 1979):

(a) *Phosphate precipitates*
These includes $AlPO_4$ or $Al(H_2PO_4)(OH)_2$ with aluminium salts, $FePO_4$ with iron salts, and $Ca_3(PO_4)_2$ with lime.
(b) *Carbon precipitates*
These are significant with lime, which forms calcium carbonate, $CaCO_3$.
(c) *Hydroxide precipitates*
With iron and aluminium salts, excess salt forms a hydroxide, $Fe(OH)_3$ or $Al(OH)_3$. With lime, magnesium hydroxide, $Mg(OH)_2$, may form.
(d) *Inert solids from chemicals*
Many chemicals supplied in dry form may contain significant amounts of inert solids. If a quicklime consists of 92% CaO, the remaining 8% may be mostly inert solids that appear in the sludge.
(e) *Polymer solids*
Polymers may be used as primary coagulants or to improve the performance of other coagulants. The polymers themselves contribute little to the total mass, but they can greatly improve clarifier efficiency, with a concomitant increase in sludge production.

1.7.3 Biofiltration sludge

Most of the dissolved organic matter and many of the otherwise non-settleable solids in waste waters applied to trickling filters are rendered settleable by adsorption and biological flocculation on the trickling-filter film. The film itself is modified by decomposition before it is sloughed off and picked up by the effluent. Destruction and loss vary with length of storage in the bed. Common limits are 30% for filters operated at low rates against 10% for filters operated at high rates. Thus secondary sedimentation normally captures 50–60% of the non-settleable suspended solids, reaching low-rate filters, whereas corresponding values for high-rate filters are 80–90%. The trickling-filter humus is generally added to primary solids for digestion. Changes in composition during digestion are greater for the high-rate humus than for the low-rate humus.

1.7.4 Activated sludge

The controlling fraction of solids in fresh, excess activated sludge is normally richer in organic matter and higher in water content than trickling-filter humus. Between 5 and 10% of the materials transferred are mineralised during formation and recirculation, depending on the proportion of solids returned and the length of aeration, which, together, govern how long the activated sludge remains in circulation. The recovery of excess sludge in secondary settling units is between 80 and 90% of the aerated, non-settleable, suspended-solid load. Excess activated sludge may be allowed to settle with primary solids, unless this practice is counter-indicated because the combined sludge responds poorly to further treatment. Digestion brings about much the same changes in composition, but not in concentration, as for plain-sedimentation primary solids. The accumulation of municipal waste water solids depends on the composition of the waste waters, the efficiency of transfer of solids, and the degree of digestion of solids incidental to the treatment.

In general, the performance of the activated sludge process is limited by the ability of the secondary clarifier to separate and concentrate the sludge from the aerator effluent (Magbanua & Bowers, 1998). This limitation is imposed by the settleability and compatibility of the activated sludge floc. The basic determinants of the settleability and compatibility of sludge are its microbial composition. Activated sludge microorganisms may be classified into two morphological types, namely, (i) filamentous microorganisms, which grow in long strands or filaments, and (ii) floc-forming or *Zoogloea*-like microorganisms, which are more spherical in shape and form. It has been observed that a well-settling sludge generally contains large and strong flocs consisting of floc-formers adhering to a backbone of filamentous bacteria. If filamentous growth is excessive, the filaments extend out of the flocs into the bulk liquid and interfere with settling and compaction (Jenkins, 1992; Kappeler & Gujer, 1992; Parker *et al.*, 1992; Sezgin *et al.*, 1978). This kind of condition, known as filamentous bulking, is the most common activated sludge separation problem with the greatest impact on the quality of effluent. An over dominance of floc-formers, on the other hand, can lead to pinpoint floc, small dispersed flocs that settle slowly, or to viscous or non-filamentous bulking, where the organisms are encased in slime capsules having a high water content and low compactibility (Eriksson & Alm, 1991; Eriksson *et al.*, 1992; Gujer & Kappeler, 1992; Jenkins, 1992; Novak *et al.*, 1993; Urbain *et al.*, 1993). Hence, a balance between the filamentous and the floc-forming organisms is necessary to obtain a well-settling sludge (Magbanua & Bowers, 1998; Matsuzawa & Mino, 1991; Metcalf & Eddy, 1991).

In tracing the history of the processing of activated sludge, Tomlinson (1982) and Albertson (1991) noted that pioneering activated sludge experiments, performed in fill-and-draw reactors and activated sludge plants with long, essentially plug-flow aeration basins, rarely suffered from bulking problems. The introduction of diffused air aeration, which substantially increased axial mixing within the aeration tank, and the increasing use of completely mixed systems, led to the increased incidence of filamentous bulking (Cetin & Surucu, 1990; Kohno *et al.*, 1991; Magbanua & Bowers, 1998). Clearly, conditions in fill-and-draw, plug-flow and compartmentalised reactors, where a spatial or temporal substrate concentration gradient exists, favoured floc-formers, while completely mixed systems, where the substrate concentration is uniformly low, favoured filaments (Chudoba *et al.*, 1973; Rensink *et al.*, 1982; Wanner, 1992).

1.7.5 Municipal waterworks sludge

The solids fractions in sludges withdrawn from settling and coagulation basins in municipal water treatment works, and in the wash water from rapid or slow filters, vary with the nature of the water treated, the amount and type of additives, and the reactions taking place during treatment. Some waterworks sludges are quite putrescible, for example, from coagulated, coloured or polluted waters. The weight of settled solids drawn from coagulation tanks and wash water settling tanks may be as little as 0.1% of the weight of mixed liquor before thickening, and as much as 2.5% after thickening. Observed values vary with the nature of the raw water and the type and concentration of the chemicals employed.

QUESTIONS

1   Discuss the objectives of sludge treatment.
2   What are the main processes for sludge treatment? Explain their purposes.
3   What are the differences between primary and secondary sludges? Define the organic content of sludge.
4   What are the main factors affecting sludge processing in a sludge treatment facility?
5   Define the moisture content of sludge. How does moisture content affect the treatment of sludge?
6   What are the differences between biological and chemical sludges?
7   What are the differences between screenings, grit and sludge?
8   What are the sources and treatment stages of sludge generation in a conventional activated sludge process?
9   Explain the following types of sludge according to source, characteristics and operating conditions:

   a)   Raw sludge
   b)   Primary sludge
   c)   Activated sludge
   d)   Aerobically stabilised sludge
   e)   Digested sludge.

Chapter 2

Pumping of sludge

2.1 INTRODUCTION

(Chou, 1958; Fleming, 1986; Guyer, 2011; MWST, 1991;
Oku *et al.*, 1990; Pergamon PATSEARCHER; WEF, 2007)

In water and wastewater treatment plants, sludge is transferred from point to point by pumping the sludge in conditions ranging from a watery sludge or scum to a thick sludge. Pumping may be intermittent, as in the case of raw sludge from settling tank to the digester, or continuous as in the return of the activated sludge to the aeration tank. A pumping system typically includes a pump, suction and discharge reservoirs, interconnecting piping, and various appurtenances (e.g., valves, flow-monitoring devices, and controls). To deliver a given volume of fluid through the system, a pump must transfer enough energy to the fluid to overcome all system energy losses.

Efficiency of sludge pumps is therefore important in the proper operation of a treatment plant. Effective sludge pumping depends on properly matching the pump, solids characteristics, and actual system head requirements. The efficiency of a sludge pump is considered subordinate to dependable, satisfactory and trouble-free operation. Sludge pumps have to be resistant to abrasion, as sludges too often contain sand and grit. Design of pumps will have to provide for sufficient clearance, as close clearance though necessary for higher efficiencies, leads to frequent stoppages and excessive wear.

2.2 SLUDGE-FLOW CHARACTERISTICS

All sludges and slurries are nothing but pseudo-homogeneous substances. Fresh plain-sedimentation solids are especially diverse in their composition; digested solids and activated sludges less so; alum and iron flocs least of all. Because many wastewater sludges are non-Newtonian fluids with plastic rather than viscous properties, their flow resistance is a function of their concentration. Hydraulics of flow are further complicated, because most sludges are thixotropic. Their plastic properties change during stirring and turbulence. Gases or air released during flow add to the difficulty of identifying probable hydraulic behaviour. As might be expected, fundamental frictional losses increase with solid content and decrease with temperature. In general, laminar or transitional flow persists at relatively high velocities, such as 0.45–1.40 m/s for thick sludges flowing in pipes 12.5–30 cm in diameter. At turbulent velocities all sludges flow more like water.

Sludges having low solid percentage (10%) can be pumped through force mains. Sludges containing solids percentage of 2 have similar hydraulic characteristics as of water. For sludges having higher solids contents of more than 2 percent, however, friction losses are from 1½ to 4 times the friction losses for water. Upon decreasing the temperature, the head losses and friction both are increase. Velocities should be kept above 2 feet per second. Grease content can leads to serious clogging, and grit will adversely affect the flow characteristics as well. Satisfactory clean-outs and long sweep turns will be used when designing the facilities of these types.

2.3   SLUDGE PIPING

The piping for sludge withdrawal must be at least 6 inches in diameter. The pipe discharge lines must have a minimum diameters of 4 inches for plants less than 0.5 million gallons per day, and 8 inches for plants larger than 1.0 million gallons per day. The short and straight pipe runs are better, while as the sharp bends and high points must be avoided. For flushing purposes, the blank flanges and valves should be provided.

2.4   SLUDGE HEAD-LOSS

The sludge head-loss takes place during the pumping of sludge mainly depends on the type, solid content, and flow velocity of sludge. It has been observed that head-losses increase with increased solid content, increased volatile content, and lower tempera-ture. When the per cent volatile matter multiplied by the per cent solid content exceeds 600, difficulties may be encountered in pumping sludge.

The head-loss in pumping un-concentrated, activated and trickling-filter sludges may be from 10 to 25% greater than for water. Primary, digested, and concentrated sludges at low velocities may exhibit a plastic flow phenomenon, in which a definite pressure is required to overcome resistance and start flow. The resistance then increases approximately with the first power of the velocity throughout the laminar range of flow, which extends to about 1.1 m/s, the so-called lower critical velocity. Above the higher critical velocity at about 1.4 m/s, the flow may be considered turbulent. In the turbulent range, the head-losses for well-digested sludge may be from two to three times the head-losses for water. The head-losses for primary and dense concentrated sludges may be considerably greater.

2.5   POWER REQUIREMENT

In order to determine the operating speed and power requirement for a centrifugal pump handling sludge, system curves need to be computed for the (1) most dense sludge anticipated, (2) for average conditions, and (3) for water. These are to be plotted on a graph of the pump curves (see Figure 2.1) for a range of available speeds. The requirement of maximum and minimum speeds of a particular pump are obtained from the intersection of the pump head-capacity curves, with the system curves at the desired capacity. Where the maximum-speed head-capacity curve intersects the system

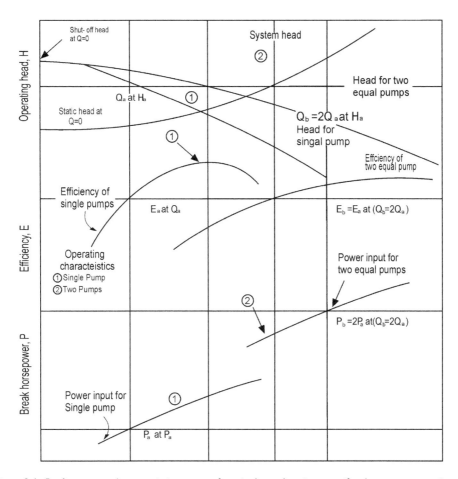

Figure 2.1 Performance characteristic curves for single and twin centrifugal pumps operating at constant speed.

curve for water determines the power required. In constructing the system curves for sludge, it is recommended that, for velocities from 0 to 1.1 m/s, the head is to be assumed constant at the figure computed for 1.1 m/s. The intersection of the pump curves with the system curve for average conditions can be used to estimate hours of operation, average speed, and power costs.

2.6 SLUDGE PUMPS
(Guyer, 2011)

Sludge pumps can be categorise as plunger (reciprocating), progressing-cavity, torque-flow, or open-propeller centrifugal pumps. Generally, primary sludges are pumped through plunger and progressing-cavity pumps, however, centrifugal pumps are considered more appropriate for the secondary sludges. In most cases, digested sludge is conveyed through centrifugal and torque-flow pumps, however, plunger and progressing-cavity pumps are used when a suction lift is involved. Plunger pumps are considered good for sludge elutriation. The standby pumps are used for pumping the primary and secondary sludges, and sludge elutriation.

2.6.1 Plunger pump

Plunger pumps of the simplex, duplex or triplex models with capacities of 150–250 lpm per plunger are available. They are suitable for suction lifts up to 3 m and are self-priming. The pump speeds should be between 40 to 50 rpm. A minimum head of 25 m is considered when designing the plunger pumps, since the accumulations of grease in sludge piping may leads to the continuous increase in head with use. The advantages of plunger pumps are as follows:

- Pulsating action tends to concentrate the sludge in the hoppers.
- Suitable for suction lifts of up to 10 feet and are self-priming.
- Low pumping rates can be used with large port openings.
- Positive delivery is provided unless some object prevents the ball check valves from seating.
- They have constant but adjustable capacity regardless of large variations in pumping head.
- Large discharge heads may be provided for.
- Heavy-solids concentrations may be pumped readily.
- Clogging easily corrected.

2.6.2 Centrifugal pumps

Centrifugal pumps are required to be equipped with variable speed drives, to suit variable capacities, since throttling the discharge to reduce the capacity with a constant drive results in frequent stoppages. The design should also consider the need for pumping sludge solids without clogging and yet avoid pumping large quantities of overlying sewage/wastewater. Presently, various modifications of centrifugal pumps are available in the market. The centrifugal pumps with non-clogging impellers are confined to bigger sizes. The scru-peller pump, a modification of the screw-feed impeller pump, is applicable specially to the pumping of sludge. It is less prone to clogging than the ordinary centrifugal pumps and eliminates some of the objections inherent in reciprocating pumps.

2.6.3 Torque flow

Torque flow pumps are very effective in conveying the sludge, since they uses fully recessed impeller. A limited size of particles can be handled mainly due to the diameter of the suction or discharge valve. A vortex is developed in the sludge by the rotating impeller thus the liquid itself becomes the main propulsive force.

2.6.4 Progressing cavity

The progressive cavity pumps are used successfully, mainly for the concentrated sludge. The main component of the pump is a singe-threaded rotor, which operates in a double threaded helix of rubber with a minimum of clearance. It is available in capacities up to 1325 liter per minute with a suction capacity of lifting up to 8.5 m. Moreover, it can pass the solids up to 2.9 cm in diameter.

2.7 PUMPS FOR DIFFERENT TYPES OF SLUDGE

Different types of sludges required to be pumped are primary, chemical, trickling-filter sludges and activated, return, elutriated and thickened/concentrated sludges. In addition, scum also requires to pump. Usually, the consistency of untreated primary sludge changes during pumping. At first, the most concentrated sludge is pumped, when most of the sludge has been pumped, the pump must handle a dilute sludge that has essentially the same hydraulic characteristics as water. This change in characteristics causes a centrifugal pump to operate farther out on its curve. The pump motor should be sized for the additional load, and a variable-speed drive should be supplied to reduce the flow under these conditions. If the pump motor is not sized for the maximum load obtainable when pumping water at top speed, it is likely to go out on overload or be damaged if the overload devices do not function or are set too high.

The application of pumps to different types of sludge can be summarised as below:

2.7.1 Primary sludge pumps

Plunger pumps may be used on primary sludge. Centrifugal pumps of the screw-feed and bladeless type, as well as torque-flow pumps may also be used for transporting primary sludge. The character of primary sludge is such that conventional non-clog pumps cannot be used satisfactorily.

2.7.2 Chemical sludge pumps

Sludge from chemical precipitation processes can usually be handled in the same manner as primary sludge.

2.7.3 Secondary sludge pumps

Non-dogging centrifugal and plunger pumps as well as airlifts and ejectors find use for secondary sludge. However, the centrifugal pump is preferred because of the advantages like greater efficiency, capacity for handling light solids concentration, uniform and smooth delivery of solids, mixing of the mass, which is being pumped, quiet and cleaner operation than the plunger pump and lower maintenance costs for continuous operation.

2.7.4 Sludge recirculation and transfer pumps

Selection of the type of pump is largely dependent on its location. Where priming is positive and no suction lift is involved, large, centrifugal pump with variable-speed drive is a good choice. The nature of solids to be handled by these pumps being similar or less concentrated than the primary sludge and since suction lifts are low, centrifugal pumps give successful results. Where a suction lift is involved, the plunger pump may be selected. Often it may be possible to combine this duty with primary or secondary sludge pumping by proper arrangement of units. Plunger pumps can be used when it becomes necessary to dewater a digester completely.

2.7.5 Bio-filter sludge pumps

Plunger pump, progressing-cavity pump, non-clog centrifugal pump, and torque-flow pump are most applicable. Sludge is usually of homogeneous character and can be easily pumped.

2.7.6 Activated sludge pumps

Non-clog or mixed-flow centrifugal pumps are preferable in this case. Sludge is dilute and contains only fine solids, so that it may be readily pumped with this type of pump, which operates at low speed because the head is low and the flocculent character of the sludge needs to be maintained.

2.7.7 Digested sludge pumps

Well-digested sludge is homogeneous, containing 5–8% solids and some quantity of gas bubbles, but may contain up to 12% solids. Poorly digested sludge may be difficult to handle. At least one positive-displacement pump should be available. However, plunger, torque-flow centrifugal type, positive displacement and progressing-cavity pumps are applicable for pumping the digested sludge.

2.7.8 Elutriation sludge pumps

Pumps that are useful for primary sludges find application for elutriation purposes also. In small plants, plunger pumps are preferred as they may often be combined with the primary sludge pumps in a common location. They are generally equipped with counters to measure the quantity of sludge pumped. In large plants, the centrifugal pump is chosen, since it is capable of smoother, quieter and cleaner operation, as well as for delivery of larger volumes of elutriated sludge. Suitable sludge meters used in conjunction with centrifugal pumps employed in elutriation permit the measurement of sludge quantities. Adjustment of washing ratios and variation of the rate of delivery to the vacuum-filters corresponding to various dewatering rates can be accomplished by variable-speed drives.

2.7.9 Thickened and concentrated sludge pumps

Plunger pumps serve this purpose better, as it can successfully accommodate the high friction-head-losses in pump discharge lines. Positive-displacement and progressing-cavity pumps too have been used successfully for dense sludges containing up to 20% solids. Because, these pumps have limited clearances, it is necessary to reduce all solids to small size.

2.7.10 Pumping of scum

Positive-displacement or progressing-cavity pumps, plunger or diaphragm, pneumatic ejectors, centrifugal pump (screw-feed, bladeless, or torque-flow types) may be used for this purpose. Scum is often pumped by the sludge pumps; values are manipulated in the scum and sludge lines to permit this. In larger plants, separate scum pumps are used.

2.8 CONTROLS

Generally, sludge pumping operation takes place at lower than the design capacity of pump. At small treatment plants, the design engineer will assess the use of a timer to allow the operator to program the pump for on-off operation. However, the use of variable speed controls should be investigated for large treatment plants (Guyer, 2011).

QUESTIONS

1   What are the key factors affecting the sludge pumping?
2   Why the head loss in sludge pumping is higher than the head loss in pumping the water?
3   Define the types of pumps and their working principal.
4   What types of pumps are used for primary sludge, secondary sludge, chemical sludge, thickened and concentrated sludge, digested sludge and scum?

Chapter 3

Treatment of sludge

3.1 INTRODUCTION

Generally, the sludges comprises 92–98% moisture, the remaining part being the putrescible organic materials. Due to the high organic content of the sludge, it requires further treatment prior to its final disposal.

The common order of unit operations employed in wastewater treatment suggests that some operations, viz. screening, sedimentation, and chemical flocculation or precipitation constitute preliminary treatment. Subsequent operations, notably those associated with trickling filtration and activated sludge treatment constitute 'secondary treatment'. Sludge treatment may include all or a combination of the following unit operations and processes: thickening, digestion, conditioning, dewatering, and incineration. Thickening is meant for the reduction of moisture content of the sludge. Digestion is a biological method of treatment and is meant for the reduction of organic content of the sludge. Conditioning improves the drainability of the digested sludge, so that dewatering may be accomplished easily by air-drying in sand drying beds or by mechanical smears. The ultimate disposal of dewatered sludge or the ash after incineration of dewatered sludge may be done on to the land or into the sea. The particular combination of unit operations to be employed for sludge treatment depends upon the quantity and characteristics of sludge. As shown in Figure 3.1, common combinations of unit operations in the category of solids concentration and stabilisation are:

(i)   Digestion of plain-sedimentation sludge followed by air drying,
(ii)   Concentration and chemical conditioning of activated sludge in advance of vacuum filtration, and
(iii)   Incineration of a mixture of trickling-filter humus and plain-sedimentation sludge after digestion, elutriation, chemical conditioning, and vacuum filtration.

The waste liquor from wastewater sludges is often putrescible and high in solids. Coagulation and concentration of the removed floc may be necessary in preparing the effluent for discharge.

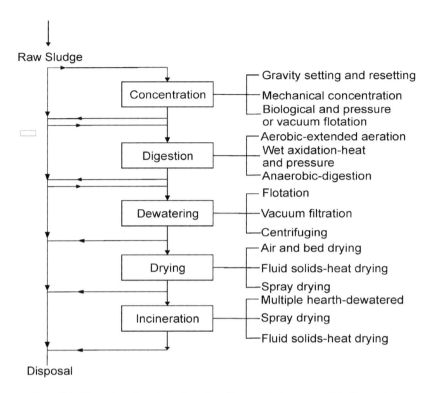

Figure 3.1 Flow chart for sludge handling (Arrows indicate possible flowpaths).

3.2 DEFINITIONS
(Coker *et al.*, 1991; Fleming, 1986)

Sludge thickening and stabilisation are the key processes in water and wastewater treatment facilities practiced before disposal of sludge. Component operations and processes include the following.

Thickening: Thickening concentrates sludge by stirring it long enough to form larger, more rapidly settling aggregates with smaller water content. An example is the thickening of activated sludge to increase its solids concentration by 3–6 fold in 8–12 hours of stirring, chlorine being added, if necessary, to impede decomposition. Displaced sludge liquor is the by-product.

Centrifuging: This operation concentrates sludge run into a centrifuge intermittently or continually to separate solids from the suspending sludge liquor, which becomes a by-product.

Chemical conditioning: This coagulates the sludge and improves its dewatering characteristics. An example is the addition of ferric chloride to wastewater sludge to be dewatered on vacuum filters.

Elutriation: Elutriation washes out the sludge substances that interfere physically or economically with chemical conditioning and vacuum filtration. An example is the reduction in the alkalinity of digested sludge in the amounts of chemicals that need to be added in advance of filtration. The elutriating water is a by-product.

Biological flotation: In this operation sludge solids are lifted to the surface by gases of decomposition. This concentrates the sludge. An example is the flotation of primary sludge in 5 days at 35°C and the withdrawal of the subnatant, which is a by-product.

Dissolved air flotation: It is similar to the biological flotation, except that compressed air is used in this operation. Dissolved air flotation units are better alternative to thicken the 'light' sludge, such as waste activated sludge.

Vacuum filtration: This withdraws moistures from a layer of sludge by suction, the sludge to be dewatered being supported on a porous medium, such as coiled springs, or cloth on screening. An example is the dewatering of chemically conditioned, activated sludge on a continuous, rotary, vacuum drum filter. A sludge paste or cake is produced. The sludge liquor removed is a by-product.

Air drying: Air drying removes moisture from sludge run on to beds of sand or other granular materials. Included moisture evaporates in the atmosphere and drains in the drying bed. An example is the air-drying of well-digested sewage sludge on sand beds, a spread able, friable sludge cake being produced. The by-product is the liquor reaching the under drains.

Heat drying: This drives off moisture by heat. The sludge can be reduced to substantial dryness. An example is the drying of vacuum- filtered, activated sludge in a continuous flash drier. If sludge is to be marketed, its moisture content must generally be reduced to less than 10%.

Sludge stabilisation: Sludge stabilisation converts raw sludge into a less offensive form that has substantially reduced numbers of pathogens and mineralised solids, and is suitable for disposal safely. Traditionally, sludge is biologically stabilised by either aerobic or anaerobic digestion process. However, there are non-biological methods also to stabilise the sludge.

Sludge digestion: It is the aerobic/anaerobic decomposition of sludge. Digestion is accompanied by gasification, liquefaction, stabilisation, destruction of colloidal structure, and concentration, consolidation, or release of moisture. The gases produced generally include, besides carbon dioxide, combustible methane and, more rarely, hydrogen.

Sludge digestion is one of the processes used to stabilise the sludge. Examples are: (i) aerobic digestion of sludge in aerated reactors under endogenous-phase conditions of respiration, and (ii) the digestion of settled solids in septic tanks (anaerobic digestion process).

Aerobic process: Aerobic process is a biological-treatment process that occurs in the presence of oxygen. Certain bacteria that can survive only in the presence of dissolved oxygen are known as obligate (i.e. restricted to a specified condition in life) aerobes.

Anaerobic process: This is a biological-treatment process that occurs in the absence of oxygen. Bacteria that can survive only in the absence of any dissolved oxygen are known as obligate anaerobes.

Denitrification: Denitrification is the biological process by which nitrate is converted to nitrogen and other gaseous end products.

Anoxic denitrification: It is the process by which nitrate nitrogen is converted biologically to nitrogen gas in the absence of oxygen.

Anoxic sludge digestion: It is nothing but the anoxic denitrification process used to stabilise the sludge. When nitrate is used rather than oxygen, the anoxic biomass stabilisation process similar to the aerobic digestion process takes place.

Dry combustion or incineration: This operation leads to the ignition and incineration of heat-dried sludge at high temperatures, alone or with added fuel. Examples are (i) the incineration of heat-dried sludge, and (ii) the burning of heat-dried sludge on the lower hearths of multiple-hearth furnace, on the upper hearths of which the sludge to be incinerated is being dried. The end product of incineration is a mineral ash. The stack gases are by-products.

Wet combustion: Wet combustion oxidises wet sludge at temperatures of about 540 of (282°C) and air pressures of 1200–1800 psig. The effluent suspension and exhaust gases are by-products.

Other unit operations: Other unit operations of sludge treatment include conditioning by heating, freezing, or physical flotation and dewatering by pressure filtration (filter pressing), etc.

3.3   METHODS OF SLUDGE TREATMENT
(MWST, 1991)

Table 3.1 enlisted the principal methods used to process and dispose of sludge. The sludge moisture is remove by thickening (concentration), conditioning, dewatering, and drying, however, aerobic and anaerobic digestion, incineration, and wet oxidation are used to stabilise the organic material in the sludge.

3.4   FLOW-SHEETS OF SLUDGE TREATMENT
(Fleming, 1986)

A generalised flow-sheet/flowchart incorporating the unit operations and processes employed in sludge treatment is presented in Figure 3.2 and 3.3. In practice, the most commonly used process flow-sheets for sludge treatment may be divided into two general categories, depending on whether or not biological treatment is involved.

Typical flow-sheets incorporating biological processing are presented in Figure 3.4. Depending on the source of the sludge, either gravity or air flotation thickeners are used. In some cases, both may be used in the same plant. Following biological digestion, any of the three methods shown (i.e. vacuum filtration, centrifugation, drying beds) may be used to dewater the sludge, the choice depending on local conditions.

Because the presence of industrial and other toxic wastes has presented problems in the operation of biological digesters, a number of plants have been designed with other means for sludge treatment. Three representative process flow-sheets without biological treatment are shown in Figure 3.5.

Table 3.1 Methods for processing and disposal of sludge.

Unit operation or treatment method	Function
(i) Preliminary Operations	
Sludge grinding	Size reduction
Sludge degritting	Grit removal
Sludge blending	Blending
Sludge storage	Storage
(ii) Thickening	
Gravity thickening	Volume reduction
Flotation thickening	Volume reduction
Centrifugal	Volume reduction
(iii) Stabilisation	
Chlorine oxidation	Stabilisation
Lime stabilization	Stabilisation
Heat treatment	Stabilisation
Anaerobic digestion	Stabilisation, mass reduction
Aerobic digestion	Stabilisation, mass reduction
(iv) Conditioning	
Chemical conditioning	Sludge conditioning
Elutriation	Leaching
Heat treatment	Sludge conditioning
(v) Disinfection	
Pasteurization	Disinfection
High pH treatment	Disinfection
Long term storage	Disinfection
Chlorination	Disinfection
Radiation	Disinfection
(vi) Dewatering	
Vacuum filter	Volume reduction
Filter press	Volume reduction
Horizontal belt filter	Volume reduction
Centrifuge	Volume reduction
Drying bed	Volume reduction
Lagoon	Storage, volume reduction
(vii) Drying	
Flash dryer	Weight reduction, volume reduction
Spray dryer	Weight reduction, volume reduction
Rotary dryer	Weight reduction, volume reduction
Multi-hearth dryer	Weight reduction, volume reduction
Oil immersion dehydration	Weight reduction, volume reduction
(viii) Composting	
Composting (Sludge only)	Volume reduction, resource recovery
Co-composting with solid wastes	Product recovery, volume reduction
(ix) Thermal Reduction	
Multiple-hearth incineration	Volume reduction, resource recovery
Fluidised-bed incineration	Volume reduction
Flash combustion	Volume reduction
Co-incineration with solid wastes	Volume reduction, resource recovery
Co-pyrolysis with solid wastes	Volume reduction, resource recovery
Wet air oxidation	Volume reduction
(x) Ultimate Disposal	
Landfill	Final disposal
Land application	Final disposal
Reclamation	Final disposal, land reclamation
Reuse	Final disposal, resource recovery

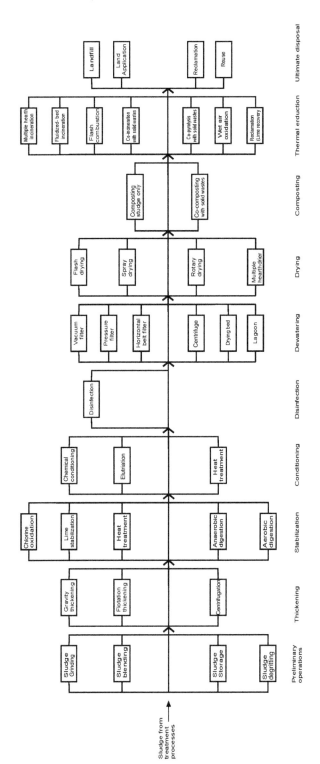

Figure 3.2   Flow chart for generalise sludge-processing and disposal.

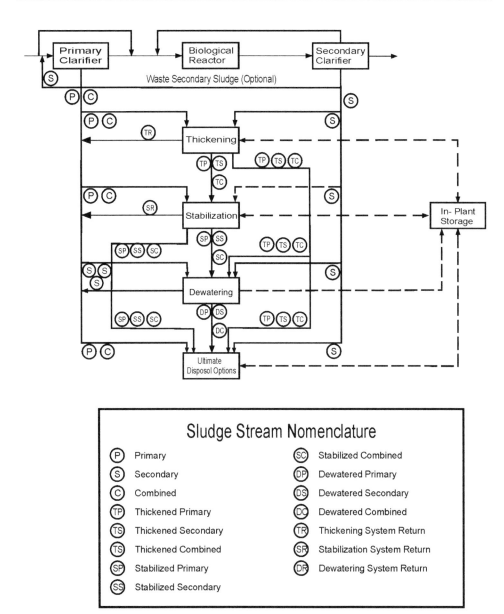

Figure 3.3 General schematic of a sludge treatment system.

3.5 PRELIMINARY TREATMENTS
(MWST, 1991; Rao & Datta, 1987; Bahadori, 2013)

Preliminary sludge treatment operations include grinding, de-gritting, blending, and storage of sludge, which helps to deliver a homogeneous feed to sludge handling

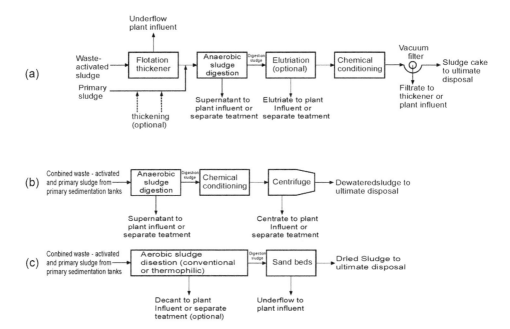

Figure 3.4 Typical sludge treatment flow chart with biological digestion and three different sludge dewatering process: (a) vacuum filteration, (b) centrifugal, and (c) drying beds.

facilities. The mixing and storage of sludge can be achieved either in a single unit designed to do both or separately in other plant facilities.

Important preliminary operations can briefly be described as below:

3.5.1   Grinding of sludge

Sludge grinding is a process in which large material contained in sludge is cut or sheared into small particles. Sludge grinders use one of two techniques: hammer mill pulverizing or cutting. The requirement of this type of unit usually depends on the specific application as mentioned below:

(a)   Sludge grinding is required before heat-treatment-process of sludge to prevent clogging of high-pressure pumps and heat exchangers.

(b)   Nozzle-disk and solid-bowl centrifuges require sludge grinding as a preceding operation/process to prevent clogging in nozzles and between disks. Nozzle-disk units may also require fine screens.

(c)   It is a prerequisite to grind sludge before chlorination, so that the chlorine contact with sludge particles can be enhanced.

(d)   Sludge grinding is required before "pumping with progressing-cavity pumps" too so that it can reduce heat and prevent clogging.

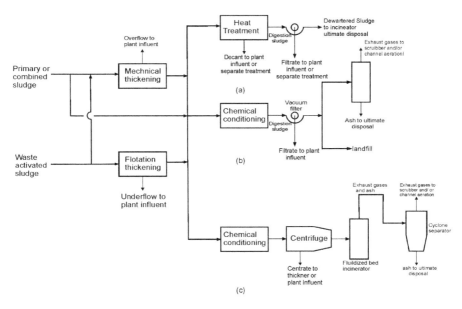

Figure 3.5 Typical nonbiological sludge treatment flow chart; (a) heat treatment with vacuum-filter dewatering; (b) multiple hearth incineration; and (c) fluidised-bed incineration.

3.5.2 De-gritting of sludge
(Bahadori, 2013)

In some plants where separate grit removal facilities are not used ahead of the primary sedimentation tanks, or where the grit removal facilities are not adequate to handle peak flows and peak grit loads, it may be necessary to remove the grit before further processing of the sludge. Where further thickening of the primary sludge is desired, it is practical to consider de-gritting the primary sludge. The most effective method of de-gritting sludge is through the application of centrifugal forces in a flowing system, to achieve separation of the grit particles from the organic sludge. Such separation is achieved through the use of hydro-clones, which have no moving parts. The sludge is applied tangentially to a cylindrical feed section, thus imparting a centrifugal force. The heavier grit particles move to the outside of the cylinder section and are discharged through a conical feed section. The organic sludge is discharged through a separate outlet.

The efficiency of the hydro-clone is affected by pressure and by the concentration of the organics in the sludge. To obtain effective grit separation, the sludge must be relatively dilute. As the sludge concentration increases, the particle size that can be removed decreases.

3.5.3 Blending of sludge
(Bahadori, 2013)

Blending of sludges is necessary to produce a uniform and homogeneous mixture of sludge. It is particularly important ahead of sludge stabilization and dewatering

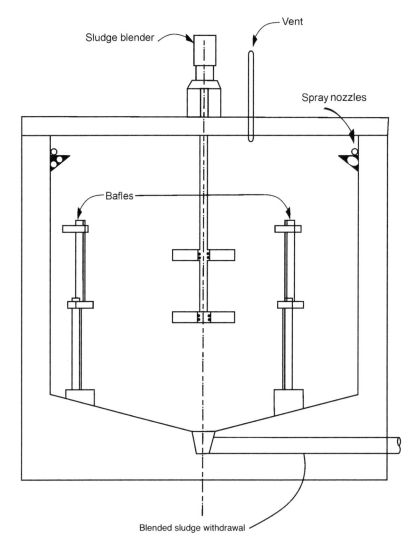

Figure 3.6 Typical sludge mixer and blender used in conjunction with sludge storage.

processes and incineration, Sludge from primary, secondary, and advanced processes can be blended in several ways, viz.:

(i)   *In primary settling tanks:* Secondary or tertiary sludges transferred to the primary settling tanks, in which settling and mixing with the primary sludge will takes palce.

(ii) *In pipes:* The careful control of sludge sources and feed rates is require to ensure the appropriate mixture.

(iii) *In Sludge-processing facilities with long retention times:* The feed sludge can be mix uniformly in Aerobic and anaerobic digesters (continuous-flow stirred-tank type).

(iv)   *In a separate blending tank:* The quality of mixed sludge can be control efficiently by this practice.

In small treatment plants ($<162\,\text{m}^3/\text{h}$), mixing is usually achieved in the primary settling tanks. At large treatment facilities, optimum efficiency is achieved by separate thickening and mixing of sludges before blending. A typical blending tank with mixing facilities is shown in Figure 3.6.

3.5.4   Storage of sludge

Sludge storage must be provided to smooth out fluctuations in the rate of sludge production and to allow sludge to accumulate during periods when subsequent sludge-processing facilities are not operating (i.e. night shifts, weekends, and periods of unscheduled equipment downtime). Sludge storage to provide a uniform feed rate is particularly important-ahead of the following processes:

• Chlorine oxidation
• Lime stabilisation
• Heat treatment
• Mechanical dewatering
• Drying, and
• Thermal reduction

Short-term sludge storage may be accomplished in wastewater settling tanks or in sludge thickening tanks. Long-term sludge storage may be accomplished in sludge-stabilisation processes with long detention times (i.e. aerobic and anaerobic digestion) or in specially designed separate tanks. In small installations, sludge is usually stored in the settling tanks and digesters. In large installations that do not use aerobic and anaerobic digestion, sludge in often stored in separate blending and storage tanks. Such tanks may be sized to retain the sludge for a period of several hours to several days. Sludge is often aerated to prevent septicity and to promote mixing. Mechanical mixing may be necessary to assure complete blending of the sludge. Chlorine and hydrogen peroxide are often used to arrest septicity and to control the doors from sludge storage and blend tanks.

QUESTIONS

1   Why sludge stabilisation is necessary before discharge into the environment or reuse?
2   What are the most common methods for reducing the sludge volume?
3   What is the difference between sludge dewatering, sludge conditioning and sludge thickening? Name the different methods used for the abovementioned purposes.

Chapter 4

Thickening of sludge

4.1 INTRODUCTION

Sludge thickening is the process used to increase the solids content of sludge by the separation and removal of a portion of the liquid phase from sludge. To illustrate, if waste activated sludge, which is typically pumped from secondary settling tanks with a content of 0.8% solids, can be thickened to a content of 4% solids, then five-fold decrease in sludge volume is achieved. Thickening is generally accomplished by physical means, including gravity settling, flotation, and centrifugation.

In view of the above, thickening is adopted for the separation of greater amount of water from sludge solids than can be attained in settling tanks. Thickening results in a saving in total unit costs compared to sludge digestion and dewatering processes without sludge thickening (Pergamon PATSEARCHER).

Three types of thickening are commonly practised, viz. (i) gravity thickening, (ii) flotation thickening, and (iii) centrifugal thickening. These are respectively based on principles of gravitation, flotation (i.e. buoyancy or negative gravitation), and centrifugation (i.e. mechanically enhanced gravitation).

4.2 OPERATIONAL PRINCIPLES

The operational principles of the various thickening processes are discussed in this section (Brechtel & Eipper, 1990; Fleming, 1986; Kondoh & Hiraoka, 1990; MWST, 1991; Turovskiy & Mathai, 2006).

4.2.1 Gravitation

Gravity thickening is based on the settling/sedimentation of suspended or floccu-lated particles (heavier than water) due to the downward force (pull) created by the gravitational field of the earth.

On the basis of the concentration and characteristics of particles, four types of gravitational settling can occur: discrete, flocculent, hindered (or zone), and com-pression. In case of sludge mass, the solids are more or less uniformly dispersed and sufficiently concentrated to enhance inter-particle forces enough to hinder the settling of neighbouring particles. During settling, as a result, the liquid tends to move up through the interstices (narrow tubular spaces) of the contacting or neighbouring particles. Consequently, the contacting particles tend to settle as a zone of "blanket", maintaining the same relative position with respect to each other. This phenomenon is known as hindered settling, which is predominant during initial phase of gravity thickening.

As settling continues, a porous structure of particles is formed at the bottom surface of the settling basin. Further settling can occur only by compression (i.e. reduction in pores volume due to squeezing out of water content) of the structure. Compression of bottom particles structure takes place by the sustaining and continuously increasing weight of top sediment particles. This "compression settling" is the prime cause for the occurrence of sludge thickening by means of gravity.

4.2.2   Flotation

Flotation is made to occur by artificially creating a field or system of upward gravitation, i.e. buoyancy, so that the particles within the system (even heavier than the liquid) can float (i.e. move upward) instead of settling down. Thus, the flotation is used to separate solid or liquid particles from a liquid phase (Kondoh & Hiraoka, 1990).

In case of flotation, the liquid-solid separation is generally brought about by introducing fine gas (usually air) bubbles into the liquid phase. The bubbles attach to the particulate matters and combined volume (of particle and gas bubbles) per unit of combined mass is significantly increased. This causes displacement of liquid in such magnitude as to generate the buoyant force enough to cause the particles to rise to the surface of liquid, from where they are skimmed off mechanically. During the process the inorganic and organic particles that are more dense than the liquid settle to the bottom of the thickener, where they are collected and mechanically scrapped into a central hopper. Thus flotation results in a unique operation of bi-directional liquid-solid separation. Flotation can be enhanced by suitable chemical additives.

The mechanism of flotation can be explained with the help of Stoke's law of sedimentation, considering the flotation as negative settling. The principal advantage of flotation over gravitational sedimentation is that, very small or light suspended particles that settle slowly can be removed more efficiently (i.e. more completely and in a shorter time) in former operation. Thus, higher loading can be permitted with flotation thickeners than gravity thickeners.

4.2.3   Centrifugation

Thickening by centrifugation involves the settling or removal of sludge particles under the influence of centrifugal forces. Centrifugation increases the settling rate or settling velocity by increasing the centrifugal force, or the artificial gravitational force. The multiple increase in the gravitational field produced by the centrifuge is directly related to the speed of rotation and the radius or diameter of the bowl (Brechtel & Eipper, 1990).

Mathemtically, $G = w^2 \cdot r / g$ (4.1)

where,
G = multiple of gravitational field (MLT)
w = speed of rotation (L/T)
r = radius of rotation (L)

If, therefore, the settling velocity of sludge solids in case of gravity thickening is given by V_g then the settling or separating velocity in case of centrifugal thickening may be represented by V_c, where

$$V_c = V_g \cdot G \qquad (4.2)$$

It may be noted that Stokes' law of gravity may be modified to incorporate the principle of centrifugation by multiplying the equation of Stokes' law [i.e. $V_g = (g/18) \cdot (P_s - P_1) \cdot (d^2/\mu \cdot u)$] with the value of G (i.e. multiple of gravitational field).

4.2.4   Gravity belt thickening

The working principle of gravity belt thickening is solid–liquid separation by coagulation-flocculation of sludge and drainage of free water from the slurry through a moving fabric-mesh belt. The thickening is dependent on sludge conditioning, usually with a cationic polymer to neutralise the negative charge of the sludge solids (Turovskiy & Mathai, 2006).

4.2.5   Rotary drum thickening

The solid–liquid separation in a rotary drum thickener is achieved by coagulation-flocculation of sludge and drainage of free water through a rotating porous media. The porous media can be a drum with wedge wires, perforations, stainless steel fabric, polyester fabric, or a combination of stainless steel and polyester fabric. The thickening is dependent on sludge conditioning, usually with a cationic polymer (Turovskiy & Mathai, 2006).

4.3   DESCRIPTIONS

Different types of thickening processes can be described as below (Coker *et al.*, 1991; MWST, 1991; Pergamon PATSEARCHER, Turovskiy & Mathai, 2006; USEPA, 2000b):

4.3.1   Gravity thickening

Gravity thickening is the most common practice for concentration of sludges. This is adopted for primary sludge, or combined primary and activated sludge, but it is not successful in dealing with activated sludge independently. It is so because most of the moisture content in activated sludge is in the form of capillary, colloidal and intercellular water which are less affected by gravity. Furthermore, gravity thickening of combined sludge is not effective when the activated sludge exceeds 40% of the total sludge weight, and other methods of thickening of activated sludge have to be considered. Gravity thickening is most effective on untreated primary sludge.

Gravity thickeners are either continuous flow or fill and draw type, with or without addition of chemicals, and with or without mechanical stirring. However, use of slowly revolving stirrers improves the efficiency. Normally, continuous flow with mechanically stirring type gravity thickeners are used.

Continuous flow gravity thickeners

Continuous flow tanks are deep circular tanks with central feed and overflow at the periphery. Dilute sludge is fed to a centre feed well. The feed sludge is allowed to settle and compact, and the thickened sludge is withdrawn from the bottom of the tank. Conventional sludge-collecting mechanisms with deep trusses or vertical pickets stir the sludge gently, thereby opening up channels for water to escape and promoting densification. The resultant continuous supernatant flow is returned to the primary settling tank. The thickened sludge that collects at the bottom of the tank is pumped to the digesters or dewatering equipment as required.

Continuous thickeners are mostly circular with diameters not exceeding 20 m and side water depths of about 3–4 m. The floor slope on these tanks generally ranges between 1:4 and 1:6, which is greater than those of conventional settling tanks. The purpose of greater slopes is to facilitate sludge collecting and to prevent the holding of sludge for too long a detention time inside the tank and thus to avoid the problem of gasification and flotation due to development of anaerobic conditions.

Concentration of the underflow solids is governed by the depth of sludge blanket up to 1 m, beyond which there is very little influence of the blanket. Concentration of underflow solids is increased with increasing sludge detention time, 24 hours being required to achieve maximum compaction. Sludge blanket depths may be varied with fluctuation in production of solids to achieve good compaction. During peak conditions, lesser detention times will have to be adopted to keep the sludge blanket depth sufficiently below the overflow weirs to prevent carry over of excessive solids.

Gravity-thickener's area and efficiency

The required cross-sectional area of a thickener can be determined performing a series of column settling tests at different concentration of solids then plotting a curve of solid flux due to gravity (SF_g) as a function of concentration of solids (c) (Figure 4.1). Now for a desired underflow concentration C_u, value of limiting solids flux (SF_L) is determined from curve (Figure 4.2) and required area (A) is calculated using the following equation (obtained from a materials balance):

$$A = (Q + Q_u)(C_i/SF_1) \tag{4.3}$$

where,
Q_u = underflow or recycle flow rate (m³/h)
$Q + Q_u$ = total volumetric flow rate to settling basin (m³/h)
C_i = influent solids concentration (kg/m³)
SF_L = limiting solids flux (kg/m³ · h)

Table 4.1 Design criteria of gravity thickeners.

Sludge Type	Influent Solids Conc. (%)	Thickening Time (h)	Thickened Solids Conc.	Dry Solids Loading kg/m² · d	Lb/ft²-d
Primary (PS)	3–6	5–8	4–8	100–200	20–40
Trickling filter (TF)	1–4	8–16	3–6	40–50	8–10
Rotating biological contractor (RBC)	1.0–3.5	8–16	2–5	35–50	7–10
WAS	0.4–1.0	5–15	2.0–3.5	25–80	5–16
PS+TF	2–6	5–10	5–9	60–100	12–20
PS+RBC	26	5–12	5–9	60–100	12–20
PS+WAS	0.6–4.0	5–15	3–7	25–200	5–40
Aerobically digested WAS	0.5–1.0	1.5–12.0	2–5	50–200	10–40
Anaerobically digested PS	4–7	20–1440	6–13	–	–
Anaerobically digested PS+WAS	2–4	20–1440	8–11	–	–

Source: USEPA, 1979.

Figures 4.1 and 4.2 are self-explanatory for the purpose of SF_L determination.

Referring to Figure 4.2 and Eqn. (4.3), if a thicker underflow concentration is required, the slope of the underflow flux line (U_b) must be reduced. This, in turn, will lower the value of the limiting flux (i.e. the limiting solids loading rate) and increase the required settling area. In that case if required settling area is not available, then either efficiency will be reduced or poorer thickening will occur. On the other hand, if the quantity of solids fed to the settling basin (i.e. solids loading rate) is greater than the limiting solids-flux value, then the solids will build up in the settling basin and, if adequate storage is not provided, ultimately overflow occurs at the top and poorer thickening will take place. Apparently, limiting solids flux (SF_L) principally governs the solids loading rate and capacity of the thickener, while underflow concentration (C_u) decides the cross-sectional area of thickener and quality of thickened sludge. Although where, SF_L and C_u are interrelated and affect each other, underflow concentration is easy to regulate, so it is used as a parameter to control the thickening process. Thus, it is necessary to ensure the provisions for:

- Regulating the quantity of dilution water needed,
- Adequate sludge pumping capacity to maintain any desired solids concentration, continuous feed and underflow pumping,
- Protection against torque overload, and
- Sludge blanket detection.

Evaluation of gravity thickening

Advantages

- Least operation skill required
- Low operating costs

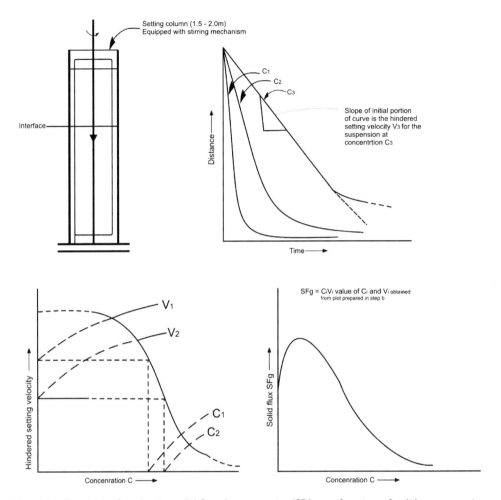

Figure 4.1 Procedure for plotting solid flux due to gravity (*SF*ₚ) as a function of solid concentration (*C*). (a) Hindered settling velocities (*V*₁) derived from column settling tests for suspension at different concentrations (*C*₁), (b) Plot of hinered settling velocities (*V*₁) obtained in step-(a) Versus corresponding concentration (*C*₁), (c) Plot of computed value of solid flux (*SF*ₚ) versus corresponding concentration (*C*).

- Minimum power consumption
- Ideal for small treatment plants
- Good for rapidly settling sludge Floating solids such as WAS and chemical
- Conditioning chemicals typically not required

Disadvantages

- Large space required
- Odor potential
- Erratic and poor solids concentration (2 to 3%) for WAS

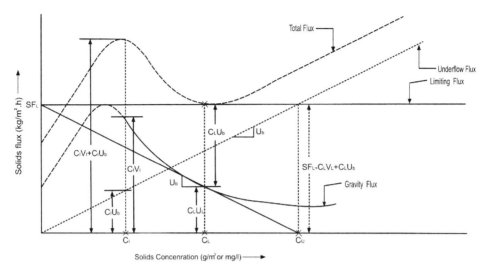

Figure 4.2 Definition sketch for the analysis of settling data using the solids-flux (SF_L) method of analysis.

4.3.2 Flotation thickening

The present practice of flotation as applied to sludge thickening is confined to the use of air as the flotation agent. Air bubbles are added or caused to form in one of the following methods to perform the flotation thickening:

- Injection of air while the liquid is under pressure, i.e. supersaturating the liquid with air, followed by release of the pressure (dissolved-air flotation).
- Aeration at atmospheric pressure (dispersed-air flotation or diffused-air-flotation).
- Saturation with air at atmospheric pressure, followed by application of vacuum to the liquid (vacuum flotation).

Further, in all these systems the degree of removal can be enhanced through the use of various chemical additives.

Flotation thickening is most efficiently used for waste sludges from suspended-growth biological treatment processes, such as the activated-sludge process or the suspended-growth nitrification process.

Dissolved-Air Flotation (DAF) units

Process mechanism: Dissolved-air flotation units are most often used to thicken the "light" sludge, such as waste activated sludge (WAS). In this process, compressed air is introduced into the pressurization tank, where it dissolves into a liquid solution-dissolved-air flotation subnatant or plant effluent – i.e. mixed with the incoming WAS (Figure 4.3).

A valve near the bottom of the central column maintains a constant back pressure. When the mixture is depressurised as it enters the DAF tank, dissolved air is released

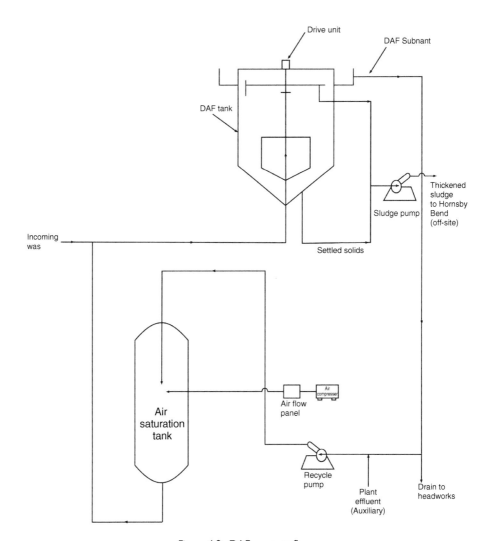

Figure 4.3 DAF process flow.

as fine bubbles that carry sludge-solids to the surface for removal. Surface skimmers move the floating sludge blanket up to a beaching plate and into a collection box. Any settled material is moved by bottom scrapers to a centre hopper. According to studies carried out by Martin and Bhattarai (1991), thickening by DAF units without polymer addition can produce solids concentration up to 3%.

Process-control: Many textbooks discussed the use of the air-to-solids ratio (ratio of kg of air to the kg of solids) as a control strategy for DAF thickeners to achieve a given degree of thickening, while little mention is made of the collector mechanism tip speed. However, practically, desired thickening range has been met while the constant air-to-solids ratio with the tip speed as the control variable (Martin & Bhattarai, 1991).

A variable-speed drive added to the collector mechanism allows the operator to slow the skimmers for a thicker sludge or increase their speed for a thinner sludge. As a general rule, a thinner sludge blanket-less than 0.3 m results in less thickening/dewatering. Also it has been found that the degree of thickening achieved depends on the initial concentration of sludge. Greater final concentrations were achieved when starting with more dilute sludges. Also, it appears that the ability to thicken waste-activated sludge will vary depending on the mean cell residence time at which the plant is being operated.

Diffused-Air Flotation (Diff.-AF) units

In air flotation systems, air bubbles are formed by introducing the gas phase directly into the liquid phase through a revolving impeller or through diffusers. Aeration alone for a short period is not particularly effective in bringing about the flotation of solids. Therefore, the efficiency of Diff.-AF units is increased by the addition of chemicals like alum and polyelectrolytes. The addition of polyelectrolytes does not increase the solids concentration but improve the solids capture from 90 to 98%.

Vacuum flotation

Vacuum flotation consists of saturating the sludges with air at atmospheric pressure, either directly in an aeration tank or by permitting air to enter on the section side of a sludge pump. A partial vacuum is applied, which causes the dissolved air to come out of solution as minute bubbles. The bubbles and the attached solid particles rise to the surface to form a scum blanket, which is removed by a skimming mechanism. Grit and other heavy solids that settle to the bottom are raked to a central sludge sump for removal. If this unit is used for grit removal and if the sludge is to be digested, the grit must be separated from the sludge in a grit classifier, before the sludge is pumped to the digesters.

The unit consists of a covered cylindrical tank, in which a partial vacuum is maintained. The tank is equipped with scum and sludge-removal mechanisms. The floating material is continuously swept to the tank periphery, automatically discharged into a scum trough, and removed from the unit to a pump also under partial vacuum. Auxiliary equipment includes an aeration tank for saturating the sludge with air, a short-period detention tank for removal of large air bubbles, vacuum pumps, and sludge and scum pumps.

Chemical additives

Chemicals are commonly used to aid the flotation process. Chemical addition has a significant effect on the performance of DAF thickener. Due to their small size, particles into the sludge may not be amenable to the flotation process, since the small size will adversely affect the proper air bubble attachment. The surface property of the particles needs to be change in order to achieve efficient flotation. The surface of sludge particles can be layered by eclectically charged cloud, which keeps the particles stable in the liquid phase. The chemical addition can neutralise the charged layer, causing the particles destabilised. Which ultimately leads the particles to coagulate, so that the air bubbles can attach to them for efficient flotation. Metal salts of aluminum and ferric, and activated silica can be used as an effective coagulant to agglomerate the particles, thus form the flocs, which can be easily entrap air bubbles. Several organic chemicals can also be used to modify the nature of air-liquid or solid-liquid interface, or both of them (Turovskiy & Mathai, 2006).

Evaluation of flotation thickening
(Turovskiy & Mathai, 2006)

The flotation thickening has the following advantages and disadvantages

Advantages

- Less space requirement than gravity thickener
- Improved solids concentration (3.5 to 5%) for waste activated sludge than gravity thickening
- Works without chemicals or with low dosages of chemicals
- No sophisticated equipments requirement

Disadvantages

- High operating costs higher than gravity thickener
- Energy intensive
- Skilled operator required
- Odor problem
- Larger are requirements than other mechanical methods
- Little storage capacity compared to a gravity thickener
- Poor performance for primary sludge
- Polymer conditioning require for higher solids concentration

Remarks

Flotation units involve additional equipment, higher operating costs, higher power requirements, more skilled maintenance and operation. However, removal of grease/ oil, solids, grit and other material as also odour control are distinct advantages.

4.3.3 Centrifugal thickening

In centrifugal thickening, centrifugal force is used to enhance the sedimentation process. Whereas, solids are settle by gravity in a gravity thickener. In a centrifuge, 500 to 3000 times of gravitational forces is applied, thus, a centrifuge perform as a highly efficient gravity thickener. Generally, the waste activated sludge is thickened by centrifuges not primary sludge, since it contains abrasive material, which can be damaging to a centrifuge (Turovskiy & Mathai, 2006).

Chiefly three types of centrifuges are used for sludge: nozzle-disc, solid bowl conveyor and imperforate (lacking the normal opening) basket. The first type is used more frequently for sludge thickening, while the other two types have special application for sludge dewatering. The *nozzle-disk centrifuge* consists of a vertically mounted unit containing a number of stacked conical disks. Each disk acts as a separate low-capacity centrifuge. The liquid flows upward between the disks towards the central shaft, becoming gradually classified. The solids are concentrated in the periphery of the bowl and are discharged through nozzles. Because of the small nozzle openings, these units must be preceded by sludge grinding and screening equipment to prevent clogging.

The extensive prescreening and de-gritting of feed sludge is require for *nozzle-disk centrifuges*. They can be used for sludges with particle sizes of 400 μm or less. *Imperforate basket centrifuges* are good for batch operation only and not continuous feed and discharge. They encounter with high bearing wear, thus need significant maintenance. Therefore, solid bowl centrifuges considered better over disk nozzles and imperforate basket centrifuges (Turovskiy & Mathai, 2006).

Solid bowl centrifuges (stated also continuous decanter scroll or helical screw conveyor centrifuges) are manufactured in two basic designs: countercurrent and concurrent. The key differences between the two are the configuration of the conveyor (scroll) toward the liquid discharge end of the machine, and the location and configuration of the solids discharge port. Sludge feed enters the bowl through a concentric tube at one end of the centrifuge. The depth of liquid in the centrifuge is determined by the discharge weir elevation relative to the bowl wall. The weir is usually changeable. As the sludge particles enter to the gravitational field, they begin to settle out on the inner surface of the rotating bowl. The lighter liquid (centrate) pools above the sludge layer and flows toward the centrate outlet ports located at the larger end of the machine. The settled sludge particles on the inner surface of the bowl are transported by the rotating conveyor (scroll) toward the opposite end (conical section) of the bowl. The main difference between a thickening and a dewatering centrifuge is in the construction of the conveyor and the conical part of the bowl. The slope of the conical part is less in a thickening centrifuge (Turovskiy & Mathai, 2006).

Figure 4.4 presents the flow diagram of the basic and auxiliary equipment of a centrifugal installation with two centrifuges. The auxiliary equipment consists of: sludge tank, sludge pump (I), screen, sludge pump (II), chemical feed pumps, chemical tank conveyor, centrifugal tank and channel for collecting the water from the centrifuged sludge or from centrifuge washing.

The performance of a centrifuge is evaluated by the thickened sludge concentration and the solids recovery or solids capture). The recovery is calculated as the thickened dry solids as a percentage of the feed dry solids. Using the commonly measured solids concentrations, the recovery (capture efficiency) is calculated using the following equation (WEF, 1998; Turovskiy & Mathai, 2006):

$$R = \frac{C_k(C_s - T_c)}{C_s(C_k - T_c)} \times 100 \tag{4.4}$$

where,
R = recovery, %
C_k = concentration of thickened (dewatered) sludge, % dry solids
C_s = concentration of feed sludge, % dry solids
T_c = concentration of centrate, % dry solids

Operational variables that affect thickening include:

- Feed flow rate
- Feed sludge characteristics, such as particle size and shape, particle density, temperature, and sludge volume index

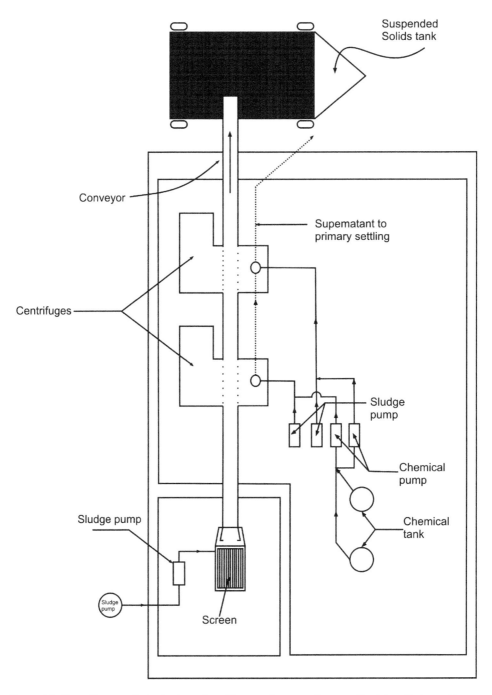

Figure 4.4 Flow diagram of the basic and auxiliary equipment of a centrifugal installation with two centrifuges.

- Rotational speed of the bowl
- Differential speed of the conveyor relative to the bowl
- Depth of the liquid pool in the bowl
- Polymer conditioning, required to enhance performance

One of most key operational factors of centrifuges is the factor of separation F, which shows how centrifugal forces are effective than sedimentation forces by the following equation:

$$F = \frac{a}{g}, \qquad a = wr, \quad \text{or} \quad F = r\frac{n}{g} \qquad\qquad (4.5)$$

where,
f = separation factor
a = speed of centrifugal force, m/s^2
g = speed of sedimentation force, m/s^2
w = angle speed of bowl (rotor), min^{-1}
r = inside radius of bowl, m
n = speed of bowl (rotor) rotation, min^{-1}

By increasing the speed of bowl (rotor) rotation, an increase in the factor of separation can be achieved. However, the high bowl rotation speed will reduce the particle sizes, which will increase the polymer dosage and reduce the efficiency of flocculation. Thus, the usual speed of centrifuges keeps between 1500 and 2500 rpm, with the factor of separation between 600 and 1600. On other hand, the thickened sludge concentration and solids recovery values remains lower at the lower values of factor of separation.

Apart from their multiple application, the main uses of solid bowl centrifuges are for thickening air or oxygen waste activated sludge. They are effective in thickening the aerobically and anaerobically digested sludge. However, the size and distribution of particles within the feed sludge affects the thickening performance considerably. The well-flocculated sludge solids cannot adhere together under the high shearing centrifugal force. Therefore, polymer addition is required to maintain the larger size and high density of flocs, which enhance the solids capturing by more than 90%.

Evaluation of centrifugal thickening

Centrifuges may offer lower overall operation and maintenance costs and can outperform conventional belt filter presses. Centrifuges require a small amount of floor space relative to their capacity. Centrifuges require minimal operator attention when operations are stable. Operators have low exposure to pathogens, aerosols, hydrogen sulfide or other odors. Centrifuges can handle higher than design loadings and the percent solids recovery can usually be maintained with the addition of a higher polymer dosage. They are particularly applicable to plants that have large volumes of activated sludge caused either by strong wastes or high hydraulic flows.

They are effective for thickening WAS to 4 to 6% solids concentration. Despite the above-discussed advantages, centrifuges have high capital cost, high power consumption and are fairly noisy. Experience operating the equipment is required to optimise performance. Special structural considerations must be taken into account. As with

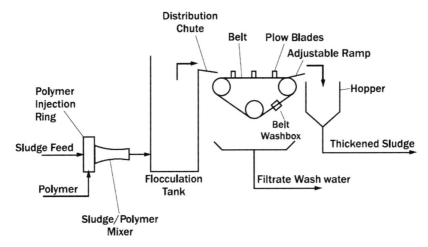

Figure 4.5 Schematic of gravity belt thickening process (Turovskiy & Mathai, 2006).

any piece of high-speed rotary equipment, the base must be stationary and level due to dynamic loading. Spare parts are expensive and internal parts are subject to abrasive wear (Turovskiy & Mathai, 2006).

4.3.4 Gravity belt thickening

The gravity belt thickener is a modified belt filter press system (Figure 4.5). The gravity belt thickeners are used for thickening the municipal (waste activated sludge, aerobically and anaerobically digested sludge) as well as industrial sludges.

In order to achieve the efficient thickening, the polymer is added into the sludge, which helps in larger floc formation. The flocculated sludge (slurry) is applied to the belt, where physical separation of solids and water takes place. The water is collected in drain vessel and transferred to a sump well. The sludge moves forward on the belt and turned over by plough blades placed on top of the belt. The sludge scraper is used to remove the thickened sludge from the belt. The belt moves to a wash-box, where the belt is washed to remove the entrapped solids. Typical hydraulic loading rates are 380 to 900 L/min (100 to 250 gpm) per meter of effective belt width. The gravity belt thickeners are available in 0.5-, 1.0-, 1.5-, 2.0-, and 3.0-m effective belt widths (Turovskiy & Mathai, 2006).

Evaluation of gravity belt thickening

Gravity belt thickening can thicken the initial sludge concentration of 0.4% with a higher than 95% solids capture efficiency. Moreover, the process includes comparatively moderate capital cost and low electricity consumption. However, it is polymer dependent process, since, 1.5 to 6 g/kg polymer addition is require on a dry weight

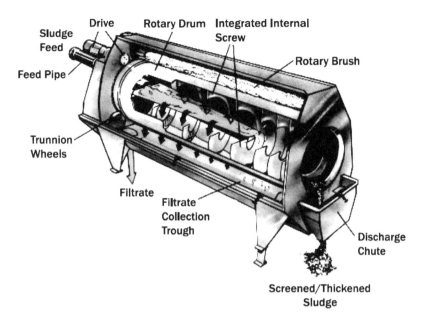

Figure 4.6 Component description of a rotary drum thickener (From Parkson Corporation, Fort
Lauderdale, FL).

basis. The odor issue and the requirement of semi-skilled staff in order to optimise
polymer feed and belt speeds (Turovskiy & Mathai, 2006).

4.3.5 Rotary drum thickening

Rotary drum thickening is achieved by an internally fed rotary drum having an integral
internal screw arrangement, which helps to move the thickened sludge out of the
drum (Figure 4.6). The drum is rotate on trunnion wheels and can be drive by a
variable-speed drive. The polymer mixed sludge fed into the drum through an inlet
pipe. The excess water is pass through the perforations of drum and collected trough a
channel. A stainless steel cover is provided for odor control and proper maintenance.
A rotary drum thickener can thicken the industrial sludges, waste activated sludge
and biologically digested sludge. They can be used mainly in small or medium scale
wastewater treatment plants with available capacity up to 1.420 m³/min (0.4 kg/m)
(Turovskiy & Mathai, 2006).

Evaluation of rotary drum thickening

Rotary drum thickening can thicken a primary sludge concentration of upto 0.5%,
with excellent solids capturing efficiency. The low energy and space requirement and
relatively low capital cost are the main advantages. The odor generation can be an
issue, which can control by proper enclosure around the unit, which can also ensure

the smooth and continuous operation of the drum under extreme weather (Turovskiy & Mathai, 2006).

4.3.6 Miscellaneous thickening

Primary clarifiers

The primary clarifiers are used to thicken the mixed sludge i.e. primary and secondary sludge. The clarifiers are designed with steeper floor slope of as high as 2.75:12, in order to reduce the depth of sludge blanket over the sludge withdrawal point. Since, the thicker sludge blanket under long retention period can leads to septic conditions and gasification due to biological activity (Turovskiy & Mathai, 2006).

Lagoons

The facultative sludge lagoons can help in additional concentrating of anaerobically digested sludge. Moreover, lagoons can help in long-term storage of sludge and continuation of anaerobic sludge stabilisation. They key advantages of lagoons are low energy requirement, no chemical conditioning require, low capital cost if land is easily available and no skilled manpower requirement. However, large area requirement, mosquito and odor problem, nitrogen rich (0.3–0.6 g/L) supernatant generation are the main drawbacks of lagoon technology.

The aerobic conditions are maintained at the surface of facultative lagoons by keeping it free from scum or biofilms. The surface mixers are arranges to provide the mixing of surface layer. Generally, the surface layer is 30–90 cm in depth and supports a dense algal population. The sources of dissolved oxygen supply to surface layer are algal photosynthesis, direct surface mixing and surface mixers. The aerobic bacteria utilised the oxygen to decompose the organic matter present in sludge; however, the solids settled down to the bottom are stabilised by the means of anaerobic decomposition. The optimum organic loading rate is 1000 kg volatile solids/hectare/day. The supernatant generated from the lagoon is returned back to the influent stage of wastewater treatment plant (Turovskiy & Mathai, 2006).

4.3.7 Emerging technologies

Recuperative thickening

In this process, digested sludges are removed from the anaerobic digestion process, thickened, and returned to the anaerobic digestion process. Recuperative thickening of 25% of the digesting solids increased solids retention time in the anaerobic digesters from 15.7 days to 24.0 days. The volatile solids reduction increased from 50% to 64%. Moreover, recuperative thickening did not affect effluent quality. This technology allows for the use of existing biosolids process equipment and does not have the additional capital costs associated with other innovative technologies that require greater capital investments (Reynolds *et al.*, 2001).

According to a full scale study at Spokane, Washington, Advanced Wastewater Treatment Plant from September 2000 to May 2001, following benefits were reported:

(1) Use of existing dissolved air flotation capacity allowed implementation with essentially no capital cost; (2) Co-thickening with waste-activated sludge showed no increase in thickening labor or power costs. Polymer use increased for thickening and decreased for dewatering.

Membrane thickening

Membrane thickening is an advanced technique mainly used for waste activated sludge. A basin with suspended biomass and a membrane system that provides a barrier for the solid-liquid separation. These membranes can be used in an aerobic environment to achieve separation of liquid from biomass. Anaerobic environments have plugged membranes too quickly in tests. Therefore, aerobic environments are needed for oxygen mixing. Thickening to over 4% solids has been reported. Flux through the membrane is reduced to half the value for membranes used in activated sludge basins. The different types of membranes are described as modular and they are of the following types: tubular, hollow-fiber, spiral wound, plate and frame, and pleated cartridge filters. Similar to membrane bioreactors (MBR) for wastewater treatment. Membranes for thickening require a smaller footprint than many established thickening technologies (Metcalf & Eddy, 2003). Membrane thickeners are in operation at several locations throughout the United States, i.e. Dundee, Michigan and Fulton County and in Georgia.

Metal screen thickening

This method provides conditioning and thickening of sludge in single basin. It employs a set of slit screens with 1-mm openings in a mixing tank. The sludge is thickened by cross-flow filtration through the screens. The system is designed to prevent clogging (which often occurs with simple screening under atmospheric pressure) with low differential pressure through the submerged screens. The preliminary findings showed a sludge treatment rate of about 200 kg solids per hour (Reynolds *et al.*, 2001).

4.4 SLUDGE-THICKENING AT SMALL PLANTS

In general, the waste sludge is transported from small treatment facilities to large scale treatment plants or lagoons for dewatering and stabilisation of sludge. In such cases, it has been found always beneficial to optimise the sludge-thickening process at each individual plant. At most plant thickening is combined with sludge storage in tanks, which are aerated with coarse bubble diffusers. Sludge thickening is achieved by stopping the aeration for some time (2–4 hours), and then drawing off the sludge liquor by a decanting device.

Increasing sludge thickenability by polymer addition has become quite popular at several small treatment plants. When the sludge storage tank is filled up and further thickening (decantation) is impossible, polymer addition to the tank gives a substantial sludge volume reduction. The polymer (type and dosage are determined by laboratory tests) is normally dosed in dry form by hand during intensive air mixing in the storage tank, followed by quiescent settling and removal of sludge supernatant. This system gives increased flexibility for the operators regarding the time of emptying the storage tank and of course reduced costs for sludge transportation and treatment. However, this system is preferable in case of activated sludge treatment plants, where aeration facilities exist with the treatment plant. In case of other type of treatment plants, gravity thickening is preferred if aeration facilities prove costly (Kohno *et al.*, 1991; Paulsrud, 1990).

4.5   BENEFITS OF SLUDGE THICKENING
(MSST, 1987; Metcalf & Eddy, 2003)

The volume reduction of sludge by sludge thickening is beneficial to subsequent treatment processes, such as digestion, dewatering and combustion, from the following standpoints:

- It permits increased loading to sludge digesters.
- It increases feed solids concentration to vacuum filters.
- Quantity of chemicals required for sludge conditioning IS reduced significantly.
- It economises on transport costs as in ocean barging in case of raw sludges.
- It saves on the amount of heat required by digesters.
- Saving on the auxiliary fuel is achieved; that may otherwise be needed when incineration of sludge is practised.
- It minimises the land requirements as well as handling costs when digested sludge has to be transported to disposal sites.

On large projects when sludge requires to be transported a significant distance, such as to a separate plant for processing, a reduction in sludge volume can result in a reduction of pipe size and pumping costs. On small projects, the requirements of a minimum practicable pipe size and minimum velocity may necessitate the pumping of significant volumes of wastewater in addition to sludge, which diminishes the value of volume reduction. Volume reduction is very desirable when liquid sludge is transported by tank trucks for direct application to land as a solid conditioner.

Sludge thickening is achieved at all wastewater treatment plants in some manner, viz. in the primary clarifiers, in sludge-digestion facilities, or in specially designed separate units. If separate units are used, the recycled flows are normally returned to the wastewater treatment facilities. In small treatment facilities (less than 162 m^3/h or 1 Mgal/d), separate sludge thickening is seldom practised. Rather, gravity thickening is accomplished in the primary settling tank or in the sludge-digestion units, or both. In larger treatment facilities, the additional costs of separate sludge thickening are often justified by the improved control over the thickening process and the higher concentrations attainable.

Separate sludge concentration can be quite beneficial to the operation of activated sludge-treatment plants, because it makes feasible the direct removal of aeration-tank mixed liquor for excess sludge wasting (the more usual practice is to waste the more concentrated return sludge). By removing a given volume of the mixed liquor each day for concentration and disposal, the sludge age or solids retention time, upon which the efficiency and operational characteristics of the activated-sludge process depend, can be closely maintained. It should be noted, however, that when this method of wasting is used, the size of the required sludge-thickening facilities will be considerably larger than those required, where sludge wasting is from the return line. Consequently, the adoption of this method is not widespread.

QUESTIONS

1   What is the purpose of sludge thickening and mention the types of sludge thickening?
2   Write three commonly used methods for sludge thickening.
3   Which sludge thickening method achieved the highest solids concentration and how much?
4   Provide an overview on the sludge thickening at small wastewater treatment plants?
5   What are the advantages of sludge thickening?
6   Write about two emerging sludge-thickening methods?
7   Which sludge shows better thickening, primary or secondary and why?
8   What are the merits and demerits of dissolved air floatation, centrifugation and gravity belt thickeners for sludge thickening?

Chapter 5

Sludge conditioning

5.1 INTRODUCTION

Sludge conditioning is a process whereby sludge solids are treated chemically or various other ways to improve the dewatering of sludge. Sludge conditioning is achieved through artificial methods, viz. by vacuum or press filters or by centrifuges. Only natural dewatering methods, viz. drying beds, sludge lagoons or land spraying do not require sludge conditioning.

Conditioning can be accomplished either by chemical methods, in which organic or inorganic flocculating chemicals are used, or physical methods using heat and freezing to change the characteristics of the sludge. Elutriation, a physical method for conditioning the digested sludge, was largely used in the past, but it is almost entirely abandoned today. Elutriation is a unit operation in which a solid-liquid mixture is intimately mixed with a liquid for the purpose of transferring certain components to the liquid. Thus, in fact, it is a physical washing operation used to reduce chemical-conditioning requirements. A typical example is the washing of digested wastewater sludge before chemical conditioning to remove certain soluble organic and inorganic components that would consume large amounts of chemicals.

Freezing and irradiation have also been investigated as sludge conditioning methods. Laboratory investigations indicate that free-zing of sludge is more effective than chemical conditioning in improving sludge filterability. Much remains to be done, however, before this method can be applied effectively. Although irradiation has been shown to be effective in improving sludge filterability, its applicability is discouraged by its high cost-benefit ratio in comparison to other available methods.

5.2 CHEMICAL CONDITIONING
(Metcalf & Eddy, 2003; Negulescu, 1985; Perry, 1973; Tenney et al., 1970; Turobskiy & Mathai, 2006)

Chemical conditioning prepares the sludge for better and more economical treatment with vacuum filters or centrifuges. Several chemicals have been used such as sulfuric acid, alum, chlorinated copperas, ferrous sulfate, and ferric chloride with or without lime, and others.

5.2.1 Introduction

The main purpose of chemical conditioning is charge neutralization of colloidal particles, which is achieved by using inorganic coagulants and organic polymers under the coagulation-flocculation process. Particle size plays an important role in sludge dewaterability, since the chemical addition enhance the size of particles and decrease the bound water. The use of chemicals to condition the sludge for dewatering is economical because of the increased yields and greater flexibility obtained. Conditioning is frequently used in advance of vacuum filtration and centrifugation. Chemicals are most easily applied and metered in the liquid form. Their dosage required for any sludge is determined in the laboratory.

Chemical conditioning can reduce sludge moisture from 90–96 per cent to 65–80 per cent, as a function of the nature of the sludge. Commonly, inorganic salts-ferric salts [e.g. $FeCl_3$, $Fe_2(SO_4)_3$] in conjunction with lime, ferrous salts or various aluminium salts [e.g. alum i.e. $Al_2(SO_3)_3 \cdot 18H_2O$] are used. Alkalinity is an important sludge characteristic that affects inorganic conditioners. Among organic chemicals polyelectrolytes (i.e. organic polymers) are to be mentioned. Polyelectrolytes are long chained high molecular weight polymerised organic coagulants.

Ferric salts are the most common coagulants used for conditioning the sludge to be dewatered by vacuum filtration (Andrews, 1975). These salts are often used in conjunction with lime to achieve the best results. Quantitatively, for example, the lime: ferric chloride ratio is typically 3:1 or 4:1 for best results. Ferrous salts are also used for sludge conditioning but not so extensively. As regards lime and hydrated limes, both the high calcium and dolomitic types can be used for sludge conditioning in conjunction with metallic salts alone. In some instances, when sludge is difficult to dewater, high dosages of lime alone may render them suitable conditioning for filtration. Salts of aluminium such as aluminium chloride ($AlCl_3$) and aluminium sulphate [$Al_2(SO_4)_3 \cdot 18H_2O$] are good flocculating agents and are widely used in Great Britain, primarily because of their cost advantage versus that of ferric salts. Regarding polymers, the cationic polymers are the most applicable to waste sludge dewatering.

5.2.2 Chemicals dosage

Before coagulants can combine with the solids fraction of the sludge, they must satisfy the coagulant demand of its liquid fraction. This is exerted by the alkalinity or bicarbonates. The alkalinity of digested sludges are quite high; in some instances 100 times those of fresh sludges. As a precipitant of bicarbonates, lime may be substituted for the portion of the coagulant that combines with liquid fraction. But it should be noted here that lime does not form floc with this fraction, but only a precipitate.

The pH range of 6.0–6.5 considered best for ferric chlorides treatment, which is further reduced to 4.5 by $FeCl_3$ addition. A higher dose of alum and ferrous sulfate is required in comparison with $FeCl_3$. The amount of chemical dosage for sludge dewatering is depend on the specific resistance of sludge. The higher the specific resistance, the higher the dosage of reagents required (Turobskiy & Mathai, 2006).

Coagulant or conditioner requirements are generally expressed as percentage ratios of the pure chemical to the weight of the solids fraction on a dry basis. Component parts are: (i) the liquid fraction requirement, approximated closely by the stoichiometry of the idealised chemical reactions, and (ii) the solids-fraction requirement, which is a matter of experience. For ferric chloride, the following equation is used to calculate the dosage.

$$P_c = [1.08 \times 10^{-4} A \cdot P/(1 - P)] + 1.6 P_v/P_f \qquad (5.1)$$

where, A is the alkalinity of the sludge moisture in mg/l of $CaCO_3$ and P_c, P, P_v and P_f are respectively the percentages of chemical ($FeCl_3$), moisture, volatile matter, and fixed solids in the sludge, all on a dry basis. The term for the solids fraction ($1.6\, P_v/P_f$) is derived from operating results for the vacuum filtration of ferric-chloride treated wastewater. Because this term is a function of the volatile-matter content of the sludge, coagulant requirements can be reduced by digesting the sludge prior to coagulating it for dewatering. By contrast, the magnitude of the term for the liquid fraction $[1.08 \times 10^{-4} A \cdot P/(1 - P)]$ is greatly magnified by the digestion. It can be reduced either by adding lime as a precipitant or by washing out a share of the alkalinity with water of low alkalinity (i.e. by elutriation).

Among the chemicals used for sludge condtioning, lime effectively enahnce the pH of the medium reduced by $FeCl_3$ and alum, improve the sludge porosity and reduce the odor by converting the sulfides into bisulfates. Moreover, lime application can enahnce the sludge stabilisation too. Despite several advatanges, chemical consitioners have some disadvantages. Some chemicals such as lime and $FeCl_3$ are corrosive in nature. Moreover, chemical conditioning enhances the volume of sludge (Turobskiy & Mathai, 2006).

5.2.3 Sludge mixing and coagulant

Intimate admixing of sludge and coagulant is essential for proper conditioning. The mixing must not break the floc after it has formed, and the detention is kept to a minimum, so that after conditioning sludge reaches the filter as soon as possible. Mixing tanks are generally of the vertical type for small plants and of the horizontal type for large plants. They are ordinarily built of welded steel and lined with rubber or other acid proof coating. A typical layout for a mixing or conditioning tank has a horizontal agitator driven by a variable speed motor to provide a shaft speed of 4–10 rpm. Overflow from the tank is adjustable to vary the detention period. Vertical cylindrical tanks with propeller mixers are also used.

5.2.4 Recent advancements

In recent years, polyelectrolytes have found increased utilization for sludge conditioning, even for pressure filtration where inorganic chemicals were traditionally preferred.

The main advantages of organic polyelectrolytes over the inorganic chemicals are easy handling, less space requirement for feed system and low dose requirement to achieve the similar degree of specific resistance reduction as of by inorganic reagents thus reduce the cost of sludge conditioning. Polymer conditioning of sludge takes place by destabiliaztion of small particles and enhancing their size by flocculation. It is commoanly practiced for sludge thickening or dewatering using with centrifuges and belt filter presses. The solution of dry polymer is prepared by using the following equipments; dry product metering, flocculent dispenser, polymer dissolving tank, storage or day tank, low-speed mixer, and solution metering pumps. The polymer dosage can be vary from 1–10 g/kg of dry solids depends on the type and dewaterability of sludge (Turobskiy & Mathai, 2006). The dosage of conditioning agent to be considered as optimal is generally assessed by laboratory tests. They involve reaching well-determined values or observing well-defined behaviour of certain characterization parameters (Chudoba *et al.*, 1973).

5.3 THERMAL CONDITIONING
(Fair *et al.*, 1968; Keey, 1972; Negulescu, 1985; Perry, 1973; Smollen, 1990)

Thermal conditioning release the water which is bound within the floc structure of the sludge and thereby improves the dewatering and thickening of the sludge.

5.3.1 Introduction
Thermal treatment of sludge means both stabilisation and conditioning process of sludge that involves heating the sludge for short periods of time under pressure. The purpose of thermal treatment is to reduce the specific resistance to filtration of the sludge. Two systems of thermal conditioning and stabilisation are known: (i) oxidative (wet air oxidation or low oxidation or zimpro), (ii) and nonoxidative (heat treatment or thermal treatment or porteous). Although apparently the oxidation process (air is added during the process) presents certain advantages, ultimately the results are no better than that of the non-oxidative process.

5.3.2 Zimpro type (low-oxidation) thermal conditioning process
Figure 5.1 illustrates the flow diagram of a low-oxidation thermal conditioning process (zimpro). The nondigested sludge separately or mixed with primary and secondary sludge is introduced to a grinder followed by a sludge storing tank and a high pressure pump (20 atm) which pushes it into the coiled pipe of a heat exchanger. The sludge inflow into the heat exchanger is connected to a high-pressure (8.4–12.6 MN/m^2) compressed air pipe. The sludge in the coil of the heat exchanger is heated by the hot sludge coming from the reactor. In the reactor the sludge is heated for 20–60 minutes, depending on the sludge characteristics, temperature and the level of hydrolysis required.

The sludge in the reactor is heated by the steam from the boiler. The temperature of the mixture in the reactor varies from 150 to 350°C. For temperatures between 150 and 250°C and detention times of the mixture in the reactor of 30–60 minutes, oxidation of organic substances is achieved up to 10–30%. Further, for equal times at temperatures higher than 250°C, the organic substances are oxidised to 90%. From the reactor the sludge passes into the body of the heat exchanger. The exit temperature of sludge from the heat exchanger is about 60°C. The sludge from the body of the heat exchanger tank passes through a separator or a thickener tank covered and provided with venting and deodorization arrangement. In the thickener tank, the sludge returns

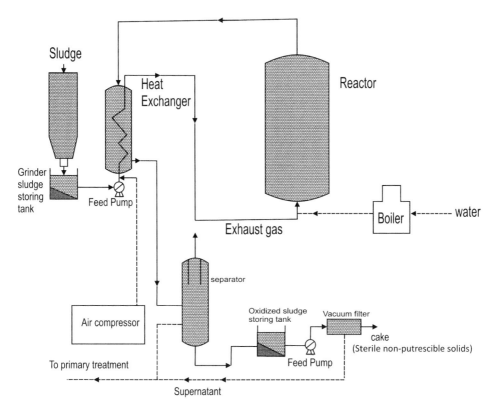

Figure 5.1 Flow diagram of a low-oxidation thermal conditioning process ("Zimpro" type).

to the atmospheric pressure and the temperature goes down to about 25°C. From the thickener tank the sludge goes into an oxidised sludge-storing tank. Gas from sludge is released and supernatant having very high BOD is sent for primary treatment. Sludge is now treated in a vacuum filter, from which being now sterile and non-putrescible it is transported and stored. The specific resistance of oxidised sludge is only 3% that of undigested sludge entering into the oxidation process. After passing through a vacuum filter, the moisture content of sludge is 60–65%.

5.3.3   Porteous type (non-oxidative) thermal conditioning process

The flow diagram of Porteous type process is illustrated in Figure 5.2. The undigested separated or mixed sludge from primary and secondary settling tanks is collected from various stages of treatment throughout the plant and passed into the raw sludge holding tank, fitted with peripheral drive stirring mechanisms; which keep the sludge at a uniform consistency. A fit pump is installed to withdraw sludge from the raw sludge tank and pass it to disintegrated sludge tank via in-line disintegrators. High-pressure raw pump pumps the disintegrated sludge directly in the concentric tube economiser heat

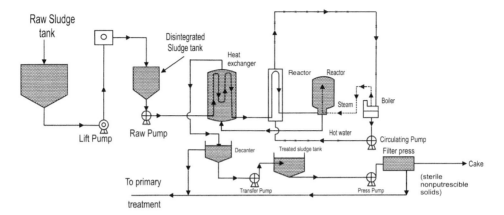

Figure 5.2 Flow diagram of a non-oxidation thermal conditioning process ("Porteous" type).

exchanger. In the heat exchanger, the sludge recovers heat from the hot treated sludge exiting from the reactor. After the heat exchanger the sludge is passed into the booster section, which is also of concentric tube design. High-pressure hot water is circulated through the annulus of the booster-heat-exchanger tubes, raising the temperature of the incoming sludge to process temperature. From the booster-heat-exchanger the sludge passes into the reactor to achieve its retention time by gradually passing to the base of the reactor. During the process plant warm-up, therefore sludge is passed to the reactor via the steam jet circulator, where steam is injected directly from high-pressure boilers to bring the plant quickly up to process temperature.

As the plant reaches process temperature, the high-pressure hot water is brought into use. This system recirculates high-pressure hot water through the booster-heat-exchanger using centrifugal pumps, and passes it through the high-pressure boiler where it receives its heat. This eliminates the need for steam during normal running to achieve a saving in overall fuel cost. Whichever form of heating is utilised, the sludge is retained in the reactor for approximately 45 minutes. After this period, it leaves the reactor and enters the heat exchanger, once again transferring its heat to the raw sludge entering the system. After the heat exchanger, the treated sludge is discharged into the decanter via the main control valve and depressurizing equipment.

In the descender, the treated sludge is settled with the aid of a peripheral drive decanting mechanism. The decantrate is drawn off via weirs at the top of the decanting tank and returned to the primary treatment. The low-pressure raw pump transfers the treated sludge having approximately 90% moisture content, into the treated sludge holding tank. The treated sludge is drawn from holding tank by a hydraulically operated flow control pump specially designed for the maintenance of constant pressure when charging filter presses. The filter press is designed to produce a low moisture content cake, this being achieved by pumping sludge under pressure into the press chamber. After a pressing cycle, which last approximately 4 h, the cake is discharged into tanks and is used as a landfill on adjacent site. Cake moisture is 50–55%, its volume representing about 8% of the volume of treated sludge.

5.3.4 Relative merits and demerits

The oxidative method (Zimpro) differs from the nonoxidative (Porteous) one by the fact that the former uses air for the sludge treatment process. The addition of air to the oxidation system produces higher levels of solubilization for a given time and temperature of reaction. This results in oxidation of a portion of the organics to CO_2 and H_2O. The level of oxidation depends on the quantity of air added, time and temperature.

A benefit of the addition of air to the process is the release of heat caused by the oxidation of carbon and hydrogen. This release of heat supports the process and reduces the auxiliary fuel requirements. Although it is possible to oxidise sufficient organics to make the process self-supporting, such a situation is not the norm for heat-conditioning. Generally, the oxidation during heat conditioning can reduce auxiliary fuel requirements by 25–45%.

Another advantage of the oxidation process is a possible reduction in tube clogging problems caused by a recombination of fibres. On the other hand, the presence of oxygen in the tubes encourages corrosion and the CO_2 end product may greatly increase the rate of chemical fouling.

5.4 FREEZE-THAW CONDITIONING
(Doe *et al.*, 1965; Logsdon & Edgerley, 1971;
Lotito *et al.*, 1990; Vesilind & Martel, 1990)

Freeze-thaw conditioning is achieved by the separation of solid and liquid fraction during ice crystal formation.

5.4.1 Introduction

Freeze conditioning of sludge is a rarely used method. It is assumed that when water transforms into ice crystals, virtually all suspended and soluble impurities are rejected, because only pure water tends to form ice crystals. In that way, sludge and other impurities are collected in frozen layers between adjacent layers of frozen water. The process is basically irreversible and during melting/thawing operation, water drains freely away from the sludge (Vesilind & Martel, 1990). The true mechanism of freeze-thaw conditioning however, has not been fully defined so far. Yet it is thought that the slow freezing of the sludge (from the outside to inside) exerts tremendous pressure that causes the chemically or biologically entrained sludge-moisture to squeeze out by forcing the solid to migrate to non-frozen areas. During the process, the pressure would continue to increase, probably accounting for the layer of clear frozen water which surrounds the solids and for the good results achieved with slow freezing. The end product of the process is a grainy material that drains readily. Nevertheless, the power costs are a major consideration in this process.

According to Doe *et al.* (1965), sludge freezing to destroy its water-binding capacity has found use in the treatment of water-works rather than the wastewater-works sludges, for which it was first developed. Recently, investigators have shown their interest to understand and control the process mechanism of sludge freeze-thaw phenomenon to condition the sludge. Few findings have been illustrated in the following sub-sections.

5.4.2   Process mechanism

The mechanism by which freeze-thaw alters the dewaterability of high solid suspensions, such as water and wastewater sludges is not yet well understood. However, Vesilind and Martel (1990) have suggested a conceptual model (as described below) for sludge freezing that helps to explain why freeze-thaw improves sludge dewaterability.

When water or wastewater sludge freezes, it is of course the water fraction of the sludge that freezes around the soil particles. When water freezes from the surface downwards, ice crystals build downward. If there are no impurities in the water, the ice front remains fairly smooth with individual water molecules being added to the ice crystals, much like bricks to a brick wall. If the water contains dissolved impurities, the ice front advances in an orderly fashion, rejecting the impurities and pushing them into a more concentrated volume, as occurs with the freezing of salt water.

The quantity of soft ice surface layer water is quite small. Tsang and Vesilind (1990) showed that surface water accounts for less than 1% of all the water in sludge. However, the freezing of this water on to the crystals has a disproportionately large effect on sludge dewaterability, since it prevents many of the smaller particles [the supracolloidal particles as defined by Karr and Keinath (1978) in the water from coming into contact and adhering to one another.

The most dramatic effect of freeze-thaw conditioning occurs for inorganic sludge such as alum water treatment sludge. The effect is irreversible, and the freeze-thawed alum sludge turns into a coffee ground-like material that drains almost without resistance, producing a perfectly clear filtrate.

Several investigators agree that freeze-thaw is most effective with smaller particles, and that some types of sludge, such as raw primary sludge, will not be nearly as effectively treated as waste alum sludge or waste-activated sludge. The freeze-thaw process does not work as well for raw primary sludge because it has a high fraction of larger particles. Cheng *et al.* (1970) showed that slow, complete freezing is absolutely essential for freeze-thaw to have a beneficial effect on dewaterability. Moreover, longer storage times, lower freezing temperatures, and a fresher sludge all improve dewaterability. Katz and Mason (1970), attempted to use rapid freezing to condition sludge and discovered that only slow freezing is effective.

5.4.3   Hypothetical model

Visual experimental evidence from the study of Vesilind and Martel (1990) indicates that if sludge with high suspended solid concentration freezes, irregular ice needles are projected into the water. The needles seek available free water molecules for growth by projecting down into the sludge, bypassing the sludge solids. As the needles thrust into the sludge, they punch aside the solids, seeking more free water molecules for continued growth. Figure 5.3 shows that as the water in concentrated slurry such as sludge begins to freeze (a), it creates a thin upper layer that sends needles into sludge (b). As ice growth continues, some sludge solids cannot be pushed in front of the ice and are trapped within the frozen mass (c and d). In time, the ice crystals dehydrate captured sludge floes (e), pushing the particles into more compact aggregates. Finally, if the temperature is low enough, surface water is also frozen compacting individual particles into tight, large solids (f).

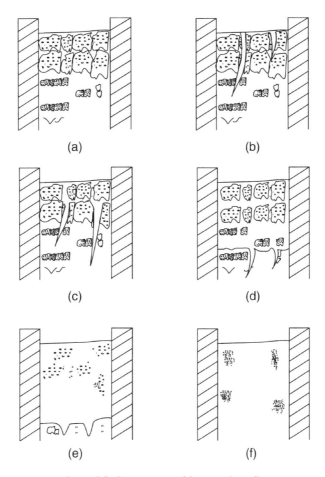

Figure 5.3 Progression of freezing (a to f).

If the water freezes too quickly, the ice crystals shoot into the sludge and trap the particles without moving him into larger, more concentrated pockets. With high freezing rates the interstitial water freezes within the floc or between the particles and is not extracted, and the individual particles are not moved closer together.

The conceptual model requires that the dewaterability of sludge can be enhanced by: (1) a slower freezing rate as opposed to rapid freezing; (2) a lower final temperature that would allow the surface water to freeze; and (3) a longer time in the frozen condition that, again, would allow the surface water to freeze.

5.4.4 Process control calculations

Freezing of sludge can be achieved either by natural freezing on open beds (where weather and seasonal conditions allow this), or by using refrigeration equipment.

The rate of natural freezing of sludge can be calculated and controlled by using the methods adopted for estimating ice thickness on lakes (Marklund, 1990).

Ice thickness can be expressed as:

$$h = k/s \tag{5.2}$$

where
h = ice thickness (cm),
k = degree-day coefficient, and
s = accumulated sum of frost days (°C day)

Further, thawing of frozen sludge can be expressed by the following equation:

$$z = k^{-1}/v \tag{5.3}$$

where,
z = actual depth of thawing (cm),
k^{-1} = degree-days coefficient, and
v = accumulated number of days (°C day) when temperature is above 0°C

The factor v is added number of non-freezing days multiplied with the mean daily temperature for each day.

In order to keep the bed free from snow, and increase the freezing velocity, the sludge should be frozen in layers of about 100 mm.

5.5 CONDITIONING PROCESS OPTIMISATION
(Campbell & Crescuolo, 1982; Campbell & Crescuolo, 1989;
Crawford, 1990; Katz & Mason, 1970)

5.5.1 Introduction

In recent years, upgrading of existing wastewater treatment plants is becoming a very important consideration. Although upgrading can take many forms, one of the most economic is to optimise the performance of the existing equipment and structures. In the realm of sludge dewatering, also an increasingly important topic, an area which has received little attention in the past is the control of the sludge conditioning process prior to dewatering (Crawford, 1990).

Organic polymers have become the primary choice of conditioning agent for sludge dewatering operations. Until now, however, there has been no simple technique of automatically controlling the addition of polymer to the sludge to allow the optimal performance of the dewatering machine. Several devices have appeared in the market in past, but none to date have been the ultimate solution to the problem (Campbell & Crescuolo, 1989). Some rely on an indirect measurement of sludge conditioning by analyzing filtrate quality, but are still subject to severe solids fouling. Another is capable of measuring the state of conditioning of the sludge on a labour intensive manual basis, but does not provide closed loop feed back control of the polymer feed pump. Yet another is applicable only to belt process, and even then, only to certain designs. In conjunction with the Wastewater Technology Centre in Burlington, Canada, Zenon

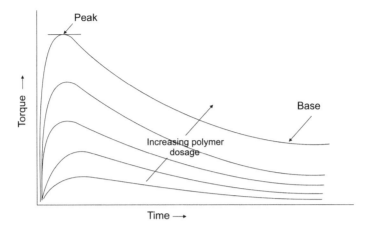

Figure 5.4 Shear response of conditioned sludge.

Water Systems Inc. has overcome all of these deficiencies by developing the new sludge conditioning controller (SCC) (Crawford, 1990).

5.5.2   Sludge conditioning controller

The Zenon SCC uses proprietary process technology, in combination with Zenon's own advanced microprocessor control technology. The SCC determines the state of sludge conditioning directly by measuring the intrinsic rheological characteristics of both the conditioned and unconditioned sludge. The rheology is the study of shear stress and shear strain behaviour of fluids. Most homogeneous materials develop shear stress in proportion to the flow rate or shear strain imposed on them. This kind of behaviour is called Newtonian (Campbell & Crescuolo, 1982). When solid particles are added to the fluid, as in wastewater sludges, the behaviour is different. Flow of fluid is inhibited by the particles until some minimum shear stress is applied. This kind of behaviour is called non-Newtonian. In addition, more the particles there are or higher the solids concentration there is, higher is the shear stress required to produce flow at the same rate.

When conditioning polymers are added to the sludge, the behaviour is different again. Fluid flow is again inhibited by the polymer chains until some minimum, but much higher, shear stress is applied. In addition, the shear stress declines with time to a lower value if the fluid flow is maintained constant. This kind of behaviour is called thixotropic.

Figure 5.4 shows schematically what the information that the SCC gathers looks like. On the graph, the torque or shear stress response of the sludge is plotted versus time. The 'peak' is defined as the highest value of torque found for a given sample, while the 'base' is the average of the value of the data near the end of the data collected. As shown, the 'peaks' get higher in magnitude as the polymer dosage increases. The lowest curve is typical of unconditioned sludges, as it has no distinct 'peak' in the data.

The SCC uses the above kinds of information, gathered from sludge samples taken in real time, to determine the state of conditioning of the sludge. It then changes the polymer flow signal to adjust the amount of polymer added to the sludge.

5.5.3 Operational tools

The hardware for an SCC consists of four main components:

1   Central control panel,
2   Local control station,
3   Sample vessel and sensor head
4   Printer

The SCC hardware is entirely automatic in operation, from taking batch samples of sludge to self-cleaning after sample analysis to adjusting the polymer pump flow rate. The only requirement for the operator is to adjust the initial polymer dosage to get the sludge dewatering operation working the way he likes it. He then presses the 'Tune' button on the Central Control Panel and the SCC takes over the control of the polymer flow rate. Although the SCC does not require any further attention, the operator can, if he wishes, adjust the 11 Auto Set point" to alter the set point that the SCC is controlling.

Figure 5.5 illustrates how the SCC fits into a typical sludge dewatering flow schematic. The sludge sampling points are as shown upstream of the polymer addition point for the unconditioned sludge and just before the belt filter press for the conditioned sludge sample.

The two samples are taken separately and independent data are collected from each. The data are transmitted to the central control panel (CCP), which makes the decision as to the change in polymer flow signal. The system therefore operates as a classic closed loop feedback control system with an additional feed forward input based on the unconditioned sludge sample.

The most vital part of any computer-controlled system is the software. For the SCC, the software controls all aspects of the operation including flushing of sample lines, filling of the sample vessel, changing of operational parameters, data collection, data analysis, data output, and the control algorithm.

5.5.4 Application of sludge conditioning controller

The application of SCC technology is very broad with respect to both the dewatering device and the material to be dewatered. In addition to all-belt press designs, the SCC can control polymer addition on belt thickeners, dissolved air flotation units, centrifuges, filter presses and vacuum filters. Although developed initially for municipal wastewater sludges, the measurement technique is equally applicable to coal refuse slurries, chemical plant organic sludges, foundry inorganic sludges and pulp and paper sludges, to name a few. In short, any dewatering or thickening operation, which uses a flocculating agent for conditioning prior to treatment, can benefit from the use of the SCC.

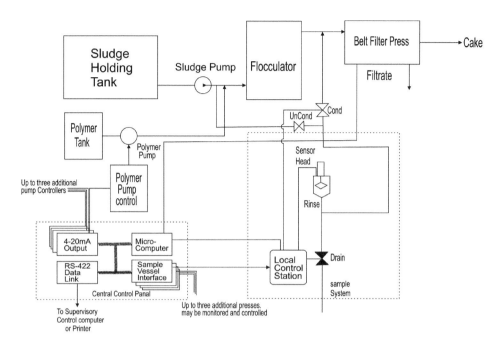

Figure 5.5 SCC block diagram.

In municipal plant operation, the range of solid concentration on which the SCC has worked successfully is from about 0.3% for waste activated sludges to about 7% for anaerobically digested sludges. In the industrial applications, the lower limit is comparable and while there is no theoretical limitation for the sensor head, the practical limitation of being able to get the sludge samples to flow into the vessel govern the upper limit. Fortunately this limitation is usually shared with the dewatering device, so that if the sludge will flow to the machine, it will flow into the SCC. As an example, in coal refuse applications, solid concentrations up to of 15–40% have been routinely sampled.

5.5.5 Process merits

There are additional benefits, which accrue with the installation of an SCC. Normally it is very difficult for the operator to adjust two variables (sludge flow rate and polymer flow rate) simultaneously. Now, with the SCC, controlling the polymer flow rate automatically, the operator can adjust the sludge flow to whatever is desirable. A future development of the SCC will be to use the solid concentration information obtained from the unconditioned sludge samples to control the sludge flow rate. By optimizing this aspect of the dewatering operation, an increase in total capacity can be obtained without resorting to the major capital investment of purchasing additional dewatering machinery. Another benefit of using the SCC is the more precise control of cake solids that it provides, because of the continuously optimised sludge conditioning. Since all forms of ultimate sludge cake disposal are cost sensitive to the cake solid concentration, this factor is very important with respect to savings and payback period.

5.6 FACTORS AFFECTING SLUDGE CONDITIONING
(Turobskiy & Mathai, 2006)

There are several factors that affect thickening or dewatering of sludge and necessitate the conditioning of sludge before further treatment.

5.6.1 Source of origin

Primary sludge, waste activated sludge, chemical sludge, and aerobically or anaerobically digested sludge. Generally, primary sludge requires low chemical dosage in comparison with biological sludge. The biological sludge generates from attached growth system require low chemical dosage than those of suspended growth system.

5.6.2 Concentration of solids

A high solids concentration will lead to effective solids conditioning in comparison with low solids concentration. High chemical dosage will require for the sludges with low solids concentration mainly due to low chemical interaction to overcome the surface charge. The higher solids concentration will leads to higher interaction of chemical and solids. Thus, a higher solids concentration will leads to improved sludge conditioning thus reducing the overdosing.

5.6.3 Particle size distribution

The main aim of sludge conditioning is to increase the particle size by addition of coagulants. The greater the number of large particles, the smaller the surface area/volume ratio. Which means lower chemical demand and lower resistance to dewatering.

5.6.4 pH and alkalinity

The coagulants decrease the pH of medium upon addition. Thus, sufficient buffering capacity is necessary of the medium in term of alkalinity. Since, the pH regulates the coagulants types, which will be existing, and the nature of charged colloidal particles. The higher alkalinity in the medium will necessitates the requirement of higher coagulant doses.

5.6.5 Surface charge and degree of hydration

Mostly, sludge solids repel rather than attract one another. This repulsion may be due to hydration or electrical effects. With hydration, a layer of water binds to the surface of the solid. This provides a buffer that prevents close approach between solids. In addition, sludge solids are negatively charged and thus tend to be mutually repulsive. Conditioning is used to overcome these effects of hydration and electrical repulsion.

5.6.6 Physical factors

Sludge storage, pumping, mixing, and treatment processes i.e. type of thickening and dewatering equipment used, are some physical factors also affects the thickening and dewatering characteristics of sludge. Long storage of sludge leads to high chemical dosage for conditioning in comparison with the fresh sludge mainly due to enhancement in degree of hydration and fine particles in sludge. Pumping process can affects the particle size due to associated shear force. The good mixing of chemicals ensure the effective coagulation and flocculation too.

QUESTIONS

1 What is the purpose of sludge conditioning?
2 Define sludge elutriation.
3 Describe the mechanism of chemical conditioning of sludge.
4 What are the controlling factors in sludge conditioning?
5 Compare the relative merits and demerits of thermal and chemical sludge conditioning.
6 Define the process mechanism of freeze-thaw conditioning.

Chapter 6

Sludge dewatering

6.1 INTRODUCTION

Sludge dewatering is a process to removes the moisture from sludge making it simpler and more economic to transport and disposal. Dewatering is mainly accomplished by drying beds or a physical process, which removes water from the sludge through presses or centrifuges. Generally, chemicals are used to improve the efficiency of processes. Sludge dewatering is necessary for one or more of the following reasons (Campbell & Crescuolo, 1982):

1   The cost of transporting sludge to the ultimate disposal site becomes substantially lower when dewatering reduces sludge volume.
2   Dewatered sludge is generally easier to handle than thickened or liquid sludge. In most cases, dewatered sludge may be shovelled, moved about with tractors filled with buckets and blades and transported by belt conveyors.
3   Dewatering is normally required prior to the heat drying and incineration of the sludge, to increase the calorific value by removal of excess moisture.
4   In some cases, removal of the excess moisture may be required to render the sludge totally odourless and non-putrescible. This is especially true for sludges stabilised by processes that create high strength recycle flows.
5   Sludge dewatering is commonly required prior to land filling to reduce leachate production at the landfill site.

Digested sludge can be dewatered either naturally (on drying beds, sludge lagoons, land spraying) or artificially (in press and vacuum filters, centrifuges etc.). Natural methods rely on natural evaporation and percolation to dewater the solids, while artificial methods use mechanically assisted physical means/devices to dewater the sludge more quickly. All dewatering devices use one or more techniques for removing moisture, viz. filtration, squeezing, capillary action, vacuum withdrawal and centrifugal settling and/or compaction.

The selection of the dewatering device is determined by the type of sludge to be dewatered and the space available. Natural methods are used especially for small amounts of sludge (i.e. at smaller plants) with enough land available and appropriate local conditions enabling the execution of constructions of drying beds/lagoons, etc. Conversely, for facilities situated on constructed sites and having appreciable amounts of sludges, mechanical/artificial dewatering devices are generally chosen (Sikora *et al.*,

1980). Natural dewatering operations do not require any pre-treatment, while the artificial methods can be applied only after a sludge conditioning (Lotito *et al.*, 1990).

Some sludges, particularly aerobically digested sludges, are not amenable to mechanical dewatering. These sludges can be dewatered on sand beds with good results. When a particular sludge must be dewatered mechanically, it is often difficult or impossible to select the optimum dewatering device without conducting bench scale or pilot plant studies.

6.2 CONTROLLING FACTORS
(Bargman & Nagano, 1958; Gale & Baskerville, 1970; Gulas *et al.*, 1979;
Hashimoto & Hiraoka, 1990; Katsiris & Katsiri, 1987; Kim, 1989;
Randall *et al.*, 1971; Tenney & Stumm, 1965; Tenney *et al.*, 1970)

Under different operating conditions, several types of dewatering equipment such as vacuum filter, belt press filter, centrifuge, filter press and screw press, have been utilised for the sludge dewatering. The performance of such dewatering equipment is critically dependent on the characteristics of the sludge and on the foregoing sludge conditioning, which is normally carried out by adding a poly-electrolyte.

Many sludge factors affecting the dewatering characteristics of sludge have been reported in the literature. These are pH, suspended solids concentration, organic content, cellulose content, particle size and its distribution, exo-cellular polymers, bound waters, etc. Further, many sludge factors determining the suitable polyelectrolyte and its dosage have been reported, viz. colloidal and low supra-colloidal particles, anionic biopolymers, biopolymers in the supernatant liquid, protein and carbohydrate particle size distribution, etc.

According to Hashimoto and Hiraoka (1990), when sewage sludge is conditioned with a cationic polyelectrolyte and dewatered by a belt-press filter, the dewatering characteristics of the sludge are affected by the following sludge factors:

1   A factor affecting the gravitational filterability of conditioned sludge is the suspended solid concentration of raw sludge.
2   A factor affecting the moisture content of dewatered sludge cake is viscosity of the sludge adjusted to 4.0% of suspended solid concentration.
3   Factors affecting the viscosity above are the intrinsic viscosity of alkaline extracts, the ratio of (VS5-Fiber A)/SS or Ash/SS or Fiber A/SS, and the charge density of sludge particles.
4   A factor affecting the extension degree of dewatered sludge cake is the charge density of sludge particles.
5   Factors affecting the amount of residual solids on the belt-filter cloth are the charge density of sludge particles and the fibrous substance content of raw sludge.

As for polyelectrolyte, the following are confirmed:

1   A highly cationised polyelectrolyte is effective to lower the moisture content, the extension degree and the amount of residual solid on the filter cloth.

2   A highly cationised polyelectrolyte is effective to increase the gravitational filterability of all mixed sewage sludges and some anaerobically digested sewage sludges, and a moderately cationised polyelectrolyte is effective to lower the gravitational filterability of the other digested sludges.

3   A factor affecting the dosage of the polyelectrolyte is anionic substances in the liquid of raw sludge.

6.3   NATURAL METHODS
(Marklund, 1990; Metcalf & Eddy, 2003; MSST, 1987; Negulescu, 1985; Smollen, 1990; Turobskiy & Mathai, 2006)

6.3.1   Sludge drying beds

Sludge drying beds are constructed on the ground and commonly characterised by the nature of the bed. If there is no danger of infecting the aquifers layer of underground water, the bed can be permeable. A waterproof bed is built only when the danger of the penetration of sludge-water (supernatant) to the aquifers layer exists.

Sludge dewatering is generally achieved by infiltration (draining) and evaporation of water from sludge. Climatic conditions and geographical locations influence the use of drying beds. Thus, in regions with low precipitations and short freezing periods drying beds are useful, since they can be used throughout the year.

Types
(Turobskiy & Mathai, 2006)

Sludge drying beds may be categorised as: (a) sand drying, (b) paved drying, (c) artificial media drying, and (d) vacuum assisted drying.

(a) Sand drying
Sand drying bed is the most common and conventional method, which is used for the population of less than 20,000. Sand drying beds are rectangular in shape, with 23–38 cm of sand (effective diameter: 0.3–0.75 mm; uniformity coefficient: <4.0) is employed over 20–46 cm of gravels (effective diameter: 3–25 mm). Sludge drying beds are covered with an enclosure in order to avoid extreme weather conditions, and to control insects and odor issues.

(b) Paved drying
Paved drying beds are rectangular in shape (6–15 m wide, 20–45 m long) and constructed by using asphalt lining or concrete covered with a 20–30 cm thick sand or gravel base. The paved drying beds can be used to accelerate the drying process, for easy removal of sludge cakes, and reducing the bed maintenance. However, the area requirement is high in comparison with conventional sand drying beds.

(c) Artificial media drying
Generally, stainless steel wedge wire or high-density polyurethane panels are used as artificial media for drying beds. The wedge wire bed is a narrow watertight rectangular

basin built-in with a false floor of wedge wire panels. Sludge dewatering can be achieved by introducing plant effluent onto the bed surface upto a depth of 2.5 cm, which works as a cushion and allow the sludge to float across the wedge wire surface. Upon filling up with sludge, the water allows the sludge to settle down and compressed against screen thus the sludge perform as filtration media.

Now, the water is permitted to infiltrate with a controlled rate. Once the water has been emptied, the sludge additionally concentrates by drainage and evaporation and thus ready for removal. The main advantages of artificial media drying beds are no clogging, continuous and fast drainage and easy maintenance. Moreover, polyurethane panels type drying beds having the benefits of low solids concentration in filtrate, dewater dilute sludge and smooth removal of sludge cakes. Artificial media drying beds are capable to dewater 2500–5000 g solids per meter square per each application.

(d) Vacuum-assisted drying

This type of beds enhances the sludge dewatering by applying the vacuum at the bottom of porous filter plates. The beds are rectangular in shape with a concrete base covered with several mm thick aggregate layer, which supports a multimedia porous filter at the top. The aggregate layer is a vacuum chamber connected to a vacuum pump. The polymer-preconditioned sludge is placed onto the bed at a rate of 9.4 L/sand to depth of 30–75 cm, and allowed to drain for approximate 1 h. After that, the vacuum system starts working with a vacuum pressure of 34–84 kPa until the cake cracks and vacuum is gone. Finally, the vacuum dried sludge is keep for air dry upto 2 days. The key advantages are short drying period, least weather independent and less area requirement.

Theoretical considerations

Dewatering on drying bed consists of initial gravity drainage and evaporation. Gravity drainage is responsible for removal of the free or gravitational water content of sludge moisture by the force of gravity. Mathematical modelling of gravitational dewatering behaviour is usually based on the specific resistance concept. By determining the basic parameters, e.g. specific resistance (R_c) at a given head loss (h_c), coefficient of compressibility (σ), initial solids content (F), initial head (H_o) and dynamic viscosity of the filtrate (μ). The total gravity drainage time in hours can be calculated by:

$$t = [1/(1+\sigma)] \cdot [(\mu \cdot R_c \cdot F)/(100 \cdot (h_c)^\sigma \cdot (\sigma + 1)] \cdot [h_c^{(\sigma+1)} - (\sigma + 1) \cdot (H_o/\sigma)h_c^\sigma$$
$$- H_o^{(\sigma+1)} + \{(\sigma + 1)/\sigma\} \cdot H_o^{(\sigma+1)}] \tag{6.1}$$

Dewatering by evaporation of liquid in a sludge layer is limited by thermal energy net input and differential vapour pressure. The thermal input raises the free kinetic energy of the chemically bonded sludge-water and the rate of evaporation. An increase in vapour in an air volume in contact with a liquid surface can continue until the rate of vaporization is balanced by condensation. Further evaporation can only be achieved by reducing the vapour level in the air.

The following empirical equation can be used to calculate evaporation from small water surfaces to air

$$\Delta m = A \cdot (\alpha / C_p)(X_v^1 - X_v) \tag{6.2}$$

where,
Δm = mass of evaporated water (kg/s)
A = water surface area (m^2)
α = heat transfer coefficient (W/m$^2 \cdot$ K)
C_p = specific heat coefficient of air (kWs/kg)
X_v^1 = vapour content at the water surface (kg H$_2$O/kg air) at constant pressure
X_v = vapour content in ambient air (kg H$_2$)/kg air)

Process mechanism and description
(Marklund, 1990; Negulescu, 1985)

Before applying the digested sludge on drying beds, sludge-solid content (i.e. initial solid content) can be as low as 0.5%, behaving as a diluted suspension. By gravity dewatering, water content is reduced by between 25 and 90%, depending on initial content. By further reducing the moisture content with evaporation the sludge acts more like a gel and finally as a solid. During this latter stage, the sludge shrinks and cracks. The evaporation can be continued to equilibrium, producing a cake as hard as dry sand or porcelain.

Reduction of sludge-moisture by evaporation involves two simultaneous processes:

(i)   Transfer of heat to evaporate liquid, and
(ii)   Transfer of mass as internal moisture into evaporated liquid.

During initial stage the rate of drying is stated to approximate that of a free water surface. With the formation of a thin solid cake, the internal resistance to moisture movement becomes significant. This makes the rate of replenishing the evaporating water lower. At specific moisture content the falling rate of internal moisture movement limits the evaporation process.

(a) Draining beds: As shown in Figure 6.1, draining beds are made of a 20 cm thick bottom layer of slag, gravel or broken stone (the grains of which range from 7 to 30 mm) and a 20 cm sand layer over the first layer (the sand grains are sized between 0.2 and 0.5 mm). For waterproof soils, the collected water is drained through drainage tubes having diameter equal to 75–150 mm, mounted at the freezing depth in stone-filled trenches. The water (supernatant) collected in drains is re-introduced in the plant ahead of the primary settling tanks.

(b) Removal of dried sludge: Withdrawal of dewatered sludge from the beds can be achieved by scrapping it off either manually or mechanically. For small platforms, the cleaning is done by shovel and removed by wheel barrow or waggon. The moisture content of sludge withdrawn from the beds ranges between 55 and 75%. Dried sludge has a coarse, cracked surface and is black or dark brown.

General considerations

In areas with an arid climate, sludge dewatering on open sand beds is commonly used. The same method has been successfully employed at industrial wastewater treatment

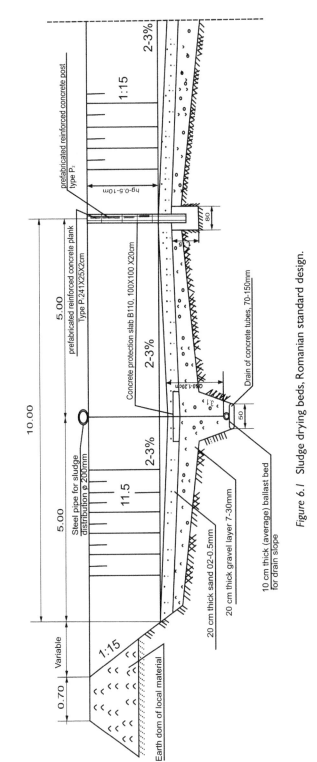

Figure 6.1 Sludge drying beds, Romanian standard design.

plant of Pali, Rajasthan (India). In cold climate zones, dewatering results tend to be unstable. In such a situation, recent developments in sludge freeze thawing technology (as described in Section 5.4) seem to enable a year round use of sand beds. In this way sand bed can be used as a sludge freeze-thaw bed during winter and as a sludge drying bed during summer.

Sludge drying beds can be used in all places where adequate land is available and dried sludge can be used for soil conditioning. In areas having greater sunshine, lower rainfall and lesser relative humidity, the drying period may be about 2 weeks, while in other areas, it could be 4 weeks or more.

6.3.2 Sludge drying lagoons

Drying lagoons may be used as a substitute for drying beds for the dewatering of digested sludge, however, the sludge is placed at depths three to four times greater than it would be in a drying bed. Lagoons are not suitable for dewatering untreated sludges, limed sludges, or sludges with a high strength supernatant, because of their odour and nuisance potential.

Drying lagoons are normally rectangular in shape, enclosed by earthen dikes 0.6 to 1.2 m high. Appurtenant equipment includes sludge feed lines, supernatant decant lines, and some type of mechanical sludge removal equipment. Unconditioned digested sludge is discharged to the lagoon in a manner suitable to accomplish an even distribution of sludge. Sludge depths usually range from 0.75 to 1.25 m. Dewatering occurs by evaporation and transpiration, of which evaporation is the most important dewatering factor. Facilities for decanting of supernatant are usually provided, and the liquid is recycled to treatment facility. Sludge is removed mechanically at moisture content of about 70%. Depending on the climate and the depth of the sludge, the time required for dewatering to a final solids content of 20 to 40% may be 3 to 12 months. Typically, sludge is pumped to the lagoon for 18 months and then the lagoon is rested for 6 months. The dewatering process depends both on soil infiltration capacity and evaporation rate. Important local factors are soil types, meteorological conditions and location in relation to residential areas (i.e. nuisance odour potential). Proper design of sludge drying lagoons requires consideration of several factors, such as precipitation, evaporation, sludge characteristics, and volume. Solids loading criterion is 35 to 38 $kg/m^3 \cdot yr$ of lagoon capacity. Per capita design criteria vary from 0.1 m^2/capita with primary digested sludge in an arid climate to 0.3 to 0.4 m^2/capita for activated sludge plants in areas where 900 mm of annual rainfall occurs.

For municipalities along the coastline, it can be quite difficult to find sites where the soil has the necessary infiltration capacity for sludge dewatering. In such cases 'artificial' lagoons may be built using proper soils transported from other sites. It is better to situate the sludge dewatering lagoons adjacent to sanitary landfills. There are two main reasons for this:

1   Most of the nuisance associated with urban waste management can be concentrated at one site in the municipality.
2   Dewatered sludge from the lagoons can be used as a top layer when terminating parts of the landfill.

The dry solids content of lagoon-dewatered sludges show great variations from place to place, but normal values are 15–30 per cent dry solids.

The major advantages include: low capital cost when land is readily available, low energy consumption, low to no chemical consumption, least operator attention and skill required. However, the main disadvantages are: large area requirement, requires stabilised sludge, design requires consideration of climatic effect, sludge cake removal is labor intensive and odor potential.

6.4   MECHANICAL METHODS

(Brown et al., 1980; Coker et al., 1991; Fair et al., 1968; Hashimoto & Hiaoka, 1990; Knocke et al., 1980; Metcalf & Eddy, 2003; MSST, 1987; Negulescu, 1985; Turobskiy & Mathai, 2006)

6.4.1   Introduction

Mechanical methods may be used to dewater raw sludge or a mixture of raw and digested sludge (in suitable ratio) preparatory to heat treatment (incineration) or before burial or landfill. Chemical conditioning is normally required prior to mechanical methods of dewatering.

Vacuum filtration is the most common mechanical method of dewatering, filter presses and centrifugation being the other methods. Digested sludge is not satisfactorily amenable to dewatering by mechanical means, because the coarser solids are rendered fine during digestion. Hence dewatering of raw primary, or a mixture of primary and secondary sludges permits slightly better yields, lower chemical requirement and lower cake moisture contents than dewatering of digested sludges. When the ratio of secondary to primary sludges increases, it becomes more and more difficult to dewater. The feed solid concentration has a great influence, the optimum being 8–10%. Beyond 10%, the sludges become too difficult to pump and lower solid concentration would demand unduly large units.

6.4.2   Vacuum filters

The function of the unit operation of vacuum filtration is to reduce the water content of sludge, whether it is untreated, digested or elutriated, so that the proportion of solid increases from the range of 5–10% to the range of 20–30%. At this higher percentage, wastewater sludge is a moist, easily handled cake. To visualise the amount of water to be removed, consider 1 Mg (1000 kg) of sludge with 5% solids (i.e. 50 kg of dry solids and 950 kg of water). After filtration to 30% solids, the 50 kg of solids would be associated with 117 kg of water [as $(50/0.30) - 50 = 117$] in 167 kg of sludge [$(50/0.30) = 167$]. Thus, 833 kg ($=950 - 117$) of water would have been extracted by the vacuum filter. This represents an 83.3% [$=(833/1000) \times 100$] reduction in the weight of sludge to be disposed off from the treatment process. Table 6.1 summarises the sludge dewatering performance of vacuum filters.

The vacuum filter consists of a cylindrical drum over which is laid a filtering medium of wool, cloth or felt, synthetic fibre or plastic or stainless steel mesh or coil

Table 6.1 Sludge Dewatering Performances of Vacuum Filters (USEPA, 1979).

Sludge type	Feed Solids (%)	Yield		Cake Solids (%)
		$kg/m^2 \cdot h$	$lb/ft^2 \cdot h$	
Raw Sludge				
Primary	4.5–9.0	20–50	4–10	25–32
WAS	2.5–4.5	5–15	1–3	12–20
Primary + Trickling Filter	4–8	15–30	3–6	20–28
Primary + WAS	3–7	12–30	2.5–6.0	18–25
Digested Sludge				
Primary	4–8	15–34	3–7	25–32
Primary + Trickling Filter	5–8	20–34	4–7	20–28
Primary + WAS	3–7	17–24	3.5–5.0	20–28

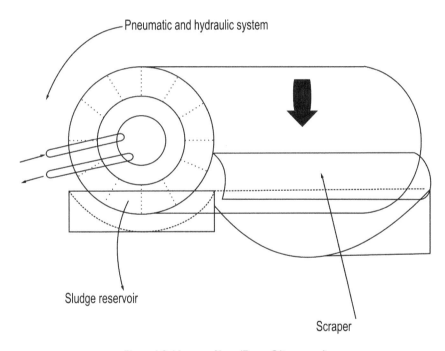

Figure 6.2 Vacuum filter (Dorr Oliver type).

springs. Vacuum filters of the rotary drum type and the spring coil types are most popular ones.

Vacuum filters of rotary drum type

These are made of a cylindrical drum (Figure 6.2) with diameters ranging between 1.50 and 2.50 m and lengths of 1–2 m and even longer. The filter medium (cloth) is placed on the drum.

Filter cloth, which can be cotton, wool, or synthetic fibre, is stretched and wired over copper mesh covering the outer surface of the revolving drum. An inner solid shell forms a compartment adjacent to the filtering surface. This space is subdivided into sections running the lengthwise direction of the filters. By arranging valves and piping suitably, each section can be placed under vacuum or pressure, as desired. About 15–40% of the filter surface is submerged in a sludge vat or sludge reservoir. The sludge in the vat is stirred by a mixer, revolving at 10–15 rpm, thus preventing the settling of sludge.

The drum revolves at about 1 rpm. When the drum passes through the sludge reservoir, a vacuum of sufficient magnitude (300–700 mm Hg) is applied to the submerged section of the drum to lift a suitable thickness of sludge on to the filter cloth. While the drum is rotating through most of the remaining arc of the circle, a drying vacuum of effective magnitude (500–700 mm Hg) draws the sludge liquor into the section and the sludge cake is formed. The cake is removed by a scraper before that part of the drum resubmerges. If necessary, slight pressure is applied to the section approaching the scraper. This lifts the cake from the cloth for easier removal. The 3–7 mm thick cake falls on to a conveyor. The sludge liquor (supernatant) separated from the solids passes through the open pores of the cloth and must be returned to the plant influent due to its high BOD value and significant amount of finely divided suspended solids.

The sludge to be filtered should be chemically conditioned previously. For fresh or digested sludges it is necessary to add ferric chloride in a dose of 2.5% of the total solids, while for excess sludge the addition is 7%. Sometimes, as much as 7–10% lime is added (Brown *et al.*, 1980). The working life of a filter medium is 1–2 months, depending on its material. Steaming, brushing and caustic solutions will clean the cloth. Binding of the cloth by finely divided sludge particles is best prevented by thorough coagulation.

The cakes resulting from the filtration of digested sludge are almost odourless. Those obtained from the filtration of undigested sludge have a foul odour, necessitating an additional treatment. If municipal wastewaters have an appreciable industrial character, they inhibit the odour and the filtration of undigested sludge can become possible and economical (Knocke *et al.*, 1980).

Spring coil type vacuum filter

Figure 6.3 illustrates a vacuum filter of the spring coil type. The coil filter uses two layers of stainless steel coil springs arranged in a corduroy fashion around the drum. These springs, which have 7–14% open area, act to support the initial solid deposits, which then serve as the filtration medium. When the two layers of springs leave the drum, the filter cake is lifted and discharged by means of positioned metal scrapers. The coils are sometimes washed before being returned to the drum.

Another type of vacuum filter is the rotary belt vacuum filter, used only rarely now. The belt is made of natural or synthetic woven cloth or metal. After leaving the drum, the belt passes over two roll system, where the cake is discharged and the cloth is washed and placed back on the drum.

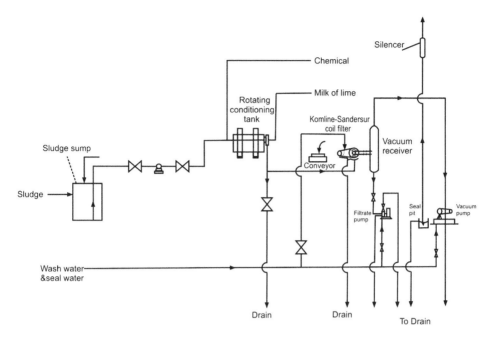

Figure 6.3 Spring-coil vacuum filter.

Appurtenances of vacuum filters

For all types of vacuum filters, besides the proper filter media, a series of appurtenances (auxiliary equipments) are necessary as illustrated in Figure 6.3.

(a) Vacuum pump: Provides the required vacuum within the aforesaid limits.

(b) Vacuum receiver: Used to separate the air from the filtrate. The receiver is designed to give retention on the air for about 3 min. It should also have sufficient volumetric capacity to allow for 4–5 min retention time on the liquid as a reservoir for the filtrate pump. The best control is achieved by providing separate receivers for pick up and dry vacuum. Lines from the filter to the receiver must not slope upward.

(c) Filtrate pumps: Along with the receiver, these pumps are generally sized by the manufacturer for the given filter and the design conditions. The filtrate pump, usually a self-priming centrifugal pump, must have a suction capacity in the same range as that provided by the vacuum pump (i.e. 300–700 mm Hg). The pumps are located nearby, generally under or directly connected to the vacuum receiver. The pump suction should always be flowing.

(d) Sludge-conditioning tank: It serves to flocculate the sludge with chemicals. Generally, the tank is constructed of corrosion-resistant materials and has a low speed agitator with a variable drive arrangement. Its design may vary depending on the chemical conditioning agent employed. Where ferric chloride and lime arc used, a detention time in the flocculator of 2–4 min is generally employed. Shorter times are used for polymers.

(e) Sludge pump: It should have a variable capacity that may be controlled locally or remotely.

(f) Wash water system: This is required to clean the belt thoroughly on each cycle (in rotary belt vacuum filter).

(g) Flow measurement device: Sludge measurement is achieved by different types of flow measurement devices.

As the materials used for the entire system for vacuum filtration are concerned, they should be corrosion proof, because these have to resist corrosive action of different chemicals. The total energy consumption of vacuum filters can be generally taken as 6.00 kWh/m^3 of sludge to be filtered.

6.4.3 Pressure filter press

In a pressure filter press, dewatering is achieved by forcing the water from the sludge under high pressure. Filtering of sludge under pressure was the first artificial mechanical procedure used to dewater the digested sludge. But due to the high consumption of coagulants in comparison with vacuum filters (usually 6–10% lime in comparison with 2.5–7% for vacuum filters) and to the numerous manual handling of filters, the utilization of filter presses was almost abandoned. However, today automation has made the utilization of filter presses attractive again. This can also be justified by the fact that the moisture content of sludge cakes is lower, on an average by 10% as compared to vacuum filters. The lower moisture content of cakes is an important advantage of this equipment.

Various types of filter presses have been used to dewater sludge. One type consists of a series of rectangular plates, recessed on both sides forming chambers that are supported face to face in a vertical position on a frame with a fixed and movable head (Figure 6.4). A filter cloth is hung or fitted over each plate. The plates are held together with sufficient force to seal them to withstand the pressure applied during the filtration process. Hydraulic rams or powered screws are used to hold the plates together.

The filtration cycle time varies from 2 to 5 h and includes the time required to fill the press, the time the press is under pressure, the time to open the press, the time required to wash and discharge the cake, and the time required to close the press. The energy consumption of filter presses is assumed to be 3 kWh/m^3 sludge introduced in filters for processing. The significant costs associated with this method are those for chemical conditioning and maintenance and replacement of filter cloths. The limited acceptance in the municipal wastewater treatment field has been due to filter media which, in past, have clogged rapidly. However, development of new types of filter media and other innovations are leading to increased acceptance of this method. Also, to reduce the amount of labour to a minimum, most modern presses are mechanised.

As for vacuum filters, the filter presses require, besides the filter press proper, a series of appurtenances (auxiliary equipments) such as, for example: feed pumps, sludge conditioning tank with chemicals, devices for cake handling and cleaning the cloth (this is cleaned periodically with an acid) etc.

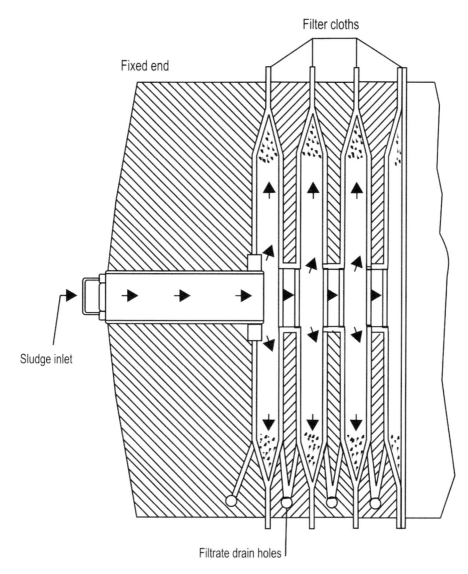

Figure 6.4 Pressure filter press.

6.4.4 Centrifugal dewatering

The centrifugation process (principle described in sections 4.2.3 and 4.3.3) is widely used in the industry for separating liquids of different density, thickening slurries, or removing solids. The process is applicable to thickening as well as dewatering of wastewater sludges. The efficiency of centrifuging is measured in terms of their effect on solid recovery and the resulting cake moisture content for a given liquid feed rate or solid rate. Basic process variables that affect the efficiency of centrifuges are the feed

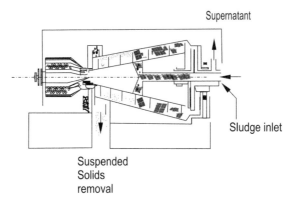

Figure 6.5 Solid bowl conveyor (centrifuge-conical).

rate, solid characteristics of particle size and density, the feed consistency, temperature and use of chemical aids.

Sludge dewatering may be accomplished by solid-bowl conveyor and basket centrifuges. Basket centrifuges may be preferred for partial dewatering at small plants. They can be used to concentrate and dewater waste activated sludge, with no chemical conditioning, at solids capture rates up to 90 per cent. To date, the use of basket centrifuges for dewatering has been limited, but combined use of basket and solid-bowl centrifuges is being investigated and developed.

Solid bowl centrifuge

(a) Equipment: The solid-bowl conveyor centrifuge usually has three types of bowl configuration: conical (see Figure 6.5), cylindrical and conical-cylindrical. The centrifuge consists of a bowl and a conveyor jointed through a planetary gear system designed to rotate the bowl and the conveyor at slightly different speeds. The helical conveyor is for pushing the sludge solids collected at the wall of the conical end. The liquid overflows at the opposite end of the bowl, which is fitted with an adjustable outlet weir or with an overflow skimmer controlling the water level within the bowl.

(b) Operation: In the solid bowl machine, sludge is fed at a constant flow rate into the rotating bowl through the hub of the conveyor, where it separates into a dense cake containing the solids deposited on the bowl wall by centrifugal force and a dilute stream called centrate (light clarified sludge liquor). The centrate contains fine, low-density solids and recycled to the untreated sludge thickener or primary clarifier. The sludge cake, which contains approximately 75–80% moisture, is discharged from the bowl by a screw feeder or conveyor into a hopper. Depending on the type of sludge, solids concentration in the cake varies from 10 to 40%. The cake then is disposed off by incineration or by hauling to a sanitary landfill.

The performance data of solid bowl centrifuges for different types of sludge are summarised in Table 6.2.

Solid-bowl centrifuges are generally suitable in the same applications as vacuum filters. Their performance is governed by the same factors that affect vacuum filters,

Table 6.2 Sludge Dewatering Performance of Solid Bowl Centrifuge (USEPA, 1979).

Type of Sludge	Feed Solids Conc. (%)	Polymer Dosage		Cake solids (%)	Solids Recovery (%)
		g/kg dry solids	lb/ton dry solids		
Primary sludge (PS)	3–7	1–3	2–6	26–36	90–97
WAS	0.5–2.5	4–8	8–16	8–20	85–94
PS + WAS	3–5	2–5	4–10	18–25	90–96
Anaerobically digested PS	4–6	2.0–7.5	2–15	25–35	92–96
Anaerobically digested PS+WAS	2–6	3–10	6–20	15–27	85–98
Aerobically digested WAS	1–3	1.5–5.0	3–10	8–12	88–91
Aerobically digested PS + WAS	1.7–4.5	3.0–5.5	6–11	11–18	92–98

viz. type and age of sludge, prior sludge processing, etc. The units can be used to dewater sludges with no prior chemical conditioning, but the solids capture and centrate quality are considerably improved when solids are conditioned with polyelectrolytes. Chemicals for conditioning are added to the sludge within the bowl of the centrifuge.

6.4.5 Filter presses and centrifuges: Relative merits

According to Coker *et al.* (1991), the belt-filter press is a simple process that operates continuously, is easy to control, and requires little operator attention. The machinery, however, includes many moving parts that require service and repair. For belt-filter presses to operate properly, belts must be washed continuously to remove residual solids to prevent binding. This continuous high-pressure wash creates a moist environment that accelerates corrosion.

Unlike the belt-filter press, the membrane press is a batch operation. Sludge is added to the unit and dewatered, then the press is opened and the cake is discharged. Because of the sticky nature of municipal sludge, an operator is essentially required to be present when the cake is discharged, to assure that all material is removed from the filter cloth. This makes membrane presses more labour intensive than the other alternatives. The batch nature of the process tends to provide inadequate conditioning or insufficient dewatering time. Control is more difficult because 2–3 hours must elapse before the operator can see the effect of a process change. The equipment, on the other hand, is simple. There are few moving parts, and those that do move operate slowly, and require minimal maintenance. The largest routine maintenance requirements are membrane and cloth replacement.

Centrifuges are a continuous process and, like belt press, the effect of a process change is immediately apparent. The centrifuge operates at a high rotational speed. If grit and similar abrasive materials are present in the sludge, scroll wear may result in high maintenance costs.

From capital standpoint, belt-filter presses are the least expensive; however, this does not present the complete picture, because it neglects sludge disposal and operations and maintenance costs. Belt-filter presses produce the wettest cake, so they incur the highest costs for landfill disposal and require the largest incinerators and the most auxiliary fuel.

6.4.6 Recent advancements

(Ensminger, 1986; Gildemeister, 1988; KHD, 1990; Lockhart, 1986; Muralidhara *et al.,* 1986; Turobskiy & Mathai, 2006)

Several new mechanical dewatering systems have been introduced in the past few years. Two of them are of more interest namely horizontal belt filters, and pressurised electro-osmotic dehydrator.

Horizontal belt filters

Four new systems (briefly described herein-after), that appears to be viable technologies, have been loosely categorised as horizontal belt filters (Metcalf & Eddy, 2003). They are the moving-screen concentrator, belt pressure filter, capillary dewatering system, and rotating-gravity concentrator. All four systems use horizontally mounted continuous belts on which the sludge is conveyed and dewatered, and all four systems appear to be designed to compete with vacuum filters. Operating complexity and energy requirements are similar. Solids capture and cake moisture content are very close to those achieved by vacuum filters.

(a) Moving screen concentrator: In the moving screen concentrator, thickened and polymer-treated sludge is distributed on two-stage variable speed moving screen (Figure 6.6). On the first screen, gravity is the major means of dewatering. When the

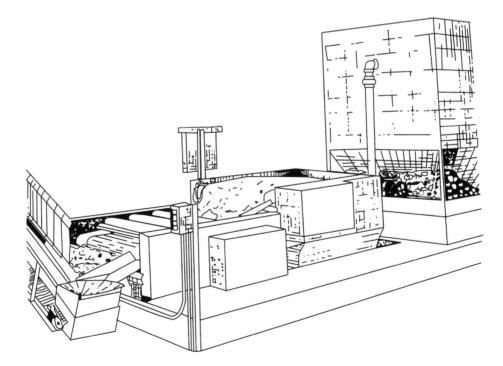

Figure 6.6 Moving screen concentration system type horizontal belt filter for dewatering sludge.

sludge is passed onto the second screen, a compression process is implemented for final dewatering. Sludge is passed under compression rollers of increasing higher pressure. At this point, relatively little free water remains in the sludge cake, and it passes out of the secondary unit and into a disposal reservoir.

(b) Belt pressure filter: As shown in Figure 6.7, the belt pressure filter consists of two continuous belts set one above the other. Conditioned sludge is fed in between the two belts. Three process zones exist. First, the sludge passes through the drainage zone, where dewatering is effected by the force of gravity. Then the sludge passes into the pressure zone, where pressure is applied to the sludge by means of rollers in contact with the top belt. Finally, the sludge is passed to the shear zone where shear forces are used to bring about the final dewatering. The dewatered sludge is then removed by a scraper.

(c) Capillary dewatering system: In the capillary dewatering system, as illustrated in Figure 6.8, chemically conditioned sludge is distributed evenly over the screen where

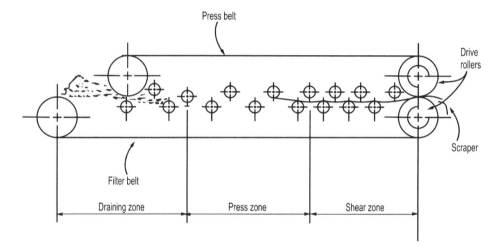

Figure 6.7 Horizontal belt pressure filter for dewatering sludge.

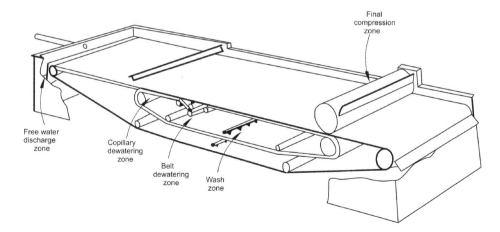

Figure 6.8 Capillary dewatering unit.

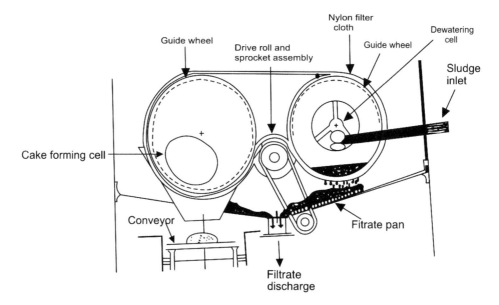

Figure 6.9 Rotating gravity concentrator.

free-water is released and the solids concentration is increased by 25%. Next, the screen carrier comes in contact with a capillary belt. Here the force for dewatering comes from the capillary action of the belt. At various stations in this zone, the filtrate is removed from the belt. Next the sludge is carried to a final compression zone, where final dewatering takes place. The sludge cake is then removed from the screen by a blade. The screen is washed and the cycle begins again.

(d) Rotating gravity concentrator: The rotating gravity concentration process consists of two independent cells formed by a fine-mesh nylon filter cloth (Figure 6.9). Dewatering occurs in the first cell, cake formation in the second. In the first cell, liquid drains from the sludge and the sludge is carried into the second cell. Here the sludge is continuously rolled into a cake of low moisture content. When the cake is large enough, excess sludge cake is discharged over the rim and on to a conveyor belt for disposal. Operation is continuous and dewatering is entirely by gravity. When more complete dewatering is required, a multi-roll press is provided. The multi-roll press consists of dual endless belts. The sludge cake from the rotating gravity concentrator is fed between the belts and graduated pressure is applied by rollers.

Pressurised electro-osmotic dehydrator (PED)

If the biologically activated sludge, which has been widely considered most difficult to dewater, were dewatered to a water content of approximately 50–60%, it could be directly composted without any additives or drying process. This has been achieved economically with the help of PED (Kondoh & Hiraoka, 1990).

(a) Process mechanism: The flow diagram of PED process is shown in Figure 6.10.

In the early stage of PED process, dewatering by pressure and electrophoresis proceed simultaneously. The particles having the negative potential are attracted to

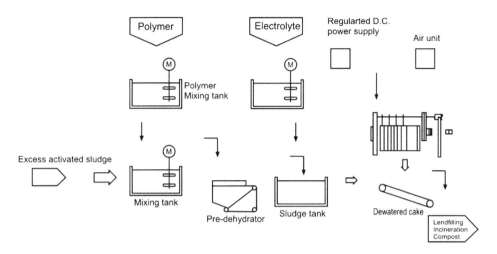

Figure 6.10 Flow diagram of PED process.

the anode, and stay apart from the negative filter surface, repelling the cathode. As a result, only few particles, having viscous matter such as bacteria and microorganisms, which usually prevent filtration, remain on the negative filter surface.

In the next stage, electro-osmosis occurs. In proportion to the negative potential of the particles, the surrounding liquid in the capillary gets the positive potential, which is known as the electric double layer in capillary tubes. Therefore, the liquid in the capillary is attracted to the cathode. Thus, the water moves smoothly through the filter cloth on the cathode since few particles, which usually cause clogging, deposit along the cathode as a result of electrophoresis. The PED process has proved its superiority and the optimum adaptability to the treatment of activated excess sludge.

(b) Remarks: Although this process is associated with high electric power consumption, its total operation cost is the least in comparison with other dehydration (dewatering) methods, viz. filter presses, belt presses, etc. due to having minimum charges of chemicals, water and landfills in case of PED.

Reed beds

The reed beds are the modified conventional drying beds with aquatic plantation onto the bottom of bed. The reed beds are rectangular in shape with top sand layer of 10–15 cm, followed by 25 cm layer of 4–6 mm gravel and the bottom layer of 25 cm of 2 cm gravel overlaid of under-drain system. A free board of 100 cm is available above the top sand layer. The reeds of *Phragmites communis* genus are planted in the middle layer on 30 cm centers. The plants help for constant water drainage from the sludge. The roots of plants are habitat of rich micro-flora, which feeds effectively on organic matter present in sludge and transform 97% of organics to water and carbon-di-oxide with a significant reduction in sludge volume. The solids loading rate designed for reeds bed is 30–60 kg per Meter Square per year. The reed beds can be operated for upto 10 years without removing the accumulated residues. The reed beds are used for the wastewater treatment plants having the capacities of less than 720 m³/h (Turobskiy & Mathai, 2006).

Quick dry™ filter bed process (Emerging Technologies for Biosolids Management, 2006)

This process includes a piping system placed on the bottom of a bed for drainage and a pre-saturating arrangement for water to entering the bed before the sludge is feed into the system. The pipes are covered with 1–1.5 cm rock under laid a top sand layer. A flocculation system (RapidFloc Mixer) including an in-line polymer preparation and injection system is also included in Quick Dry process. The rapid gravity drainage takes place in the process with additional water removal by solar evaporation process. The saturation of filter bed forces out the trapped air and permit the sludge to flow uniformly across the bed surface to accomplish the throughout distribution. Upon opening the underdrain, a vacuum effect is formed which leads the sludge dewatering as approximately 90% water can be withdrawal in 12 hours. The findings revealed that a sludge cake upto 0% dry can be produced. The sludge can be dewaters more than 50% within 7 days with a space requirement approximate 30% of conventional sludge drying beds (Evans, 2006).

Screw press (Emerging Technologies for Biosolids Management, 2006)

Mixed sludge (primary and secondary) having a solid concentration of 1–2% is transferred to a flocculation tank, where polymer is added in feed sludge and mixed in a static inline mixer. The flocculated sludge overspills into an inclined screw (~20°) rotating inside a stainless steel, wedge wire screen (200 micron). As the sludge is move forward, the filtrate flows out through the screen. The frictional pressure at the sludge/screen interface combined with improved pressure produced by the outlet restriction yields the sludge cake. The dewatered sludge (20–25% solids) drops on a conveyor or directly into a collection bin. The screw press setup is fabricated of stainless steel and is fully covered, which helps to protect the unit from corrosion, odor control and providing good working condition. Moreover, the unit operation is fully automated thus reducing the operational costs if compare with conventional methods (Atherton et al., 2005).

GeoTube® container (Emerging Technologies for Biosolids Management, 2006)

Geotube® brand geotextile tubes are comprised of high-strength polypropylene fabric. The tube is filled by polymer-conditioned sludge and allows the water to filter trough the tube wall, while fine-particle material is retain by the tube. The volume reduction within the container permits for repetitive filling and dewatering cycles. After the final cycle, retained fine particle materials continue to consolidate by drying since remaining water vapor outflows through the geotextile. The dried sludge is removed from the tube when retained solids achieve dryness aims. Geotube® container are advantageous in terms of enhanced dewatering, effective odor control, low suspended solids in effluent and economic operation.

6.5   SLUDGE DEWATERING AT SMALL PLANTS
(Crawford, 1990; Gale & Baskerville, 1970; Gulas *et al.*, 1979; Katsiris & Katsiri, 1987; Marklund, 1990; Paulsrud, 1990; Randall *et al.*, 1971; Smollen, 1990)

6.5.1   Introduction

Sludge at small wastewater treatment plants is often limited. In most cases the treatment consists of one or two sludge storage units, with a submerged aerator and a decanting device. For final dewatering and treatment, the sludge has to be transported to and processed in a central facility at a larger plant. This process has the disadvantage regarding considerable transportation cost to transport non-dewatered sludge, often over vast distances. Co-processing of sludge from small and large plants also prohibits its reuse in farming, as more polluted sludges usually cannot be separated from less polluted ones in the treatment chain.

Further, installation of dewatering equipment like centrifuges, belt presses and filter presses is very expensive for small plants. In view of all of the above reasons, any of the following systems can be used in order to comply with the sludge dewatering recommendations (Marklund, 1990):

1   If a larger wastewater treatment plant exists within a reasonable distance, then transportation of liquid sludge from a smaller to a larger plant may be preferred. At a larger plant dewatering by mechanical devices is generally opted.
2   Use sludge drying beds if sufficient land is available economically and also if climate permits (see section 6.3.1).
3   Transportation of liquid sludge-to-sludge lagoons for dewatering (see section 6.3.2).
4   Using recently developed mobile dewatering units (see section 6.5.2).

6.5.2   Mobile dewatering systems

In recent years mobile dewatering trucks have attained increasing interest for dewatering management of small sludges, particularly of septic tank sludge (septage). The mobile systems are based on the principle that septage pumped from one septic tank is dewatered while moving the truck to the next septic tank. The sludge liquor is collected on the truck and poured back into the next septic tank after having it emptied first. The sludge cake is also stored on the truck and is normally disposed off at the end of the day or the next morning. The following mobile systems that dewater the sludge have recently been developed.

Hamstern system

Figure 6.11 indicates the major components of the mobile Hamstern dewatering unit. The unit consists of a special vacuum filter, which differs from conventional design. The tanks include a holding tank for conditioned septage, a tank for filtrate, sludge cake container and a container for dry lime.

The lime conditioning system used by Hamstem, both improves sludge filterability and gives a lime stabilised sludge ready for disposal to agricultural land. The lime

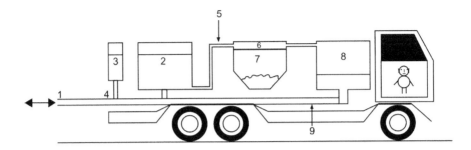

1. House
2. Holding tank, conditioned septage
3. Lime container
4. Lime pump
5. Sludge feed

6. Dewatering unit
7. Cake container
8. Flltrate collection tank
9. Flltrate feed back pipe

Figure 6.11 The different components of the Hamstern mobile dewatering unit.

dosage is normally within the range of 200–400 g/kg dry solids, depending on the sludge type. For sludges from biological and chemical treatment plants, cake solids content will be 15–20% and with filtrate suspended solids concentrations of 500–1000 mg/l. Due to high alkalinity of the filtrate there should be a buffering system before re-circulating this stream to the inlet of small wastewater treatment plants. The capacity of the Hamstem unit is 6–10 m^3 sludge per hour.

Moos-KSA system

The whole unit consists of two main parts; the truck chassis and a rectangular container. Within the container there is a storage (vacuum) tank for unconditioned sludge, a polymer tank, a dewatering tank and a filtrate tank. The rectangular dewatering tank has a drainage system in both sidewalls, covered by a filter fabric, and also a double wall construction with filter cloth along the centre of the tank. Polymer conditioned sludge is pumped into the dewatering tank, where the solids settle to the bottom and the supernatant is filtered by gravity through the filter walls. The sludge cake is removed by tilting the whole container and opening a gate in the back wall. A schematic process procedure for septage handling is shown in Figure 6.12.

With a polymer consumption of 3–5 g polymer/kg dry solids, biological and/or chemical sludges can be dewatered to about 15% dry solids. The filtrate will be low in suspended solids (100–300 mg/l), and can be recirculate to the plant inlet without disturbing plant performance. The capacity has shown to be about 30 m^3/pd for wastewater treatment plant sludges.

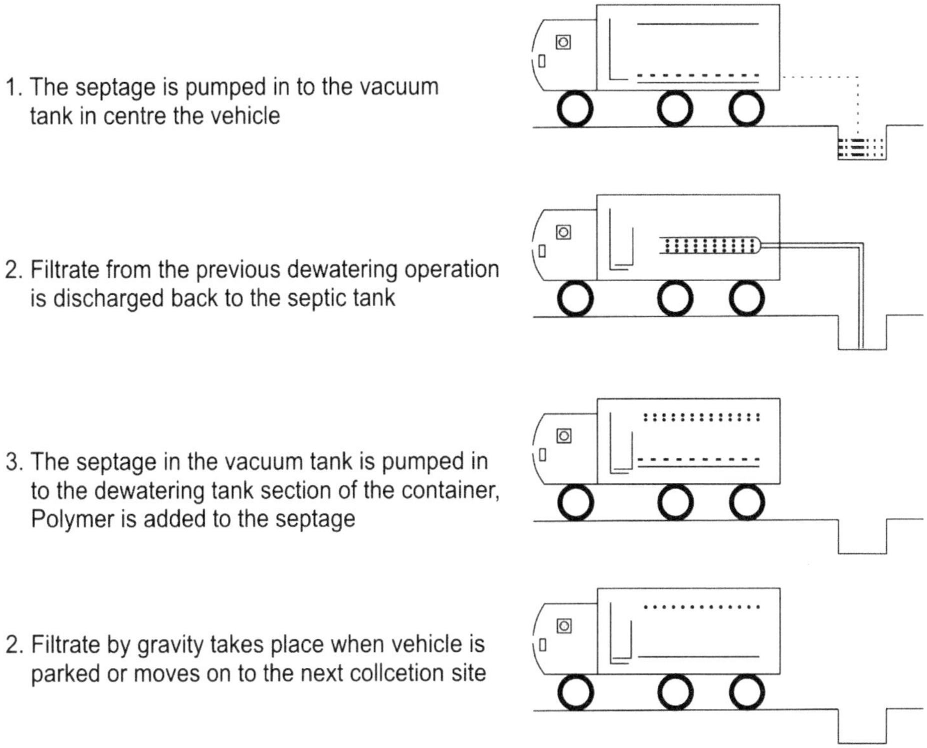

1. The septage is pumped in to the vacuum tank in centre the vehicle

2. Filtrate from the previous dewatering operation is discharged back to the septic tank

3. The septage in the vacuum tank is pumped in to the dewatering tank section of the container, Polymer is added to the septage

2. Filtrate by gravity takes place when vehicle is parked or moves on to the next collcetion site

Figure 6.12 Schematic procedure for septage dewatering with the mobile Moos-KSA unit.

QUESTIONS

1 What is the significance of sludge dewatering?
2 What the governing factors in sludge dewatering?
3 What are the natural dewatering methods? Write the merits and demerits of natural sludge dewatering methods.
4 Compare the natural and mechanical dewatering methods.
5 What are the emerging sludge dewatering methods?
6 How sludge dewatering practiced at small wastewater treatment plants?

Chapter 7

Biological stabilisation of sludge

7.1 INTRODUCTION

Sludge solids contain a large fraction of volatile solids or rich in organic content, which makes it easily decomposable, putrescible or unstable. Hence, sludge stabilisation is necessary, otherwise it will be full of nuisance value or a potential case for putrefaction. Thus, sludges are stabilised to:

- Reduce pathogens,
- Eliminate offensive odours, and
- Inhibit, reduce, or eliminate its potential for putrefaction.

The success in achieving these objectives is related to the effects of the stabilisation operation or process on the volatile or organic fraction of the sludge, to convert them into minerals. Survival of pathogens, release of odour, and putrefaction occur when microorganisms are allowed to flourish in the organic fraction of the sludge to digest and hence to stabilise the sludge biologically. There are two means to eliminate these nuisance conditions through stabilisation. They are:

(i)   Biological means, viz. aerobic-digestion, anaerobic-digestion, anoxic-digestion, composting, etc.; known as sludge-digestion.
(ii) Non-biological methods, e.g. chlorine oxidation, lime stabilisation, chemical-fixation, heat treatment, etc.; known as physico-chemical stabilisation of sludge.

7.2 SLUDGE DIGESTION

7.2.1 Objective of sludge digestion

The principal objective of sludge digestion is to subject the organic matter present in the settled/thickened sludge of the primary and final sedimentation tanks to anaerobic or aerobic decomposition, so as to stabilise the sludge as well as make it innocuous and amenable to dewatering on sand beds or mechanical filters before final disposal on land, lagoon or the sea. Regardless of the process used, sludge digestion brings about a reduction in its volume.

It has been observed that digested sludge has better dewaterability than that of raw sludge. Thus, in addition to the decomposition and stabilisation of biodegradable organic content of the raw sludge, sludge-digestion improves the dewatering characteristics of sludge. The mechanism through which dewater-ability of sludge is improved by sludge digestion, has been studied by Baier and Zwiefelhofer (1991), which is explained as below:

7.2.2   Sludge digestion: An aid for dewaterability

Baier and Zwiefelhofer (1991) explain the phenomenon by which sludge digestion facilitates the dewatering process to render an extra advantage, in addition to reduce the putrescibility of sludge. By definition, the aim of mechanical dewatering of sewage sludge is the separation of water from solid material. In digested sludge, water – the liquid phase is present in different forms: e.g. free water and intracellular water. Free-water, the major part of the liquid phase, can be easily separated from the solids by taking advantage of the difference in specific weight, or by compressing the solid phase and thus forcing free water to escape.

Main hindrances to the escape of intracellular water are non-uniform particle size distribution and non-optimal operation of sludge-thickeners. Sludge with a large amount of large and very small particles will not release free-water very efficiently, because inter-particle volume is high and water-release channels are clogged easily.

Conceptual background

Any process that makes particle size distribution uniform, either by eliminating small or breaking up big particles, will enhance free-water release. Anaerobic digestion accomplishes much of this by hydrolyzing smaller particles. In breaking up bigger particle material, anaerobic hydrolysis is slowed, and longer digester residence time is needed. Aerobic thermophilic pre-treatment deals very efficiently with this problem; at elevated temperatures, the hydrolysis rate of aerobic organisms is an order of magnitude higher when compared to anaerobic mesophiles. By combining the fast aerobic thermophilic hydrolysis with anaerobic stabilisation, sludge with a uniform particle size distribution and an enhanced free-water release is obtained. With the chosen pre-treatment reactor, including external recirculation and air injection, the mechanical treatment of sludge by exposing it repeatedly to higher shear forces effectively supports big-particle breakdown. The combined effect of high mechanical shear forces and elevated temperature results in death and lysis of microbial cells present in raw sludge. Cells exposed to high temperatures show a weakening of cell wall structure and an increase in membrane permeability. As a result, intracellular water is set free. The exposure of weakened cells to mechanical stress, as occurs in the low-pressure gas injector, supports these effects.

7.3   ANAEROBIC DIGESTION PROCESS

7.3.1   Introduction

Anaerobic digestion is a conventional method used for stabilisation of sludge. It comprises the decomposition of organic and inorganic matter in the absence of molecular oxygen. The major applications have been, and remain today, in the stabilisation of concentrated sludges (particularly mixture of primary and secondary sludges).

In the anaerobic digestion process, the organic content of sludges, is biologically hydrolyzed, liquefied and gasified to methane (CH_4) and carbon dioxide (CO_2) under anaerobic condition. However, not all of the organic matter is quickly decomposed. Lignin and other cellulosic substances are examples of resistant components. They remain substantially unaltered even during prolonged digestion. Therefore, an equational relationship of sludge digestion will read as follows (Fair *et al.*, 1968):

Organic matter + Bacteria (new cells) + Residual, resistant, organic matter

$$+ CH_4 + CO_2 + H_2O \tag{7.1}$$

7.3.2   Process mechanism
(Andrew, 1975; Chung & Neethling, 1990; Duarte & Anderson, 1982; Fair *et al.*, 1968; Metcalf & Eddy, 2003; Mitsdorffer *et al.*, 1990; Negulescu, 1985)

The driving force in sludge digestion is the consumption of foodstuffs in the sludge by living organisms. The biological conversion of organic matter of sludges in anaerobic digestion process is thought to occur in either two or three steps. In the three-step sequence, the first step involves the enzyme-mediated transformation (liquefaction) of higher-weight molecular compounds into compounds suitable for use as a source of energy and cell carbon. The second step involves the bacterial conversion of all compounds resulting from the first step into identifiable lower-molecular weight intermediate compounds. The third step (gasification) involves the bacterial conversion of the intermediate compounds into simpler end products, principally methane and carbon dioxide.

As shown in Figure 7.1, in two-step sequence, the microorganisms responsible for the decomposition of the organic matter are commonly divided into two groups:

1   Acid-producing bacteria (acid formers), and
2   Methane-producing bacteria (methane formers).

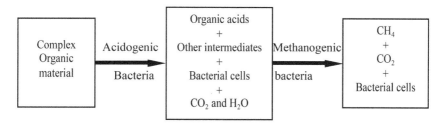

Figure 7.1 Mechanism of anaerobic sludge digestion with two-step sequence.

Thus, for simple understanding, the whole digestion process can be seen into two phases: liquefaction and gasification, as enumerated below:

Liquefaction

During this phase of digestion the decomposable solids, required mainly for bacterial food, include sugar, starch, cellulose and soluble compounds of nitrogen (i.e. nitrites and nitrates). Anaerobic bacteria use oxygen from the organic matters and the soluble compounds of nitrogen. The decomposition products are volatile organic acids (acetic, propionic and butyric acid), carbonic acid and gases, especially carbon dioxide (CO_2), Hydrogen (H_2), as well as hydrogen sulphide (H_2S) and limited amounts of methane (CH_4).

The following equations may represent the reactions of liquefaction:

$$2C_2H_5OH + CO_2 = 2CH_3COOH + CH_4 \tag{7.2}$$

$$CO + H_2O = CO_2 + H_2 \tag{7.3}$$

$$S^{2-} + 2H^+ = H_2S \tag{7.4}$$

The digestion at a pH = 5–6 lasts for about two weeks at a temperature of 15°C; a longer period (about 3 months) follows, during which a slightly decreased acidity is recorded. Formation of CO_2 and H_2S decreases during this period, and the odour from sludge decomposition becomes extremely strong. At the end of the acid period, called the "mature period" and lasting for about 6 months (at 15°C), the pH value rises to 6.8–7.0. Maintaining its grey colour, the sludge becomes sticky and foamy giving rise to gases, which are subsequently removed. And a scum is formed on the surface.

Gasification

The methane digestion that follows has an alkaline character. The most resistant substances including organic acids and proteins are converted into gaseous products during this digestion. Nitrogen is converted into ammonia and, since the fatty acids (acetic, propionic and butyric acids) are also decomposed, now the process becomes completely alkaline. The fatty acids, formed previously during the acid digestion, are decomposed into carbon dioxide and methane. The freed hydrogen acts upon CO_2 forming CH_4. The pH value becomes constant at 7, even if small amounts of acids or alkalis occur in the digestion area. Enzymes (digestion agents) and a large number of bacteria crowd up in the sludge and cooperate to the completion of the converting process. After about one month, the sludge is stabilised, its colour becomes dark grey and it has a tarry odour. Only tomato pips and hair remain un-decomposed (Hills & Dykstra, 1980). The gasification stage of digestion ceases at temperatures of 10–12 DC. When this stage is over, sludge can be removed out of the digestion space and dewatered. If dewatering is achieved on drying beds, the aerobic bacteria intervene again to mineralise the last organic matters.

Although there are numerous specific adaptations with regard to the specific mechanisms involved in the formation of methane, methanogenic bacteria appear to be capable of using the following three categories of substrates:

1 The normal and isoalcohols containing from one to five carbon atoms (methanol, ethanol, propanol, butanol, pentanol), e.g.

$$2C_2H_5OH \rightarrow 3CH_4 + CO_2 \tag{7.5}$$

2 The lower fatty acids containing six or fewer carbon atoms (formic, acetic, propionic, butyric, valeric, caproic), e.g.

$$CH_3COOH \rightarrow CH_4 + CO_2 \tag{7.6}$$

$$4C_2H_5COOH + 2H_2O \rightarrow 4CH_3COOH + CO_2 + 3CH_4 \tag{7.7}$$

3 Three inorganic gases (hydrogen, carbon monoxide, and carbon dioxide), as given below:

$$4H_2 + CO_2 \rightarrow CH_4 + 2H_2O \tag{7.8}$$

$$CO + 2H_2 \rightarrow CH_4 + H_2O \tag{7.9}$$

$$CO_2 + 8H^+ \rightarrow CH_4 + 2H_2O \tag{7.10}$$

In addition, many other groups of anaerobic and facultative bacteria use the various inorganic ions present in the sludge. Desulphovibrio is responsible for the reduction of the sulphate ion SO_4^{2-} to the sulphide ion S^{2-}, while other bacteria reduce nitrates NO_3^- to nitrogen gas N_2 (i.e. denitrification).

7.3.3 Process microbiology

The anaerobic sludge digestion ecosystem is complex, consisting of many different bacteria growing in a close synergistic environment. The microbiology and biochemistry of the process have been studied for several years and revealed a complex picture with the following major components.

1 Hydrolysis of complex, high molecular weight solid particles is carried out by hydrogenotrophic bacteria, to make complex organic compounds accessible to bacterial degradation.
2 Acid-producing organisms convert complex organic compounds to organic acids, primarily acetic, butyric, and propinic acid. These organisms use a variety of simple organic compounds as substrate.
3 Acetogenic bacteria convert larger organic acids into acetic acid, which is subsequently converted to methane and carbon dioxide by acetoclastic bacteria.
4 A second group of methane producing organisms use hydrogen (produced as a by-product in several of the afore-mentioned reactions) and carbon dioxide to produce methane.

Since all these reactions proceed simultaneously, the activity and number of individual species cannot easily be measured. For example, methane production rate measurements will reflect the activity of all methanogens, acetoclastic plus hydrogenotrophic bacteria.

7.3.4   Process kinetics
(Hosh & Pohland, 1974; Negulescu, 1985)

The curve presented in Figure 7.2 is an important pictorial representation of the kinetics of the anaerobic digestion process. The curve expresses gas production as a function of time, for unaugmented sludge sample at 20°C. It is noted that the gas production rate is lower at the beginning, gradually increasing and decreasing again when gas production comes closer to the limit value. The behaviour of total organic solids during anaerobic digestion depends on the accumulation of enzymatic products. If the feeding of fresh stage in the fermentation space is continuous (daily feeding), the enzymatic products of reaction accumulate and promote the process. The dotted line in Figure 7.2 defines, on the abscissa, a lag period which in continuous digestion disappears, and the rate of digestion can be stepped up appreciably. Seeding and mixing can lead to an increase in the digestion rate.

The anaerobic digestion of wastewater sludges can be measured in terms of the reduction in their volatile matter content, or their generation of gas. Formulation of the gasification has been expressed by various equations, resulting from the processing of experimental data and general relationships determined by Streeter and Phelps as discussed by Paulsrud and Eikum (1975).

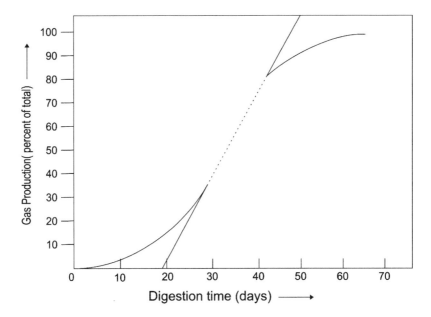

Figure 7.2 Gas production versus sludge digestion time.

Two autocatalytic equations are used, namely,

$$dy/dt = K_1(L - y) + K_2(L - y)^2 \qquad (7.11)$$

and

$$dy/dt = K_1(L - y) + K_2 \cdot y(L - Y) \qquad (7.12)$$

The logistic curve is characterised by the following relationship

$$(100 \cdot y)/L = 100/[1 + m \cdot \exp(nt)] \qquad (7.13)$$

where,
y = amount of gas produced in time t_1
L = saturation value to which gasification is asymptotic, and
K_1, K_2, m, n = coefficients.

The coefficients of logistic-curve equation are readily determined from the coordinates of three equally spaced points on a straight-line plot on logistic paper, as:

$$L = [2 \cdot y_o \cdot y_1 - y_1^2, (y_o + Y_2)]/(y_o y_2 - Y_1^2) \qquad (7.14)$$

$$m = (1 - y_o)/y_o^2 \qquad (7.15)$$

$$n = (1/t_1)L_n[y_o(L - y_1)]/y_1(L - y_o)] \qquad (7.16)$$

For $L = 100$ and $y = 20$, 50, and 80 in Figure 7.2 at $t_o = 15$, 20 and 25 days, respectively, and conveniently replaced, in turn, by time units 0, 1 and 2:

$$y = 100/[1 + 4\exp(-1.386t)] \qquad (7.17)$$

7.3.5   Process controlling factors
(Fair et al., 1968, Negulescu, 1985; Tezel et al., 2011; Turobskiy & Mathai, 2006)

Temperature

The temperature affects the physicochemical properties of sludge, biochemical reaction rates, and gas transfer rates in a digester. Overall, it has a profound effect on the digester performance. Reaction velocity constant increases with temperature in accordance with the Van't Hoff-Arrhenius relationship. The influence of temperature can be noted by the inspection of the digestion results, the time required for gasification to reach a useful degree of completion, such as 90%. The relationship connecting time and temperature is:

$$t/t_o = \exp[C_t \cdot (T - T_o)] = Q_t^{(T - T_o)} \qquad (7.18)$$

where,
t = time required to reach technically complete digestion of sludge at a temperature of T degrees

t_o = time required at a reference temperature of T_o degrees,
C_t, Q_t = slopes read from straight-line plots of Eqn. (7.18)

Fair and Moore (1937) observed that digestion exhibit two significant response ranges to temperatures:

1   A range of responses to moderate temperatures, in which the common moderate-temperature-loving (mesophilic) saprophytes and methane formers are active, and
2   Range of responses to high temperatures, in which heat-loving (thermophilic) organisms are responsible for digestion. The intermediate upswing and drop-off of the mesophilic range may appear due to the thermal death point of normal saprophytes.

At lower temperatures between 25 and 15°C, the rate of hydrolysis and acido-genesis as well as methanogenesis drops. However, below 15°C, biogas production decreases not only because of the reduced methanogenesis rate but also as a result of enhanced gas dissolution at the lower temperatures. At higher temperatures, sludge constituents become more soluble and the sludge surface area increases. As a result, the hydrolysis rate increases resulting in high organic acid production. In treatment plants, sludge is always digested at moderate temperatures (30–33°C); for this purpose, the heating of sludge is permanently required.

It is important that a stable operating temperature should be maintained in the digester. Sharp and frequent fluctuations in temperature affect the bacterial activity, especially methanogens. Process failure can occur at temperature changes greater than 1°C/d. Changes in digester temperature greater than 0.6°C/d should be avoided.

Solids retention time

Generally, SRTs of 15–20 days for mesophilic and 8–12 days thermophilic digestion, are applied. SRTs lower than 8 days are insufficient for a stable digestion and result in very low VS destruction and accumulation of volatile fatty acids in the digester.

pH and alkalinity

Each group of microorganisms involved in the anaerobic digestion process has a different range of optimum pH. Acceptable enzymatic activity of hydrolytic and acidogenic bacteria occurs above pH 5.0. Methane-producing bacteria are extremely sensitive to pH. The optimum pH for methanogens ranges between 6.7 and 7.5. Therefore, single-stage anaerobic digesters are generally functioned in a pH range of 6.8–7.2.

Volatile acids produced in the acid-forming phase tend to reduce the pH. The pH of the digester then increases as soon as the acids are consumed and alkalinity is produced in the form of bicarbonate. To ensure that sufficient buffering capacity is available to compensate sudden increases in the VFAs content, alkalinity (HCO_3) should be greater than 1000 mg/L. The best way to increase pH and buffering capacity in a digester is by the addition of sodium bicarbonate. Lime will also increase bicarbonate alkalinity but may react with bicarbonate to form insoluble calcium carbonate, which promotes scale formation.

Toxicity materials

The heavy metals, ammonia, sulfides, and some inorganic materials are of toxic to anaerobes. Toxic conditions normally occur from overfeeding, excessive addition of chemicals, or from industrial wastewater contributions with excessive toxic materials to the plant influent.

The ammonia nitrogen concentration of greater than 1000 mg/L considered highly toxic. Generally, ammonia is a outcome of the anaerobic breakdown of urea and proteins. High level of sulfides i.e. >200 mg/L in sludge may also cause the toxicity to methanogenic population. Sulfide precipitation as iron sulfide by adding the iron salts can be a good option.

OLR and the sludge feed composition

These are two main parameters affecting digester performance. The OLR is usually expressed in terms of VS. High OLRs cause problems in digester mixing and homogenous heat transfer. Therefore, the typical OLR range for anaerobic digesters is between 1.6 and 4.8 kg VS/m^3/d.

Typically, sludge fed to digesters is a mixture of primary and secondary sludge. The primary sludge is discrete and rich in lipids, whereas the secondary sludge is more polymeric and protein rich. As a result, the hydrolysis of primary sludge is faster than that of secondary sludge and is more biodegradable. In general, primary sludge is blended with thickened secondary sludge and digested together to improve VS and COD destruction. Increasing the primary sludge fraction in the feed sludge, as well as the SRT, generally increases the overall VS destruction.

Redox potential

Low redox potential below −200 mV is necessary for efficient anaerobic digestion and especially methane formation. To keep a low redox potential, oxygen, sulfate, and nitrogen oxides should be eliminated from the feed sludge.

Nutrients

Nutrients must also be present in sufficient quantities to facilitate cell growth, and thus ensure efficient digestion. The nutrients required for all microorganisms are N, P, and S. However, the nutrient requirements for anaerobic microorganisms are low compared to aerobic ones because the growth yield of microorganisms under anaerobic conditions is lower. A nutrient ratio C:N:P:S of 500–1000:15–20:5:3 or an organic matter ratio COD:N:P:S of 800:5:1:0.5 is sufficient for efficient anaerobic digestion. These nutrients are usually in sufficient levels in the feed sludge.

Sludge heating

In temperate climates where yearly average temperatures range around 15°C, the sludge must be permanently heated to create optimum digestion conditions which are at

30–33°C, and to avoid too long a digestion time corresponding to the low temperatures. The heat supplied to heat up sludge digestion tanks must be sufficient:

1   To raise the temperature of incoming sludge to that of the tank,
2   To offset the heat lost from the tank through its walls, floor and roof, and
3   To compensate for heat lost in piping and other structures between the source of heat and the tank.

Seeding and mixing of sludge

Even if sludge digestion starts to develop under normal conditions, it is still necessary to add and create an intimate contact between fresh and old sludge, to enable a good development of the process by avoiding acidification. In this respect, the fresh sludge must be continuously mixed with the older digested sludge, to equalise the quality of sludge in the tank and to put the mature anaerobic bacteria from the old, well-digested sludge into intimate contact with the fresh sludge to decompose it uniformly. If the mixture fails to take place, then the sludge in the tank tends to stratify, the digested sludge falling to the bottom. When the sludge is properly mixed, the temperature becomes uniform and the surface scum and foam are destroyed.

Volatile suspended solids
(Hosh & Pohland, 1974)

Since successful operation of this biological process depends on maintaining a careful balance in the system ecology, it is important to be able to measure the variable biomass in the system. Bacterial concentrations in wastewater treatment processes are often estimated based on the volatile suspended solids (VSS) concentration. However, since the primary sludge fed to the anaerobic sludge digester contains large amounts of volatile particulate material, VSS measurements, include both the available substrate concentration and the bacterial cell mass. To interpret VSS data from anaerobic sludge digesters, one must account for the various particulate fractions included in the VSS measurements, viz. VSS originally present in the particulate feed, VSS produced during digestion after bacterial growth, VSS destroyed during digestion, and inert residual volatile solids that accumulate after cell death.

Due to the limitations of VSS as an indicator of active biomass, several alternative methods targeting specific constituents in microbial cells have been examined, including amino nitrogen content, ribonucleic acid content, adenosine triphosphate (ATP) concentration, phosphate activity, and dehydrogenase activity (DHA). Their potential usefulness in wastewater treatment process as superior indicators of microbial activity has been well recognised. It is important to recognise that these measurements reflect bacterial activity and not necessarily bacterial mass. They are often correlated with other bacterial activity indicators, such as oxygen uptake rates in aerobic systems, or gas production rates in anaerobic systems.

7.3.6 Process description
(Cetin & Surucu, 1990; Fair *et al.*, 1968; Metcalf & Eddy, 2003;
Negulescu, 1985; Turobskiy & Mathai, 2006)

Anaerobic digestion process is carried out in an airtight reactor. Sludges are introduced continuously or intermittently and retained in the reactor for varying periods depending upon the operating temperature range. The stabilised sludge, which is withdrawn continuously or intermittently from the process, is non-putrescible, and its pathogen content is greatly reduced.

Two types of digesters are commonly used: low-rate or conventional sludge digesters, and high-rate or continuous sludge digesters. A combination of basic processes involved in these two digesters has been used to develop the two-stage digesters.

Low-rate or conventional digestion

It is also known as non-continuous sludge digestion. In this process, the contents of the digester are usually unheated and unmixed. The digestion is carried out in a single state process, where the functions of digestion, sludge thickening, and supernatant formation are carried out simultaneously in one tank, as shown in Figure 7.3.

Operationally, untreated sludge is added in the zone at mid-depth of the tank, where the sludge is actively digesting and the gas is being released. In this zone the untreated sludges mix with digesting solids and are seeded and buffered by them. As gas rises to the surface, it lifts sludge particles and other materials, such as grease, oil, and fats, ultimately giving rise to the formation of a scum layer. Obviously, due to the bio-flotation, some degree of mixing of sludge contents may occur by the interchange of solids between the sludge and scum zones.

(a) Demerits of conventional digestion: In this type of digestion process, much of the digester volume is wasted and sometimes acidification takes place in the top and middle layers. This can lead to areas of low and high pH in the system, which restricts optimum biological activity. Further, chemicals added for pH control are not dispersed throughout the tank and their effectiveness is limited. Grease breakdown is

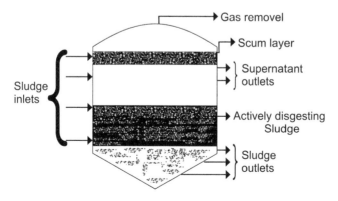

Figure 7.3 Typical anaerobic digester with conventional standard rate single-stage process.

poor, because the grease tends to float to the top of the digester, while the methane bacteria are confined to the lower level. The scum accumulation at the surface of supernatant seeks additional mechanism for its destruction. Methane bacteria are removed with the digested sludge and not recycled to the top, where they are required, resulting in poor operation.

High-rate sludge digestion

In high-rate digestion (Figure 7.4), the sludge is more or less continuously added and vigorously mixed, either mechanically or by recirculating a portion of the digestion gases through a compressor. The digester is heated to maintain maximum activity in the mesophilic region. Thus, mixing, heating, thickening and constant feeding are the vital features of high-rate digestion, which collectively can make an identical situation. High rate digestion is particularly useful for upgrading conventional digester units to accommodate additional loads where land is scarce (Tanghe *et al.*, 1994).

The heating mainly helps to enhance the microbial activity, which ultimately leads to high rate of digestion and subsequent gas production. High rate anaerobic digesters works at the mesophilic temperature of 35°C. The external heat exchangers are used for heating purpose mainly due to easy maintenance and flexible operation. Supporting mixing in anaerobic digester helps to provide better contact between substrate and inoculum, to avoid the thermal stratification, decreasing scum accumulation, and diluting toxic compounds. Thickening helps to decrease the volume of sludge thus the volume of anaerobic digester. The sludge is feed to high rate digesters constantly or at routine intervals in order to maintain the steady state conditions in the digester. Which can help to cope up with shock loading conditions (Turobskiy & Mathai, 2006).

Merits of high-rate digestion

1   Because of good mixing, there is no stratification and hence loss of capacity does not arise due to supernatant or scum, or dead pockets.
2   By adopting continuous addition of raw sludge and resorting to pre-thickening of the raw sludge to a solid content of 6%, the detention times could be reduced to

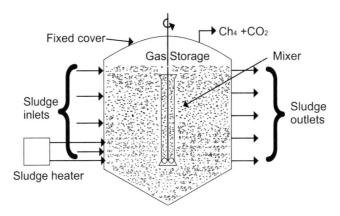

Figure 7.4 Anaerobic sludge digester with high-rate, continuous-flow stirred tank, single-stage process.

10–15 days. (Thickening beyond 6% affects the fluidity of the sludge, hampering its movement in pipes and mixing in the digester).

However, there is no supernatant separation in the high-rate digester, and only 50% destruction of volatile solids (VS) is achieved and given off as gases.

Thermophilic digestion

Thermophilic anaerobic digester operates in the range of 50 to 57°C and the group of microorganisms, called thermophilic bacteria. Generally, thermophilic anaerobic digestion has several advantages over mesophilic digestion listing as:

- Rapid rate of reaction, which ultimately enhances the hydrolysis of volatile solids and resulting in higher biogas production.
- High degree of pathogens removal

However, the disadvantages of thermophilic anaerobic digestion include:

- Higher energy demands for heating
- Poor-quality supernatant
- Higher odor potential
- More sensitive to temperature fluctuations than mesophilic process
- Poor dewaterability

Two-stage anaerobic sludge digestion

As shown in Figure 7.5, these units are designed to have their space divided into two tanks.

During the first stage (first tank or digestion tank), the sludge is heated and kept in the uniform digestion activity. For this the sludge mass is continuously stirred, to mix old and fresh sludges intimately, to avoid any zonal differentiation (i.e. stratification). Gas is usually collected during this stage only. Here, after 5 days, about 67% of the total gas is generated, while after about 14 days nearly 90% of the gas is produced,

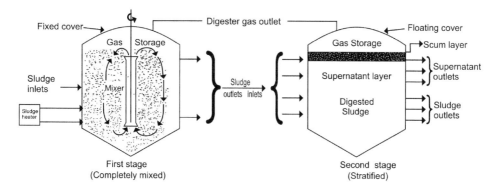

Figure 7.5 Schematic diagram of anaerobic sludge digester with two-stage process.

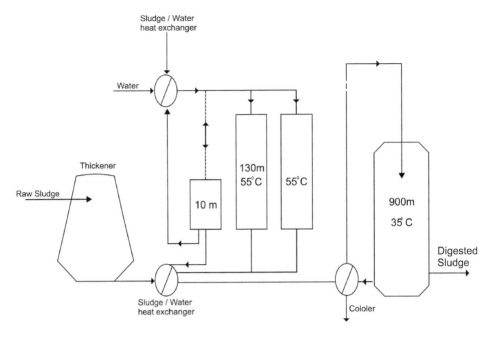

Figure 7.6 Two-stage thermophilic/mesophilic digestion.

which is the limit at which digestion is practically over. Gas may then be stored in the gasholder.

The second tank is used for storage and gravity thickening of digested sludge and to produce a dense sludge and relatively clear supernatant. Frequently, both tanks may be made identical, in which case either one may be the primary. But the second tank may also be an open and unheated tank, or a sludge lagoon. Second tanks may have fixed roof or floating covers used to capture gases, but the amount of gas is very low.

In the second stage, digestion goes on at a slower rate with no artificial help such as heating, stirring, etc. There is a zonal differentiation of digester activity but with little, if any, scum formation.

Supernatant is removed periodically for conservation of space. The oldest, densest and most stable sludge is produced and removed from this tank. Additional storage capacity needed for the monsoon periods or for winter seasons also be provided in the second stage digester.

As per study of Mitsdorffer *et al.* (1990), the stabilisation and disinfection of sewage sludge can be achieved more satisfactorily using the two-stage anaerobic (thermophilic and mesophilic) digestion. In 1987, he conducted his study on a real plant in Germany. Figure 7.6 shows the schematic diagram of the two-stage digestion system. He obtained the following results:

(a) Stabilisation: With the temperature ranging between 50 and 55°C, and the hydraulic retention time (HRT) being about 5 days, safe operation of the thermophilic pre-digesting stage is achieved, leading to stable production of biogas

(60% CH_4). In total a specific gas production of 610 l/kg volatile suspended solids (VSS) and a degree of 66% reduction of VSS respectively, is achieved by thermophilic plus mesophilic digestion. Bypassing the first stage leads to a significant decrease of stabilisation efficiency.

(b) Disinfection: Health risks which may arise from agricultural use of digested sludge can be eliminated completely during the thermophilic stage of digestion. Enterobacteria and salmonellae which are taken as indicators of health risk are reduced several orders in magnitude, which is within the acceptable range.

Dairy-sludge treatment (Special application of anaerobic digestion)

As per Karpati (1989), in case of dairy-waste treatment, if alum is used in the precipitation stage, anaerobic digestion is proposed for treatment of the sludge produced. In this case, flotation with polyelectrolyte addition can be used for phase separation.

If the precipitation stage is carried out with slaked lime and ferrous sulphate, flocculation and settling of the sludge without polyelectrolyte addition is a more favourable possibility. The use of lime-containing sludge is very advantageous, especially in the case of sour soils (i.e., cold wet land). Its use increases the pH of the soil. This hinders the uptake of heavy metals and decreases the phytotoxic effect of aluminium but increases the zinc concentration. However, the overall heavy metal content of the sludge is low, and agricultural application of the sludge is recommended. The lime and plant nutrient (nitrogen, phosphorus, sulphur, and organic compounds) contents mean that the sludge can be used advantageously for the amelioration of sour soils. The remaining impurities in the preheated effluent can be decomposed easily in a municipal sewage treatment plant.

Biotransformation and degradation of nonylphenol mono- and diethoxylates in anaerobic sludge environment

Now a days, nonylphenol ethoxylates (NPEOs) are extensively used as surfactants in various industrial products. NPEOs are a mixture of polyethoxylated monoalkylphenols, predominantly para-substituted and are used in the manufacturing of paints, detergents, inks, and pesticides. Surfactants are common water pollutants because of their use in aqueous solutions, which are discharged into the environment in the form of wastewater from treatment plants or sludge stored in landfills. Degradation products of alkylphenol polyethoxylates, i.e., nonylphenol (NP), have the potential to be bioaccumulated, thereby becoming toxic to aquatic and soil microorganisms (Ejlertsson et al., 1999; Lewis, 1991).

The partial degradation of NPEOs can proceed aerobically as well as anaerobically. Although the metabolic pathways are not completely understood, it is believed that biotransformation commences at the hydrophilic part of the molecule and the C-2 units (ethylene glycol) are removed one at a time giving rise to nonylphenol mono and diethoxylates (NPEOl-2) (Swisher, 1987). Complete degradation of NPEOl-2 may occur under aerobic conditions, but they have been reported to be more persistent in anaerobic environments (Ekelund et al., 1990; Jones & Westmoreland, 1998). Under aerobic conditions carboxylated metabolites may be formed (Ahel et al., 1994a; Jones & Westmoreland, 1998). Furthermore, because NPEOs with one or two ethoxy groups are less hydrophilic than polyethoxylated NPEOs, they are subjected to non-biological elimination by absorption to hydrophobic sludge constituents and organic matter, among other materials.

Recently, Ejlertsson *et al*. (1999) have conducted a study to qualitatively investigate the anaerobic biotransformation and degradation of nonylphenol mono- and diethoxylates by microorganisms derived from (i) an anaerobic sludge digester treating wastewater from a pulp plant and industrial wastewater containing NPEOs as a pollutant, (ii) a landfill site where the very same sludge is deposited, and (iii) a municipal waste land fill, the latter acting as a reference source. It was observed that in the samples with 10 and 100 per cent anaerobic digester sludge amended with 2 mg/litre, NP levels increased steadily during the experiment, at the end of which the NP level was 57 per cent in the 100 per cent digester sludge sample and 31 per cent in the 10 per cent sludge sample. Transformation of NPEOl-2 also occurred in samples amended with 60 mg/litre for 100 per cent sludge, in which NPEO-2 was transformed to NPEO-l. However, no such transformation occurred in the diluted sludge inoculum.

Other researchers have also reported earlier that the mass flow of short-chained NPEOs in sewage treatment plants as well as in natural aquatic environments depends not only on microbial activity but also on physico-chemical processes (Ahel *et al*., 1994b, 1994c; Oman & Hynning, 1993; Schnurer *et al*., 1994). It has also been observed that the factors influencing the degradability of alkyl ethoxylates include the structure of their hydrophobic region, the length of their hydrophilic unit, and the incorporation of other glycol molecules in the hydrophilic region (Ejlertsson & Svensson, 1996; Holt *et al*., 1992; Wahlberg *et al*., 1990). Surfactants containing straight-chain alcohol ethoxylates can be transformed quickly and completely under anaerobic conditions, whereas branched alcohol ethoxylates often are more difficult for the organisms to degrade (Kravetz *et al*., 1991). Information on the anaerobic degradation of alcohol ethoxylates is limited. Salanitro and Diaz (1995) showed, for example, that straight-chain alcohol ethoxylate gave a methane yield that was 84 per cent of the theoretically calculated methane value.

Transformation of β-hexachlorocyclohexane (β-HCH) under anaerobic conditions by methanogenic granular sludges

In many countries, the use of lindane, the p-isomer of hexachlorocyclohexane (HCH), is still made as an insecticide for use in agriculture and forestry, which leads to many environmental problems (Hernandez *et al*., 1991; Johri *et al*., 1996). Consequently, since the early 1970s, the microbiological transformation of HCH has been extensively studied, both under aerobic and anaerobic conditions (Middeidorp *et al*., 1996; Sahu *et al*., 1995). In most cases, the 13-isomer of HCH was found to be least susceptible to microbial dechlorination. This is probably due to the spatial (all equatorial) arrangement of the chlorine atoms as was postulated by Beurskens *et al*. (1991). In soils contaminated with HCH, rapid degradation of the a- and y-isomer of HCH was found under aerobic conditions by the microbial population naturally present (Nagasawa *et al*., 1993; Sahu *et al*., 1995). Further, in anaerobic (flooded) soils, a- and y-HCH are almost always transformed, whereas the 13-isomer is mostly found to be persistent under methanogenic conditions (Sahu *et al*., 1992; Senoo & Wada, 1990; Van Eekert *et al*., 1998a).

In their study, Van Eekert *et al.* (1998a) has described the anaerobic dechlorination of 13-HCH by granular methanogenic sludge from upflow anaerobic sludge blanket (UASB) reactors fed with different primary substrates. The granular sludge from UASB reactors has high biomass content and consists of mainly acetogenic and methanogenic bacteria (Buser & Muller, 1995; Holliger *et al.*, 1993; Van Eekert *et al.*, 1998b). These aerobic bacteria contain large amounts of corrinoids and other cofactors such as F_{430} etc. (Florencio *et al.*, 1993; Schanke & Wackett, 1992; Scholetn & Stams, 1995). These factors are part enzymes, which catalyze important pathways in anaerobic bacteria, like the acetyl-CoA pathways and the last step in methane formation. The results had shown that β-HCH transforming bacteria were present in different anaerobic environments. This finding may be of importance for the application of anaerobic bioremediation on sites contaminated with HCH isomers.

7.3.7   Process control

For efficient functioning of an anaerobic sludge digestion system, both the non-methanogenic and methanogenic bacteria are to be kept in dynamic equilibrium. To establish and maintain such a state, the reactor contents should be devoid of dissolved oxygen and free from inhibitory concentrations of such constituents as heavy metals and sulphides to favour the growth of methane formers. Further, the pH of the aqueous environment should range from 6.6 to 7.6. Sufficient alkalinity should be present to ensure that the pH will not drop below 6.2, because the digestion-rate controlling methane formers cannot function below this point. A sufficient amount of nutrients (generally in the ratio equal to BOD:N:P = 100:2.5:0.5), which is low due to low growth rate of methane-formers, must also be maintained. Temperature is another important environmental parameter. The optimum ranges are 30–38°C for mesophilic and 49–57°C for thermophilic sludge-digestion. In addition to this, for optimization of the digestion process it is made continuous, i.e. feeding and desludging is accomplished simultaneously, to have permanent seeding and buffering capacity for maintaining alkaline conditions. Moreover, by constant stirring of the digesting sludge, the process is made uniform, to avoid any stratification and to put fuel value of gas in in-plant uses, e.g. for plant heating or power production, etc.

Digestion is greatly expedited by thermal optimization of the process, which can be achieved by adopting one of the following heating systems:

- Heating the incoming sludge with steam or in a counter current heat exchanger powerful enough to take care of all heat requirement.
- Heating the tank through fixed or moving hot-water circulating coils.
- Burning the sludge gas in the sludge digester.
    Introducing steam, hot water, or hot sludge liquor at the bottom of the digester and withdrawing condensate or water through supernatant removing pipes.

Sludge heating in the heat exchangers is the most widely used method. When external heaters are installed, the sludge is pumped at high velocity through the tubes, while water circulates at high velocity around the tubes. This promotes high turbulence on both sides of the heat-transfer surface and results in higher heat-transfer coefficients and better heat transfer. Another advantage of external heaters is that untreated cold sludge on its way into the digesters can be warmed, intimately blended and seeded with sludge liquor before entering the tank.

Digestion is continued long enough to produce an end product that can be disposed of economically and without nuisance either immediately or after further treatment by dewatering, drying, or incineration. Complete digestion will also enable to have maximum amount of gas (rich in fuel value), which can be utilised subsequently. The fuel value of sludge gas can be put to use in various ways, both within treatment plants and for non-plant purposes. The following are a few common ways to utilise the fuel value of sludge gas:

Plant heating: digesters, incinerators, building and hot-water supply;
Plant power production: pumping, air and gas compression, and operation of other mechanical equipment;
Minor plant uses: gas supply to the plant laboratory for gas burners and refrigerators; and Motor fuel for municipal cars and trucks.

However, collection, storage and utilization of sludge gas are economically justified only when the sludge treatment plants are large and the skilled attendance is ensured.

7.3.8 Merits and demerits

Merits: The energy rich methane gas can be recover, which can be further utilise for electricity or heat generation. Low sludge production, well stabilised and odor free sludge and rich in nutrients (Nitrogen and phosphorus) are other advantages. Moreover, high rate of pathogen destruction can be achieved in thermophilic anaerobic digesters.
Demerits: High capital cost dues to large reactor size including the arrangements for feeding, mixing and heating. Slow digestion rate, process sensitivity against toxic compounds, temperature, pH, oxygen intrusion, high ammonia and sulfide concentration are other major issues. The supernatant of anaerobic digesters is rich in organics, nitrogen and phosphorus, thus require an additional treatment before disposing.

7.3.9 Operational issues

It is necessary to maintain stable operating conditions in an anaerobic digester due to its complex bio-chemical processes. Such as, the high and frequent variation in operating temperature can disturb the metabolic processes of anaerobic bacteria. The uncontrolled operating conditions can lead to digester upset and failure.

Reactor functioning

In order to assess the performance of anaerobic digester following process stages should be monitored (WEF, 1998):

- Feed sludge: pH, temperature, alkalinity, TS, VS.
- Digester content: Temperature, alkalinity, TS, VS, volatile acids.
- Digested sludge: alkalinity, TS, VS, volatile acids.

- Biogas produced: percentage of methane, carbon dioxide, and hydrogen sulfide.
- Supernatant: pH, BOD, COD, TS, total nitrogen, ammonia nitrogen, phosphorus.

There are several indicators, which can reflect the performance of anaerobic digesters. The reduced production of methane and a corresponding increase in the CO_2 content can be a result of increase in volatile acid concentration, and decreases in the alkalinity and pH of the digester content indicate imbalanced anaerobic treatment operation. Several other factors like sudden change in temperature, organic loading, composition of feed sludge, or toxic loading, or a combination of these can also hamper the anaerobic activity. At the normal anaerobic digestion condition of pH between 6.6 and 7.4 and CO_2 content of 30 to 40%, the bicarbonate alkalinity will be in the range 1000 to 5000 mg/L as $CaCO_3$. The concentration of bicarbonate alkalinity in the digester content should be about 3000 mg/L as $CaCO_3$.

Odor control

Hydrogen sulfide, the main source of odor at anaerobic technology based systems, is a highly corrosive gas, which needs to be remove from the biogas by scrubbing before use in engine generators or boilers.

Supernatant

The supernatant of anaerobic digesters is rich in solids, organics, nitrogen and phosphorus, thus require an additional treatment before disposing. However, when the supernatant is recycle back to the wastewater treatment plant inlet point, it can enhance the organic load onto the wastewater treatment process. Thus, physical (adsorption), chemical (coagulation-flocculation, precipitation) or biological treatment of supernatant is necessary before recycled back to wastewater treatment stream.

Struvite

Struvite (magnesium ammonium phosphate) is a precipitate formed in anaerobic digesters under the excess accumulation of ammonium and phosphate ions as a result of anaerobic digestion of sludge. Thus, a magnesium ammonium phosphate supersaturated condition takes place in the reactor. Struvite formation leads to the scale depositing and chocking of pipes, thus cause the maintenance problems in digesters. Acid washing can be an effective but costly treatment. Iron salts addition can be a good a good option, so that the phosphorus can be precipitate down (Turobskiy & Mathai, 2006).

Cleaning of digester

A frequent cleaning of digesters is require in order removing floating scum and deposited grit, which will ultimately increase the active volume of anaerobic digesters. The effective mixing in digesters and arrangement of multiple sludge withdrawal pipes can improve the condition and reduce the regular cleaning of digesters.

7.4 AEROBIC DIGESTION PROCESS

7.4.1 Introduction

Aerobic digestion is a biological method of sludge treatment that designed to stabilise and reduces the total mass of organic waste by biologically destroying volatile solids in the presence of oxygen, which is achieved through endogenous respiration of the sludge microorganisms, mainly bacteria (Katz & Mason, 1970). It is a suspended growth biological process that is similar to the extended aeration-type activated sludge process. The process produces a stable product, reduces mass and volume, and reduces pathogenic organisms (Turobskiy & Mathai, 2006).

This process may be used to treat only the following:

1   Waste activated sludge
2   Mixtures of waste activated sludge or trickling filter sludge and primary sludge
3   Waste sludge from activated sludge treatment plants designed without primary settling

7.4.2 Mechanism
(Fair *et al.*, 1968; Hartman *et al.*, 1979; Mc Clintock *et al.*, 1988; Metcalf & Eddy, 2003; Messenger *et al.*, 1990; Turobskiy & Mathai, 2006)

Aerobic digestion is tried in the endogenous respiration phase of aerobic bacteria. As the supply of available substrate (food) is depleted, microorganisms begin to consume their own protoplasm to obtain energy for cell maintenance reactions. This phenomenon of obtaining energy from cell tissue, known as endogenous respiration, is the major reaction in aerobic digestion. The cell tissue is aerobically oxidised to CO_2, H_2O and NH_3, which is subsequently oxidised to nitrate. Oxides of nitrogen and sulphur are also formed. These are immediately converted into soluble salts. In actuality, only about 65 to 80% of the cell tissue can be oxidised. The remaining 20 to 35% is composed of inert components and organic compounds that are not biodegradable. Some non-biodegradable materials are hemicellulose and cellulose employed in the cell walls, which require several months for degradation. Thus, in the design of aerobic digestion the rate of this decomposition can be assumed to be negligible (Datar & Bhargava, 1988; Ganczarczyk *et al.*, 1980). The material that remains after completion of the digestion process is at such a low-energy state that it is essentially biologically stable. The bio-oxidation of biomass results in the reduction of the volume of residual solids requiring disposal. The first step of the aerobic digestion process, oxidation of biodegradable matter, can be shown by the equation 7.20:

$$\text{Organic matter} + O_2 \xrightarrow{\text{bacteria}} CO_2 + H_2O + \text{Cellular Material} \qquad (7.20)$$

The next step of endogenous respiration can be shown by equation 7.21:

$$C_5H_7O_2N \text{ (cell mass)} + 5O_2 \rightarrow 4CO_2 + H_2O + NH_4HCO_3 \qquad (7.21)$$

In biomass decomposition, oxygen is used to oxidise cell mass to carbon dioxide and water. This reaction also produces ammonium bicarbonate. The ammonia released in this process, combines with some of the CO_2 that is produced to form ammonium bicarbonate.

Equation 7.22 showing the complete nitrification/denitrification step:

$$2C_5H_7O_2N + 11.5O_2 \rightarrow 10CO_2 + 7H_2O + N_2 \tag{7.22}$$

The complete nitrification–denitrification shows a balanced stoichiometric equation. If all the ammonia released is nitrified and denitrified, there is balance. Biomass plus oxygen is converted to carbon dioxide, nitrogen gas, and water.

Where activated or trickling filter sludge is mixed with primary sludge and the combination is to be aerobically digested, there will be both direct oxidation of the organic matter in the primary sludge and endogenous oxidation of the cell tissue. However, the overall reaction can shift to a lengthy phase of direct oxidation of biodegradable matter. Therefore, longer detention times are required to accommodate the metabolism and cellular growth that must occur before endogenous respiration conditions are achieved.

Theoretically, about 1.5 kg of oxygen is required per kilogram of active biomass (1.5 lb/lb) in non-nitrifying systems, whereas about 2 kg of oxygen per kilogram of active biomass (2 lb/lb) is required in nitrifying systems. In a system with complete nitrification–denitrification, it provides the opportunity to (1) reduce the oxygen requirements (a 17% reduction), (2) avoid alkalinity depletion because the alkalinity produced in denitrification is used to offset the existing alkalinity that is required for nitrification, and (3) and nitrogen is removed. The oxygen requirements for mixed primary and activated sludge digestion are substantially higher than those are required only for activated sludge digestion, because of the longer time required to oxidise the organic matter in primary sludge (Turobskiy & Mathai, 2006).

Aerobic digesters can be operated as batch or continuous flow reactors as shown in Figure 7.7. In general, primary and waste activated sludge goes to the aerobic digester either directly from the primary and secondary clarifier, respectively, or after preliminary concentration in a sludge thickener. It had been a common practice to pump the WAS directly from the clarifiers into the aerobic digester; however, it required larger tanks to fulfill the SRT requirement. Now a days, WAS is thickened before transfer to the digester, thus decreasing the requirement of large size digester.

7.4.3 Process kinetics
(Hills & Dykstra, 1980; Hosh & Pohland, 1974)

The growth phase during which cells use their own cell material and surrounding dead cells for food is termed as the endogenous respiration phase. The rate of endogenous decay is represented by the following equation of first order biochemical reaction:

$$dx/dt = -K_d \cdot x \tag{7.23}$$

where
dx = change in biodegradable cell material in time interval dt,
dt = time interval,
K_d = degradation constant, and
x = concentration of biodegradable cell material at time t.

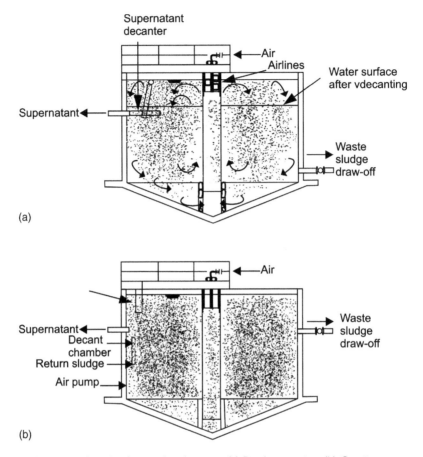

Figure 7.7 Operational modes for aerobic digesters: (a) Batch operation, (b) Continuous operation.

Since the biodegradable cell material, i.e. the active biomass is difficult to determine directly, volatile solids (VS) of the activated sludge, which are easy to determine, describe the performance of aerobic digester. The coefficient K_d, can be calculated on either a volatile suspended solids basis (K_{vs}) or a total suspended solids basis (K_{ts}). The biochemical oxygen demand (BOD) of the sludge, which is a measure of oxygen required for oxidation of biodegradable (carbonaceous) organic matter can be used as another variable representing directly the active biomass.

Substituting BOD (L) of sludge for biodegradable cell material (x), Eqn. (7.23) becomes:

$$dL/dt = -K_b \cdot L \tag{7.24}$$

where,
dL = change in BOD of activated sludge during time interval dt,
dt = time interval,

K_b = degradation constant for BOD reduction, and
L = concentration of BOD at time t.
   Integrating Eqn. (7.24) between definite limits

$$\int_{L_o}^{L} dL/L = - \int_{o}^{t} K_b \cdot dt \tag{7.25}$$

where,
L_o = concentration of BOD, mg/l, at time $t = 0$

$$\text{or} \quad [\ln(L)]_{L_o}^{L} = -k_b[t]_o^t \tag{7.26}$$

$$\text{or} \quad [\ln(L/L_o)] = -k_b t \tag{7.27}$$

$$\text{or} \quad L = L_o e^{-K} \tag{7.28}$$

Chemical oxygen demand (COD) reduction during the aerobic digestion process is mainly due to the oxidation of organic material present in the sludge and, therefore, can also be assumed as a first order reaction kinetics.
   Mathematically:

$$dc/dt = -K_c \cdot C \tag{7.29}$$

where,
C = COD concentration at time t, (mg/l),
dc = change in COD concentration in time interval dt, (mg/l),
k_c = degradation constant for COD reduction reaction.
   This equation can be integrated between finite limits ($t = 0$, t and $c = c_o$, c), and reduced to the following relation:

$$\ln(C/C_o) = -K_c \cdot t \tag{7.30}$$

$$C = C_o e^{-K} \tag{7.31}$$

7.4.4   Process controlling factors

Factors that must be considered in designing aerobic digesters include hydraulic detention time, process loading criteria, oxygen or air requirements, energy requirements for mixing, environmental conditions, and process operation (Bishop & Farmer, 1978; Metcalf & Eddy, 2003; Messenger et al., 1990; Negulescu, 1985; Turobskiy & Mathai, 2006).

Feed sludge characteristics

The aerobic digestion process is best suited for stabilizing biological solids such as WAS, because the process keeps the system predominantly in the endogenous respiration phase. If a mixture of primary sludge and biological sludge is digested, a longer detention time is required to oxidise the excess organic matter in primary sludge before the endogenous respiration can be achieved. The concentration of solids in feed sludge to an aerobic digester is important in the design and operation of a digester. Advantages of higher feed solids concentration include longer SRTs, smaller digester volume requirements, easier process control (less or no decanting in batch-operated systems), and increased levels of volatile solids destruction. However, higher solids concentrations require higher oxygen input levels per digester volume.

Hydraulic retention time

The amount of volatile solids (VS) in sludge is reduced more or less linearly up to a value of about 40 per cent at a hydraulic detention time of about 10–12 days. Although VS removal continues with increasing detention time, the rate of removal is reduced considerably. Depending on the temperature, the maximum reduction ranges between 45% and 70%.

Oxygen requirements

The oxygen requirements that must be satisfied during aerobic digestion are those of the cell tissue and, with mixed sludges. It is due to biochemical oxygen demand in 5 days (BOD_5) in the primary sludge. The oxygen requirement for the complete oxidation of cell tissue, computed using Eqn. (7.28) is equal to 7 mol/mol of cells or about 2 kg/kg of cells.

$$C_5H_7NO_2 + 7O_2 \rightarrow 5CO_2 + NO_3^- + 3H_2O + H^+ \tag{7.32}$$

The oxygen requirement for the complete oxidation of the BOD_5 contained in primary sludge varies from about 1.7 to 1.9 kg/kg destroyed. Considering fraction of O_2 in air equal to 23.2 per cent and density of air as 1.201 kg/m^3 and assuming suitable oxygen transfer efficiency (say p %), the air requirement can be calculated as below:

$$\text{Air requirement } (\text{m}^3/\text{d}) = O_2 \text{ requirement } (\text{kg/d})/[1.201 \times 0.232 \times (P/100)] \tag{7.33}$$

On the basis of operating experience, it has been found that if the dissolved-oxygen concentration in the digester is maintained at 1–2 mg/L and the detention time is greater than 10 days, the sludge dewaters well (Drier & Obma, 1963).

Mixing

The contents of the aerobic digester need to be well mixed to ensure intimate aeration to meet the oxygen requirement of sludge solution. Generally, mechanical aerators require 20–40 kW/10^3 m^3. In diffused air mixing, air supply rates of 1.2 to 2.4 m^3/m$^3 \cdot$ h have been reported; the higher values are recommended for sludges of high solids concentrations. If polymers are used in the pre-thickening process, especially for centrifugal thickening, a greater amount of unit energy may be required for mixing. In cases where the air-mixing requirement exceeds the oxygen transfer requirement supplemental mechanical mixing should be considered rather than overdesigning the oxygen transfer system.

pH and temperature

The pH value and temperature play an important role in the operation of aerobic digesters. It has been observed that the operation of aerobic digester is temperature-dependent, lower temperatures retard the process whereas higher temperatures speed it up. Especially, at temperature below 20°C, provided hydraulic detention times are of the order of 15 d. As the hydraulic detention time is increased to about 60 d, the effect of temperature is negligible. In the past, aerobic digestion has normally been conducted in unheated tanks similar to those used in the activated-sludge process. However, as understanding of the thermophilic aerobic digestion process increases, it is anticipated that more use will be made of well-insulated or even partially heated tanks.

Process operation

Depending on the buffering capacity of the system, the pH may drop to a rather low value ±5.5 at long hydraulic detention times. Reasons advanced for this include the increased presence of nitrate ions in solutions and the lowering of the buffering capacity due to air stripping. Although pH does not seem to inhibit the process, but the pH requires to be checked periodically and adjusted, if found to be excessively low.

Aerobic digesters need to be equipped with decanting facilities, so that they may also be used to thicken the digested solids before discharging them to subsequent dewatering facilities or sludge drying beds. If the digester is operated so that the incoming sludge is used to displace supernatant and the solids are allowed to build up, the mean cell residence time will not be equal to the hydraulic residence time.

Volatile suspended solids
(Anderson & Mavinic, 1984; Benefield & Randall, 1978;
McClintock et al., 1988; Messenger et al., 1990)

The use of VSS as a parameter for the assessment of aerobic digestion efficiency is limited. Some authors have suggested that in order to obtain a more representative idea of the reduction in actual biodegradable microbial material, the VSS values should be corrected for the loss of non-biodegradable material (measured as non-volatile residue or fixed suspended solids; FSS). This loss can occur through such processes as solubilization of FSS over time. To overcome this problem in differentiating between the various forms of suspended solids, some authors have suggested to use total suspended solids (TSS) in any assessment of digester performance. In this way, reduction in both volatile and fixed suspended solids would be accounted for.

The overall TSS reduction can be correlated with odour nuisance for a given type of sludge. However, the changing quality of the raw sludge fed into an aerobic digester will often make this parameter difficult to use during practical operation.

Moreover, most of the aerobic digesters are operated in a semi-continuous fashion. The air supply is usually shut off and the supernatant is withdrawn from the unit before the raw sludge is pumped into the digester. This type of operation will increase the solids concentration in the unit. In order to calculate the solids reduction, a complete mass balance has to be undertaken. This is time consuming and quite often difficult to do.

Biochemical oxygen demand (BOD) is used as a measure of the quantity of oxygen required for oxidation of biodegradable organic matter present in the system by the aerobic biochemical action. Oxygen demand of wastewaters and sludges is exerted by three classes of materials:

(i)   Carbonaceous organic materials usable as a source of food by aerobic organisms,
(ii)   Oxidizable nitrogen derived from nitrate, ammonia and organic nitrogen compounds, which serve as a food for specific bacteria (e.g. nitrosomonas and nitrobacter), and
(iii) Chemically reducing compounds, e.g. ferrous ion (Fe^{2+}), sulphites (SO_3^{3-}) and sulphides SO^{2-}, which are oxidised by dissolved oxygen.

A standard BOD test represents oxygen demand in 5 days at $20°C$ due to carbonaceous organic materials and denoted by BOD or BOD_5. The BOD of a sludge sample can, therefore, be looked upon as representing biodegradable organic matter present in the sample.

Chemical oxygen demand (COD) corresponds to the amount of oxygen required to oxidise nearly all organic matter present in a sample of sewage or sludge by dichromate oxidation in an acid solution. However, it also includes oxygen required to oxidise inorganic matter, if present in the sample, and oxidizable by dichromate oxidation in an acid solution.

7.4.5 Process description

At present two proven variations of the aerobic process are in practice. These are as follows:

- Conventional aerobic digestion, and
- Pure-oxygen aerobic digestion

A third variation, thermophilic aerobic digestion, is also under intense investigation. Furthermore, upgrading of available variations are also being developed and used. As a result a fourth variation, mesophilic and thermophilic aerobic digestion, has come into the picture.

Conventional aerobic digestion

It is the aerobic digestion process of sludge accomplished with air and the most popular one at present. In this system oxygen-fraction of the air is utilised in oxidation of organic matter and the biomass. Digested sludge has excellent dewatering characteristics and significant fertilizer value.

In this process, the sludge is aerated for an extended period of time, with a minimum food to micro-organisms (F/M) ratio, in an open and unheated tank using conventional air-diffusers or surface aeration equipment. The process may be operated in a continuous or batch mode, later being preferred in smaller plants.

Pure oxygen aerobic digestion

Pure oxygen aerobic digestion is a modification of the conventional aerobic digestion process, in which pure oxygen is used in lieu of air. The digested sludge is similar to sludge from conventional air aerobic digestion. This modification is an emerging technology, because O_2 transfer-efficiency is higher in this process.

In case a covered aeration tank is used in this variation, a high-purity-oxygen atmosphere is maintained above the liquid surface and oxygen is transferred into the sludge via mechanical aerators. On the other hand, if an open aeration tank is used, oxygen is introduced to the bottom of liquid-sludge by a special diffuser that produces minute bubbles of oxygen. The pressure of diffusing oxygen and depth of sludge suspension is so adjusted that the bubbles totally dissolve before reaching air-liquid interface.

Pure-oxygen aerobic digestion can be used only by large installations, when the cost of oxygen-generation equipment instrumentation is offset by the saving obtained by reduced reactor volumes, and lower energy requirements for oxygen-dissolution equipment.

Thermophilic aerobic digestion

Thermophilic aerobic digestion (Figure 7.8) represents a refinement of both the conventional air and pure-oxygen aerobic digestion. It has been shown in large-scale pilot plant studies that thermophilic aerobic digestion (temperature range 45–65°C) can be used to achieve high removals of the biodegradable fraction (up to 70%) at very short detention times (3–4 days). Advantageously, thermophilic digestion can be achieved

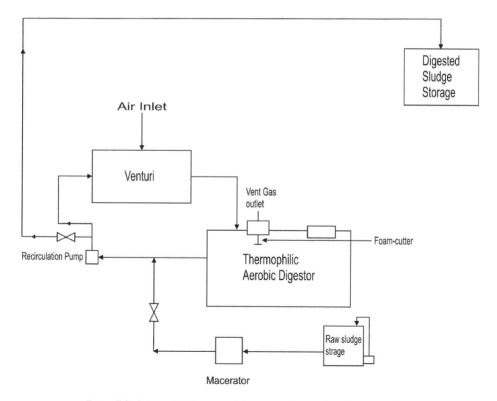

Figure 7.8 Schematic diagram of thermophilic aerobic digestion plant.

with a simple aeration system. Moreover, because of the high operating temperatures, the digested sludge is pasteurised as well.

As the aerobic process is exothermic, the thermophilic digestion, without much external heat input, can be achieved by using the heat released during microbial oxidation of organic matter to heat the sludge (Jewell & Kabrick, 1980). It has been estimated that more than 25 kcal/l of heat energy is released in the aerobic digestion of primary and secondary sludges (between 2 and 5% solids). This quantity of heat is sufficient to heat up wet sludges containing from 95 to 97% water, up to a temperature of 45°C (i.e. thermophilic range), provided that sufficiently high oxygen-transfer efficiencies can be obtained to avoid air in oxygen stripping of the heat (Cummings & Jewell, 1977). However, studies of Datar and Bhargava (1984, 1988) shows that thermophilic conditions are unfavourable for aerobic digestion of activated sludges.

7.4.6 Merits and demerits
(Turobskiy & Mathai, 2006)

The major advantaged of aerobic digestion in comparison of anaerobic digestion are

- Lower capital cost
- Odorless end product
- Easier to operate
- Lower BOD, TSS, and ammonia nitrogen in supernatant liquor
- No odor problems

Major disadvantages of aerobic digestion in comparison with anaerobic digestion are:

- Operating cost is higher, in terms of the power consumption for supplying the oxygen.
- Methane gas, a energy rich by-product, is not produced.
- The performance is affected by the concentration of solids, type of sludge, location, and type of mixing–aeration system

7.4.7 Operational issues
(Turobskiy & Mathai, 2006)

Operational issues include pH decrease, foaming problems, supernatant characteristic, and sludge dewaterability.

pH decrease

The drop in pH is caused by acid formation that occurs during nitrification; the drop in alkalinity is caused by lowering of the buffering capacity of the sludge due to air stripping. It has been observed that the system will acclimate and perform well at a pH as low as 5.5. However, filamentous growth may occur at low pH values. If the feed sludge has low alkalinity and the pH in the digester continues to drop below 5.5, provisions may have to be made to increase the alkalinity by adding chemicals.

Foaming problems

Foaming may occur in aerobic digesters during warm weather periods, due mainly to high organic loading rates. Growth of filamentous organisms can also cause foaming problems. The main control measures are to chlorinate the digester feed to destroy filamentous organisms, turning off the aeration equipment to create transient anaerobic conditions. Water sprays are normally used to control the foaming problems.

Supernatant characteristics

The aerobic digesters produce a better-quality supernatant if compare with anaerobic digesters. As the supernatant is generally returned to the head end of the treatment plant, the actual loading to the aeration basins from the supernatant is represented by soluble BOD, which is typically less than the organic strength of the wastewater. The suspended solids do not exert a high load to the aeration basin because the solids are in the endogenous stage of respiration.

Dewaterability

Belt filter press dewatering of aerobically digested sludge produces cake with 14 to 22% solids. Dewaterability of aerobically digested sludge is also affected by the degree of mixing provided during the digestion process because the high degree of mixing destroys the structure of the solids floc.

7.5   PROCESS ADVANCEMENTS

Over the period, several developments in conventional methods and innovative methods were reported by the researchers. This section is all about the process advancements in biological sludge stabilisation methods in past years.

7.5.1   Airlift autothermal biodigester

When cost-effective operation of sludge digestion process is desired, it is necessary to maximise the rate of oxygen transfer for cell growth, while minimizing the power requirement (Rich, 1982). One of the more promising contactors designed for high mass-transfer rates with low power consumption is the airlift tower (Figure 7.9). The airlift tower is a bubble column divided into two sections. The gas is diffused into one section, causing the dispersion density in that section to be lower than the density in the unsparged section. The resulting pressure difference causes liquid circulation. An upward velocity is observed in the sparged section and a downward liquid velocity is observed in the unsparged section. The downward liquid flow entrains some gas, therefore, mass transfer can take place in all parts of the tower.

As shown in Figure 7.9, circulation of liquid is initiated by compressed air into the riser section. The air bubbles travel upwards and create a partial voidage in this portion of the riser. This causes the column of liquid in the down corner (subject to its static head pressure) to move downwards. The cyclic circulation pattern is thus started. The down corner air is then fed and engulfed downward along with the liquid.

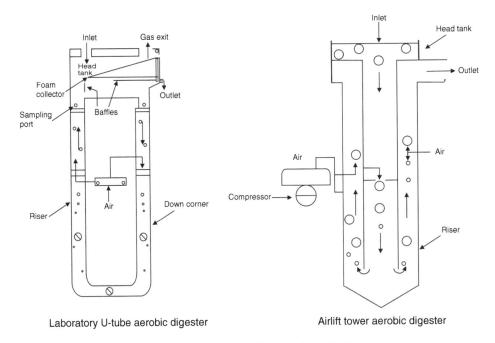

Laboratory U-tube aerobic digester   Airlift tower aerobic digester

Figure 7.9 Sectional diagram of U-shaped aerobic digesters.

As these bubbles descend, they shrink in size due to increasing static head pressure and their dissolution in the liquid phase. At the bottom, almost complete shrinkage and dissolution of air bubbles occur and, thus, maximum mass transfer of O_2 takes place. Bubbles are again released in the riser section due to the progressive decrease of static head pressure. At the top, where the bubbles enter the head tank, they partially disengage and exit through the opening provided in the head tank cover. A water heating jacket around the shaft allows the desired temperature to be maintained.

Tran and Tyagi (1990) conducted laboratory as well as pilot plant scale studies on airlift bio-reactor under mesophilic and thermophilic ranges of aerobic sludge digestion. The following conclusions were drawn by them:

1   In the airlift reactor, an eight times higher VSS loading rate is achieved compared to the recommended VSS loading in conventional systems.
2   Nitrification is almost completely inhibited above 40°C.
3   The high volatile loading rate and the high per cent VSS reduction capability of the airlift bio-digester minimise the reactor volume requirements.
4   The high oxygen transfer efficiency minimises the aeration energy costs.
5   The airlift bio-reactor produces stabilized sludge in significantly shorter retention time than conventional digesters. Sludge stabilisation is achieved in less than two days at 59°C.

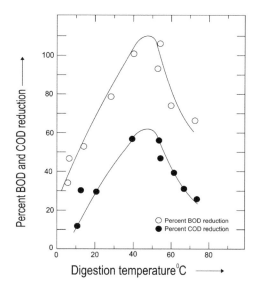

Figure 7.10 Plots of BOD and COD reduction versus digestion temperature.

These factors make the airlift autothermal biodigester an extremely efficient system with specific advantages in the digestion of high solid content sludges.

7.5.2 Mesophilic aerobic digestion

Mesophilic aerobic digestion is an auto-heating digestion process that includes sludge thickening followed by two or three stages of treatment in aerobic reactors. Sludge has to be pre-thickened to 4 to 5% solids in order to achieve heat balance in the process. Mesophilic aerobic digestion can qualify as a process to significantly reduce pathogens; however, compared to conventional aerobic digestion, use of this process is limited, due to high capital and operating costs (Turobskiy & Mathai, 2006).

BOD/COD reduction

As shown in Figure 7.10, Datar and Bhargava (1984, 1988) found that at optimum digestion temperature of 30–35°C yielding maximum destruction of activated sludge (in minimum digestion period) the BOD reduction ranges from 70 to 92.5% of its initial value. The BOD that remained at the end of the digestion period constituted fixed BOD contributed by chemically reducing compounds, which were oxidised by dissolved oxygen. At digestion temperature lower than 30°C the non-biodegradable BOD and COD concentrations were found to increase with reducing temperatures, probably because of reducing efficiency of aerobic bacteria (in utilizing their own cell materials) with decreasing temperature. It is also observed that all digestion temperature higher than 35°C, he nonbiodegadable BOD and COD concentration increase with increasing temperatures, because of reducing efficiencies of aerobic bacteria with increasing temperatures.

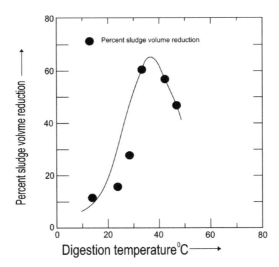

Figure 7.11 Plots of sludge volume reduction versus digestion temperature.

Reduction in sludge volume

During the endogenous respiration phase (aerobic digestion) the sludge volume reduction results from conversion of a substantial part of the sludge by oxidation into volatile products (viz. CO_2, NH_3, H_2). Figure 7.11 depicts the relation between sludge volume reduction and digestion temperature. It is seen that the reduction in sludge volume increased with increasing temperature from 9.0 percent at 5°C to the maximum value of 62.0 percent at 35°C and then reduced with increasing temperatures, reaching a minimum value of 40 per cent around 45°C. At digestion temperatures higher than 45°C no sludge volume reduction was observed.

Practical application

According to the above-mentioned study of Datar and Bhargava (1984, 1988), the digestion temperatures around 30–35°C are seen to be optimum for aerobic digestion of activated sludge from overall considerations of minimum digestion time, maximum reduction in BOD, COD and sludge volume and minimum supernatant BOD. The practical application of this finding is that the aerobic digester's temperature can be maintained at or near the optimum level of 30–35°C during summer season due to continuous aeration. During winter season, however, the digestion temperature is bound to reduce below 30°C, thereby dropping the efficiency of the process rapidly with decreasing temperatures. In that case the digestion temperature can be maintained at or near the optimum level by using (preheated) warm air for aeration.

Further, in another study, Datar and Bhargava (1984, 1988) made the following observations:

(i)   Aerobic digestion of activated sludge showed less destruction of biomass, in terms of reductions in initial TS, VS, BOD and COD concentrations in thermophilic digestion temperature range than in mesophilic digestion temperature range.

(ii)   In 40–45°C aerobic digestion temperature range, 40–50% sludge volume reduction of activated sludge was observed with turbid supernatants. At higher digestion temperatures, however, no sludge volume reduction was observed.

(iii) A considerable amount of cell material were solubilised during thermophilic aerobic digestion, thereby increasing supernatant BOD values to high levels.

(iv)   The thermophilic conditions are unfavourable for aerobic digestion of activated sludges.

7.5.3 Anoxic digestion

The removal of nitrogen in the form of nitrate by conversion to nitrogen gas can be accomplished biologically under anoxic (without oxygen) conditions. The process is known as denitrification. In this process, the facultative anaerobic bacteria obtain energy for growth. from the conversion of nitrate to nitrogen gas, but require an external source of carbon for cell synthesis. Nitrified effluents are usually low in carbonaceous matter and so methanol is commonly used as carbon source, but industrial wastes that are poor in nutrients have also been used. Overall energy reaction of this process may be given as:

$$6NO_3^- + 5CH_3OH \text{ (methanol)} \rightarrow 5CO_2 + 3N_2 + 7H_2O + 6OH^- \text{ (alkalinity)} \quad (7.34)$$

On the basis of experimental laboratory studies, McCarty *et al.* (1969) developed the following empirical equation to describe the overall identification reaction:

$$NO_3^- + 1.08CH_3OH \text{ (methanol)} + H^+ \rightarrow 0.065C_5H_7NO_2 \text{ (new cell)}$$
$$+0.47N_2 + 0.76CO_2 + 2.44H_2O \quad (7.35)$$

Process control

Because of a close relationship between the endogenous nitrate respiration (ENR) uptake rate and volatile suspended solids (VSS) destruction rate, the anoxic sludge digestion can be monitored and controlled by measuring NO_3^- consumption rate, as the endogenous decay rate (i.e. ENR uptake rate) of organisms depends on the availability of nitrate (Mavinic & Koers, 1982). This provides an easier and more reliable alternative than the VSS determination for which sampling often introduces errors. We may see here that use of ENR uptake rate (in terms of measuring nitrate consumption rate) is analogous to the use of endogenous oxygen uptake rate (in terms of measuring BOD removal rate) as a control parameter in the aerobic sludge digestion. Mineralization of organic-N provides alkalinity (OH^-) just like the anoxic case. However, under favourable conditions some NH_4^+ is nitrified to nitrate, resulting in alkalinity consumption.

Nitrification:

$$NH_4^+ + 2O_2 - NO_3^- + H_2O + 2H^+ \quad (7.36)$$

Alkalinity consumption:

$$OH^- + 2H^+ - H_2O + H^+ \qquad\qquad (7.37)$$

Thus, the decrease in pH [i.e. production of H^+ ions in Eqn. (7.32)] during aerobic digestion depends on the degree of nitrification, which in turn, is limited by the availability of alkalinity (as consumption of H^+ ions shift the equilibrium of Eqn. (7.31) to right side). Under certain conditions (e.g. in pH controlled digester or during initial period of the batch sludge digestion), the NH; ions may be completely oxidised to nitrate.

7.5.4   Dual digestion

Hao and Kim (1990) and, Kim and Hao (1990) studied the pre-anoxic and aerobic sludge digestion system (Figure 7.12) and found that the dual-digestion system has several advantages over conventional aerobic digestion process. Dual digestion includes aerobic thermophilic digestion as a first stage followed by mesophilic anaerobic digestion in a second stage. As a result power requirement is reduced significantly (McClintock *et al.*, 1988). Secondly, the system produces an adequate buffer capacity because of endogenous nitrate respiration (ENR) in the anoxic digester. This can result in a complete nitrification in the aerobic digester. Consequently, the aerobic digester's

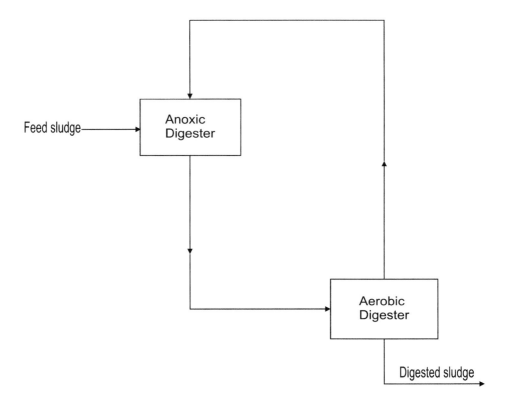

Figure 7.12 Schematic diagram of pre-anoxic and aerobic sludge digestion system.

supernatant nitrogen contains mostly nitrate to be recycled to anoxic reactor, to regulate ENR. Further, the high pH and alkalinity facilitate the aerobic decay as reported by Anderson and Mavinic (1984). Generally, residence time in the aerobic digester is typically 18 to 24 hours at the temperature range 55 to 65°C, however, residence time in the anaerobic digester is about 10 days.

For practical applications, the anoxic sludge digestion must be incorporated with the aerobic digestion, in either a semi-batch or continuous mode. For semi-batch operation, the aeration period can be incorporated into the aerobic cycle, which generates nitrate, to re-cycle into anoxic reactor, which undergoes the ENR and provides alkalinity (and high pH) for the subsequent aerobic nitrification in aerobic reactor.

Advantages of dual digestion are (1) increased levels of pathogen reduction, (2) improved volatile solids reduction, (3) increased methane gas production in the anaerobic reactor, (4) fewer odors produced by the stabilised sludge, and (5) one-third less tankage required than for a single-stage anaerobic digester (Turobskiy & Mathai, 2006).

7.5.5 Biological denitrification

Nielsen (1996) showed that nitrate was reduced in activated sludge from a wastewater treatment plant when ferrous iron was added. It was not clear whether the process was mainly chemical or biological, but since an activated sludge system provides perfect environmental conditions for the development of microbial communities with a capability to oxidise Fe(II) with nitrate or nitrite, it was hypothesised that it was mainly biological. A modern wastewater treatment plant with biological nitrogen and phosphorous removal provides periods when the microbial communities are simultaneously exposed to ferrous iron and nitrate or nitrite in the absence of oxygen (Harremoes et al., 1991; Wanner, 1994). Nitrate is produced by nitrification in aerobic periods, and ferrous iron is produced by microbial Fe(III) reduction under anaerobic conditions (Nealson & Saffarini, 1994; Nielsen et al., 1997; Straub et al., 1996). Thus, in the subsequent denitrification period, nitrate and Fe(II) are present together. Furthermore, iron is often present in large quantities, because it is added to enhance chemical phosphorus removal (Heron et al., 1994; Frolund et al., 1996; Rasmussen & Nielsen, 1996).

Recently, Nielsen and Nielsen (1998) performed a study to investigate whether oxidation of ferrous iron by nitrate in activated sludge is mainly biological or chemical, whether, it is a significant process in activated sludge, and also to get more information about the stoichiometry and kinetics of the process. It was found that a biological reduction of nitrate takes place in activated sludge concomitantly with the oxidation of ferrous iron to ferric iron. This process was found to be predominantly biological and present in different types of activated sludge treatment plants with variable rates.

This study has confirmed the suggestions given by Straub et al. (1996) and Nealson and Saffarini (1994) that microbial Fe(II)-dependent nitrate reduction can be of significance in the environment. The process can be important, particularly in activated sludge treatment plants, with biological nitrogen and phosphorous removal. The reaction is of interest in relation to at least three processes in a typical treatment plant with nutrient removal. First of all, the process may be involved quantitatively in the denitrification process, in addition to the organic matter-dependent nitrate reduction. In the investigation of treatment plants with an active Fe(II)-dependent nitrate removal, the rates ranged from 25 to 68% of the maximum nitrate removal rate that was observed in the plants when lactate was added. It means that this process, if the conditions are optimal, can be almost as important for the denitrification process as the organic matter-dependent denitrification. The extent will depend on the concentration of Fe(II), nitrate, and organic matter, but the dependence on these factors is not well described so far and should be studied in greater detail. Second, the chemical phosphorous precipitation in treatment plants relies on the presence of Fe(III), and for this reason, it is important that Fe(II) can be oxidised also under anoxic conditions, promoting an improved P removal (APHA, 2005; Ganaye et al., 1996). Finally, Fe(III) possesses better flocculation properties than Fe(II) (Caccavo et al., 1996; Henze et al., 1995). Therefore, Fe(II) added under anaerobic conditions (which is not unusual in many treatment plants for P-removal) or produced by microbial Fe(III) reduction can be reoxidised in the presence of nitrate (e.g., in a denitrification tank), preventing deflocculation of the microbial flocs.

In a nutshell, Nielsen and Nielsen (1998) found in their study that a biological reduction of nitrate and nitrite take place in activated sludge concomitantly with the oxidation of ferrous iron to ferric iron. The highest activity was found in plants with biological nitrogen and phosphorus removal. The highest Fe(II)-dependent nitrate removal rate was found to be 0.31 mmol NO^{-3} (g VSS)$^{-1}$ h^{-1}, which corresponds to 68% of the maximum dissimilatory nitrate reduction rate in the presence of lactate. The Fe(II)-dependent, nitrate removal rate was strongly pH-dependent, with a maximal rate at pH value of 8, which is almost four times the rate at a pH value of 6. The main product of Fe(II)-dependent nitrate removal was most probably dinitrogen, as no accumulation of ammonia, nitrous oxide, or nitrite could be observed. The process may be of significance in the activated sludge treatment plant with regard to nitrate removal and with regard to the oxidation of Fe(II) to Fe(III), which influences the chemical phosphorous removal and the flocculation properties of the sludge.

7.5.6 Temperature Phased Anaerobic Digestion (TPAD)

Two phase anaerobic sludge digestion or temperature-phased anaerobic digestion (TPAD) treatment combining a thermophilic step (55°C) with a short retention time and mesophilic step (35°C) with a longer retention time is known to provide pathogen destruction and effective organic matter removal by reducing the volatile solids (Han et al., 1997; Huyard et al., 2000).

The two stages of treatment comprise: hydrolysis and acidogenesis in one reactor and methanogenesis in second reactor. The research findings show that enhanced sludge digestion can be achieved by optimizing the both stages individually. The first reactor is known as acid-phase digester and the hydrolysis and acidogenesis is achieved at a retention period 1 to 2 days. This reactor can be operated either at mesophilic or thermophilic temperature. pH in the reactor maintains between 5.5 and 6.5 and no methane generation occurs at this stage. The second reactor is methane-phase digester, is designed for approximate 10 days retention period and operates at mesophilic temperature range (Turovskiy & Mathai, 2006).

It has also been stated that a two-stage process is better than a single-stage process because it separates faster acidogenesis reactions in the first-stage from the slower methanogenesis reactions in the second-stage (Solera *et al.*, 2002). Such kinetic separation of different steps in anaerobic digestion enhances the overall process by enhancing the individual steps. The TPAD treatment improved management of waste sludge in terms of volatile solids reduction and gas production compared to the single-stage mesophilic treatment (Han & Dague, 1997). The main advantages of two-phase anaerobic digestions are enhanced biogas generation with high methane fraction, high volatile solids reduction and high pathogen reduction.

7.6 SLUDGE COMPOSTING
(Fair *et al.*, 1968; Fleming, 1986; Metcalf & Eddy, 2003; MSST, 1987; Negulescu, 1985; Turobskiy & Mathai, 2006)

7.6.1 Fundamentals of composting

Introduction

Composting is a process in which organic material undergoes biological degradation to a stable end product. Sludge that has been composted properly is a sanitary, nuisance-free, humus-like material with high fertilizing capacity. Approximately 20–30% of the volatile solids are converted to carbon dioxide and water.

It can be shown by following formula:

$$C_6H_{12}O_6 + 6O_2 = 6CO_2 + 6H_2O + 674\,\text{kcal} \tag{7.38}$$

Composting can also be of anaerobic type, which can be described by following formula:

$$C_6H_{12}O_6 = 2C_2H_5OH + 2CO_2 + 27\,\text{kcal} \tag{7.39}$$

The aerobic method produces a stable and higher caloric content, and more rapid process than the anaerobic process.

In addition, because the sludge is essentially pasteurised, the composted sludge may be used as a soil conditioner. Although the process works well, the primary problem has been the lack of a market for the stabilised end product. Composting at present, need further research work to produce such end product, which can prove competent among conventional fertilizers, so that, it can be sold easily, and overall process may prove cost-effective.

Composting microbiology

Composting microbiology represents the collective activity of a sequence of mixed populations of bacteria, actinomycetes, and fungi. The bacteria played a key role in decomposition of a major portion of the organic matter. Composting follows in three successive phases: the mesophilic (40°C), thermophilic (70°C), and curing phases. In curing phase, the microbial activity is reduced, and the composting process is completed. The acid producers bacteria are responsible to metabolise the proteins, sugars and carbohydrates in the mesophilic phase. The thermophilic bacteria metabolise the lipid, fats and proteins during thermophilic phase and are also one of the main cause of heat production. The fungi and actinomycetes are found in both the mesophilic and thermophilic stages and causes the decomposition of cellulosic and complex organic compounds (Turobskiy & Mathai, 2006).

Merits and demerits of composting

The main merits of sludge composting:

- Nutrient rich compost can be used as a fertilizer
- Compost enhances the water retaining capacity (sand soil), water infiltration (clay soil) and aeration of soils.
- Composting techniques are relatively simple and easy in operation (windrow and aerated static pile composting), requires less space and effective in odor control (In-vessel composting)

The main demerits of sludge composting are as follows:

- Large area requirement and odor issue (Windrow and aerated static pile composting).
- Sensitive to environmental conditions: ambient temperatures and weather conditions (windrow and aerated static pile composting).
- High operation and maintenance requirement: Little flexible & maintenance intensive (In-vessel reactors).

7.6.2 Process description
(Turobskiy & Mathai, 2006)

Most composting operations consist of three basic steps:

1 Preparation of the wastes to be composted,
2 Decomposition of the prepared wastes, and
3 Preparation and marketing of the product.

The preparation steps are as follows:
Receiving, sorting, separation, size reduction and addition of moisture (as it permits the passing of air throughout the compost layer) and nutrients (carbon) are part of preparation step. In order to control the porosity and moisture, a bulking agent (wood chips, sawdust, rice hulls, wood ash) is added to the dewatered sludge upto an extent to achieve final solids content of 40–50% in prepared waste mixture. Addition of bulking agent also enhances the porosity of waste, which subsequently increase the aeration in waste. Moreover, bulking agent is considered a good source of carbon, which helps to enhance the decomposition of waste. The best range of carbon-to-nitrogen ratio is 25:1 to 35:1 (Turobskiy & Mathai, 2006).

Next step is high-rate decomposition step, in which the prepared waste mixture is aerate by manual or mechanical turning and, aeration by using a blower or by using the both ways. Moreover, the thermophilic temperature ranges of 40 to 70°C needs to be achieved, which also confirms the reduction of moisture content (evaporation) and destruction of pathogens. To accomplish the decomposition step, several techniques have been developed. In windrow composting (or natural composting) the prepared wastes are placed in windrow/heaps in an open field. The windrows are turned once or twice a week for a composting period of about 5 weeks. The turning helps in movement of air and moisture reduction. Generally, the width of 2.0–4.5 m and the height of 1–2 m is considered good for a windrow (Turobskiy & Mathai, 2006). The material is usually cured for an additional 2–4 weeks to ensure stabilisation. Inside the heaps, the temperature rises spontaneously to about 70°C as a result of the decay process. During digestion the water content is diminished and germs are killed; thus, plague germs are killed at 45°C, tuberculosis germs at 55°C, Taenia germs at 65°C, etc.

As an alternative to windrow composting, several mechanical systems have been developed, including the aerated pile process. By controlling the operation carefully in a mechanical system, it is possible to produce humus within 5–7 days. A grid of aeration piping system is used for forced aeration in an aerated pile system. The blower or fan is used for aeration of pile. Generally, the height of piles are keep from 2 to 2.5 m. The top thick layer of pile is composed of 15–20 cm of wood chips. The forced aeration helps to maintain a good oxygen and temperature conditions in pile (Turobskiy & Mathai, 2006). Often the composted material is removed, screened and cured for an additional period of 3–4 weeks.

In-vessel composting take place in a covered reactor with air forced through the reactor. It generates a well-stabilised end product in fewer period of 10–21 days, since, the process capability of controlling the environmental conditions like as air flow, oxygen concentration and temperature. Generally, in-vessel composting systems are of two types: plug flow and agitate type reactors. There is two types of plug flow reactors: horizontal and vertical plug flow reactors. In a horizontal plug flow reactor, the air is provided through a blower by using floor-mounted diffusers. The reactor is constructed of concrete with typical dimensions of 5 m width, 3 m depth and 20 m length. The prepared waste is feed through the inlet point and drives to the outlet of reactor using a hydraulically operated ram. In a vertical plug flow type reactor, the prepared waste is feed into the reactor from top. The waste material is aerated rather than mixing and agitation. After the bottom layer of compost is removed, the waste material in vessel moves forward toward the bottom outlet. In agitated in-vessel reactor, the periodic mechanical mixing is provides to the waste material. The reactors are rectangular shaped bins with open-topped and enclosed in a structure. The aeration is provided from the bottom and the depth of compost bed is maintained to 2–3 m (Turobskiy & Mathai, 2006). Once the compost has been cured, it is ready for the third step, which is product preparation and marketing. This step may include fine grinding, blending with various additives, granulating, bagging, storing, shipping and in some cases, direct marketing.

Sometimes, the first-step-mixture is introduced into a bio-stabilizing drum, where it remains for about one day at a temperature of 120°C. After that, the resulting matter is commuted and heaped up to a height of 1.50 m and after a few days of anaerobic digestion it is used as fertilizer. An enclosed mechanical compost system is usually preferred to open composting in excessively humid or cold areas, because conditions can be better controlled.

7.6.3   Process influencing parameters
(Turobskiy & Mathai, 2006)

The key factors that affect the biochemical reactions in composting are moisture, temperature, pH, nutrient concentration, and oxygen supply.

Moisture

Decomposition of organic matter depends on moisture. Less than 40% moisture may limit the rate of decomposition. The optimum moisture content is 50 to 60%. Dewatered municipal sludges are usually too wet (65 to 82% moisture content) for composting. However, mixing the dewatered sludge with a dry bulking material can reduce the moisture content of the sludge cake.

Temperature

The efficient composting can be achieved at the temperature range of 50 to 65°C. However, a temperature higher than 65°C can cease the biological activity. The composting temperature can be influenced by the shape and size of piling, aeration rate, moisture content, atmospheric conditions and nutrient availability.

pH

The pH ranges from 6 to 9 considered best for composting. However, the bacterial and fungal communities grow well at pH ranges from 6 to 7.5 and 5.5 to 8, respectively.

Nutrients

A carbon to nitrogen ratio ranges from 25:1 to 35:1 is considered best for composting. Low C/N ratios increase the nitrogen loss by volatilization as ammonia, which ultimately leads to loss of nutrient value and odor problem of ammonia. High C/N ratio causes to increasingly lengthier composting period. In some cases, bulking agents needs to be add into the sludge in order to improve the C/N ratio by providing the supplemental carbon.

Oxygen supply

An oxygen level between 5 to 15% by volume of gas mass considered good for composting. However, oxygen concentrations greater than 15% will leads to a decrease in temperature due to high rate of airflow.

7.6.4   Operational consideration
(Turobskiy & Mathai, 2006)

Detention period

In 1970, U.S. Department of Agriculture established the active composting period for aerated piles to be 21 days followed by 30 days of curing period without aeration for the composting mix of dewatered wastewater sludge and wood chips. However, the recommended period for in-vessel composting is as little as 14 days for active composting, although most horizontal agitated systems use 21 days.

Aeration and temperature control

In order to control the temperature upsurge, adequate aeration needs to be facilitated. The increase in aeration rate will decrease the temperature of system as well as moisture too. The aeration rate of 34 $m^3/Mg \cdot h$ considered best to facilitate enough oxygen for biological activity, sufficient drying, and suitable temperature for pathogen removal. The temperature needs to be controlled below 70°C, since a higher temperature may halt the microbial activity.

Screening

To recover and recycle the bulking agents from compost can save upto 80% of the cost of new bulking agent. An efficient screening can be achieve for the compost having moisture content greater than 50%.

Odor control

Organics decomposition leads to the production of odorous compounds like sulfides, ammonia, fatty acids, and mercaptans. However, good operational practices i.e. suitable mixing, proper collection and treatment of leachate can helps to reduce the odor problem at compositing plants. Moreover, bio-filters and wet scrubbers can also be used to as effective odor controlling methods.

Safety and health issues

The composting facilities must be equipped with good ventilation system, workers should wear the mask on site as well as good housekeeping like dust suppression must be performed for healthy working conditions.

7.6.5 Co-composting options

Sludge may be composted either separately, as has been discussed, or in combination with wood chips or other solid wastes (co-composting).

Co-composting with wood chips

Co-composting of sludge with wood chips usually requires that the sludge be dewatered initially. In addition, it must be blended with a bulking material, such as wood chips. The most attractive of the woodchip co-composting processes appears to be the aerated pile process (Figure 7.13). In this process, sludge is first mixed with wood chips. The mixed material is then placed in a pile and covered with a 300 mm layer of screened compost for insulation and odour control. Oxygen is supplied by forced aeration. After about 21 days, plus 2 days of drying, the wood chips are post stored and recycled. After curing for an additional 30 days, the compost is ready for product preparation and marketing.

  Feed sludge may consist of digested or untreated sludges. Digested sludge is reported to compost more slowly than untreated sludge, particularly during wet, cold periods, possibly because of the lack of sufficient digestible energy material for rapid biological oxidation. On the other hand, compost systems using untreated sludges are

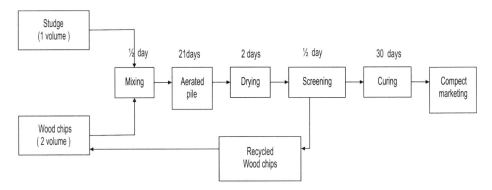

Figure 7.13 Schematic diagram of aerated pile method for co-composting of sludge.

often more susceptible to odour problems. Odour has been a problem with the aerated pile process.

Co-composting with solid wastes

Co-composting of sludges and municipal solid wastes usually does not require sludge dewatering. Feed sludges may have a solid content ranging from 5 to 12%. A mixture of solid wastes to sludge in the ratio of 2:1 is recommended. In fact, any amount of sludge can be mixed with solid wastes for composting, provided that the sludge is dewatered adequately. The solid wastes should be pre-sorted and pulverised in a hammer mill prior to mixing.

7.6.6   Refinery sludge composting

Economical and environmental considerations have forced oil industries to substantially reduce the amount of hydrocarbon waste material produced at their facilities, either by improving plant designs and operating procedures or by installing end-of-the-line remediation processes. Bioremediation is one approach to convert industrial organic waste into water, carbon dioxide and biomass, which is suitable for disposal through traditional waste treatment facilities. Due to significantly lower economic costs, biorermediation has been extensively studied and applied to petroleum wastewaters (Atlas, 1991; Milne *et al.*, 1998). In general, it has been found that with proper design procedures, all petroleum waste organics (except for the large polyaromatic compounds found in asphaltenes and resins) can be completely metabolised during the bioremediation process. Unfortunately, refinery sludge normally consists of a large quantity of asphalteries and/or resins making them difficult constituents for the bioremediation process; but some reports have indicated quite large organic conversions of these materials (Vail, 1991; Venketeswaran *et al.*, 1995).

Sludge produced in the oil industry can contain up to 80% oil and 40% solids. There have been a wide range of bioremediation processes applied to petroleum sludge clean-ups (Prince & Sambasivam, 1993). Due to economics and simplicity, land farming has traditionally been the biological treatment method chosen to dispose of most oily sludge (Persson & Welander, 1994; Wilson & Jones, 1993). In this procedure the hydrocarbon wastes are mixed with land and sometimes additional nutrients to a depth of less than 0.6m. The subsequent mechanisms involved in the disappearance to some extent degradation by sunlight and leaching. On the negative side, land farming often requires a large surface area and usually takes a long time to complete. Uncertainty about the long-term effects of oily sludge on the ecosystem of the land has made it increasingly difficult to obtain governmental permission to use land farming process. Composting of petroleum wastes has, therefore, received increasing attention as a potential substitute technology for land farming (Dhuldhoya et al., 1996; Jack et al., 1994; Nordrum, 1992; Potter & Glasser, 1995; Riggle, 1995). Advantages of composting include the use of less land area and faster biodegradation rates. In addition, the final compost may sometimes be suitable as a saleable soil amendment, thus avoiding landfilling of the final wastes.

Most wastes including oily sludge, requires a bulking agent as the main structural component of the compost pile. Fienstein et al. (1987) reviewed the guidelines of Environmental Protection Agency (EPA), USA, for selecting appropriate bulking agents and compared them to several actual applications. The guidelines ensure that the final compost heap has adequate porosity and the correct amount of moisture, so that the composting process proceeds correctly. The bulking agent enhances air permeability and acts as an immobilization surface for the active microbial community. In addition, the bulking agent often reduces the toxicity level of hydrocarbons and helps to maintain high moisture content within the compost. Wood chips, chopped wheat or barley straw, shredded tires and peat moss are among the numerous materials, which have been used as bulking agents. Some of the commercially available bulking agents also contain active microbial flora and other nutrients suitable for toxic wastes. This makes it unnecessary for the user to experiment with fertilizer and inoculant preparations, when trying to compost a new industrial waste material.

The inoculum or source of microorganisms is also an important issue for any successful bioremediation process. Commercial preparations of pure or mixed microbial cultures have generally not given significant bioremediation enhancements, as compared to indigenous microorganisms (Venosa et al., 1992). However, it appears that commercial preparations that include large supplies of nutrients and co-substrates have proven to be of significant benefit to the bioremediation process (Leavitt & Brown, 1994). Also, it has recently been shown that application of high concentrations of estimated, mixed hydrocarbon degrading bacteria enhanced the removal rate of hydrocarbons from soil by over 20 % (Vecchioli et al., 1990).

Recently, Milne et al. (Milne et al., 1998) undertook preliminary experiments to determine suitable ingredients for the successful composting of a heavy oil refinery sludge using a range of bulking agents, moisture contents, nutrients, co-substrates and inocula. Afterwards, the best ingredients were chosen for a detailed study of the CO_2 evolution rate and total petroleum hydrocarbon (TPH) reductions versus time during the composting process. The initial TPH content of the refinery sludge was found to be 30% (mainly C_{15}–C_{30}), two-thirds of which were saturates and aromatics that were thought to be biodegradable. The results show a significantly higher TPH reduction using solv-II (55% reduction) as a bulking agent/nutrient source, compared to heat-treated peat moss and barley straw (both at 30% reduction). The CO_2 evolution was high with solv-II and barley straw, indicating the presence of high microbial activities in those composters. Thus, it was observed that both the solv-II and contaminated soil from the heavy oil refinery are good sources for active microorganisms to biodegrade heavy oil sludge. Both heat-treated peat moss and solv-II were found to be the best bulking agents for activating the biodegradation process on oily sludge. It was also shown that no compost formulation could operate with greater than 250,000 ppm of oily sludge. Further, the product solv-II shows the highest TPH reduction (55%) during composting of a Western Canadian heavy oil refinery sludge. The final product was found to be suitable for direct land disposal at an industrial site, according to the most recent Canadian environmental guidelines (CCME, 1991), whereas, the original toxic sludge would not be disposed of by land application. Nevertheless, it was finally concluded that more research needs to be performed to determine whether an improved moisture content or added co-substrates mixed with heat-treated peat moss would make this bulking material as effective as solv-II in biodegrading heavy oil refinery sludge.

QUESTIONS

1   Describe the sludge digestion? How to achieve sludge digestion by two different methods?
2   What are the differences between aerobic and anaerobic sludge digestion?
3   What are the criteria to choose either aerobic or anaerobic digestion?
4   Describe the anaerobic digestion of sludge and different reactions involved within the process?
5   What are the benefits of feeding a higher substrate concentration in aerobic and anaerobic reactors? What restricts the upper limits of solids concentration?
6   What are the keys factors in determining the retention time and operating temperature relationships in sludge stabilisation?
7   Name the four different phases of anaerobic digestion of sludge. Explain each phase in brief.
8   What are the main parameters need to control and check during anaerobic digestion of sludge?
9   What are the possible reasons or factors, which causing the disturbance in anaerobic digestion of sludge?
10   What is the ratio of CH_4 and CO_2 in biogas generated from anaerobic digestion of sludge?
11   What are the main causes of sludge bulking during sludge digestion? How to control the sludge bulking?
12   What are the dangers connected with a high hydrogen sulphide content in the digester gas? How can hydrogen sulphide be removed from digester gas?

Chapter 8

Non-biological sludge stabilisation

8.1 INTRODUCTION

The non-biological sludge stabilisation comprises the physical and chemical methods, which are used to achieve oxidation of volatile substances present in sludge and to inhibit its potential for putrefaction. The main physical and chemical methods include heat treatment, lime stabilisation, chlorine oxidation, etc. (Ekelund *et al.*, 1990; Metcalf & Eddy, 2003). Moreover, several new developments are also taking place in this field.

8.2 THERMAL TREATMENT
(Keey, 1972; Metcalf & Eddy, 2003; Negulescu, 1985)

Under thermal treatment, sludge is heated in a pressure vessel to temperatures up to 260°C at pressures up to 2.75 MN/m^2 for short periods of time. Thermal treatment essentially serves as both a stabilisation process and conditioning process. [Sludge is conditioned to improve its dewatering characteristics. For more information please refer to section 5.3]. It conditions the sludge by rendering the solids capable of being dewatered without the use of chemicals. When the sludge is subjected to high temperatures and pressures, the molecular activity of sludge particles increases resulting in the hydrolysis of the encapsulated water solid matrix and dying of the biological cells. Obviously, the thermal activity releases bound water and results in the coagulation of solids. Thus, the thermal treatment coagulates the solids, breaks down the gel structure, and reduces the water affinity of sludge solids. As a result, the sludge is sterilised, practically deodorised and is dewatered readily on vacuum filters or filter presses without the addition of chemicals. In addition, hydrolysis of proteinaceous materials occurs, resulting in cell destruction and release of soluble organic compounds and ammonia nitrogen.

The thermal treatment process is most applicable to biological sludges that may be difficult to stabilise or condition by other means. The high capital costs of equipment generally limit its use to large plants (more than 720 m^3/h) or facilities where space may be limited. Processes used for thermal treatment include the low pressure Zimpro and the Porteous types as described in sections 5.3.2 and 5.3.3. A major disadvantage associated with heat treatment results from the very high strength of the supernatant and filtrate. The recycled liquor is composed mostly of organic acids, sugars, polysaccharides, amino-acids, ammonia, etc. Thus, the recycled liquor is highly polluted and contains a high proportion of non-biodegradable matter. As a result, separate treatment of the recycle flow may be required.

8.3 ALKALINE STABILISATION
(Paulsrud & Eikum, 1975; Metcalf & Eddy, 2003; Negulescu, 1985; Tenney et al., 1970; Turobskiy & Mathai, 2006; USEPA, 2000a)

The alkaline stabilisation of sludge is achieved by using the hydrated lime, quicklime (calcium oxide), flyash, lime and cement kiln dust, and carbide lime. Quicklime is commonly used because it has a high heat of hydrolysis and can significantly enhance pathogen destruction. Fly ash, lime kiln dust, or cement kiln dust are often used for alkaline stabilisation because of their availability and relatively low cost. In the lime stabilisation process, lime is added to untreated sludge in sufficient quantity to raise the pH to 12 or higher. The high pH creates an environment that is not favourable to the survival of microorganisms. Consequently, the sludge will not putrefy, create odours, or pose a health hazard, so long as the pH is maintained at this level.

Lime addition to untreated sludge has been practised for many years as a conditioning process to facilitate dewatering. However, the use of lime as a stabilizing agent has only recently gained recognition. Lime stabilisation requires more time per unit weight of sludge processed than that necessary for dewatering. The higher lime dose in the range of 5 to 15% of the amount necessary for initial pH elevation is required to maintain the elevated pH because of slow reactions that continue to occur between lime and both atmospheric carbon dioxide and sludge solids. In addition, sufficient contact time must be provided before dewatering to affect a high level of pathogen kill. Lime treatment at a pH higher than 12 for a period of 3 hours has been reported to achieve pathogen reduction beyond that attainable with anaerobic digestion. Because lime stabilisation does not destroy the organics necessary for bacterial growth, the sludge must be disposed off before the pH drops significantly, or it can become re-infected and putrefy.

8.3.1 Design criteria
The design criteria for alkaline stabilisation facility for sludge are as follows:

- Solids concentration of feed sludge.
- Required reuse criteria i.e. Class A and Class B, which govern the alkaline dosage and reaction time
- Odor management facility
- Storing space

The alkaline stabilisation facility require the following setup:

- Sludge feeding equipment
- Lime storeroom

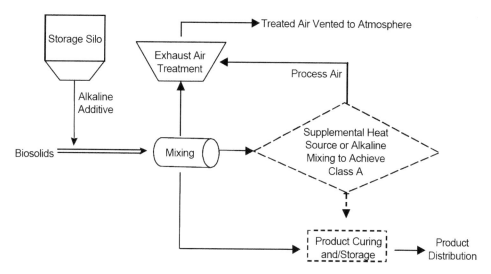

Figure 8.1 Flow diagram of a typical alkaline stabilisation operation (Source: Parsons Engineering Science, Inc., 1999).

- Lime transfer device
- Blender
- Dust and odor controlling device

Figure 8.1 presents a typical flow diagram for alkaline stabilisation.

8.3.2 Performance

Alkaline sludge stabilisation methods are flexible in operation and reliable, and often used to produce Class A or Class B solids. The beneficial effects of the alkaline process can be summarised in following subheadings (Turobskiy & Mathai, 2006):

Odor reduction

Sludge treatment with lime provides a low pH condition into the sludge, which ceases the microbial growths that are responsible for odorous gases generation. Sulfides are the primarily responsible for odor generation in term of hydrogen sulfide. A pH above 7 reduces the total sulfides in the form of hydrogen sulfides by 50%, which further reduces to zero at pH 9 mainly due to conversion in nonvolatile ionized forms.

Pathogen reduction

The alkaline treatment achieves significant pathogen reduction, provided that a sufficiently high pH is maintained for an adequate period of time. The liquid lime stabilisation of raw sludge reduced total coliform, fecal coliform, and fecal streptococci concentrations reduced by more than 99.9% (U.S. EPA, 1979). The number of

Salmonella and *Pseudomonas Eruiginosa* were reduced below the level of detection. There are few studies about the removal of viruses during lime stabilisation, which showed a higher virus removal at pH 12 (U.S. EPA, 1982). The helminth ova could not survive during alkaline treatment and a temperature of 50°C to 70°C and reaction time of few hours to some seconds, respectively.

Dewatering characteristics

Lime is considered a good conditioning agent to enhance the sludge dewaterability. Moreover, a little higher dose than those required for sludge dewatering would helps in sludge stabilisation too. However, high lime dose leads to the scaling problem.

8.3.3 Process variants
(Turobskiy & Mathai, 2006)

Sludge stabilisation by alkaline method can be achieved by following three ways: (i) liquid lime stabilisation, (ii) dry lime stabilisation, and (iii) advanced alkaline stabilisation.

Liquid line stabilisation

Lime slurry is mixed with liquid sludge for two hours reaction time in order to produce class B biosolids. However, liquid sludge stabilisation requires high quantity of lime than that require for sludge dewatering only. High quantity of lime is necessary to achieve the keep the pH level above 12 for two hours in order to stop the bacterial regrowth.

Dry lime stabilisation

In dry lime stabilisation, dry lime (hydrated lime or quicklime) is mixed with dewatered sludge cake to raise the pH of the medium. Quicklime is less expensive than hydrated lime and is easier to handle in large-scale facilities using lime in excess of 3 to 4 tons/day. Additionally, the heat generated from the exothermic reaction of quicklime and the water in sludge cake can enhance pathogen destruction. Significant advantages of dry lime stabilisation include no additional water added to sludge in the form of lime slurry, no special requirements for sludge dewatering equipment, and no lime-related abrasion and scaling problems to the dewatering equipment.

Advanced alkaline stabilisation

The main advantages of advanced alkaline stabilisation are the production of class A biosolids, low odor potential, low capital cost and easy operation. However, the drawback involves the high operation cost, larger volume of solids generation to transportation and the solids are not suitable for alkaline soils. Upon mixing with sludge quicklime releases the heat as a result of exothermic reaction. Per kilogram of 100% quicklime generates almost 15,300 cal/g · mol of heat and achieved a reaction temperature of more than 70°C for more than 30 minutes.

In a different process, the lime kiln dust is mixed with the sludge to attain the pH values of 12 or higher for a minimum period of 7 days. The sludge is dried for minimum 30 days, till a solids concentration of 65% is accomplished.

8.3.4 Operation and maintenance

Alkaline sludge stabilisation methods are relatively simple and easy to operate. The staff require for maintenance, instrumentation and operation of heavy machines. Since alkaline material is corrosive in nature, high maintenance of equipments is require. In order to ensure the consistency and homogeneity of product, the accurate design and operation of mixing device is require (USEPA, 2000a).

8.3.5 Treatment cost
(USEPA, 2000a)

The cost estimation of alkaline sludge stabilisation can be carry out by considering the following:

- Procurement and installation of equipments
- Biosolids drying and storing facility
- Transportation facility
- Operation and maintenance of equipments
- Chemicals
- Workforce
- Odor controlling device and chemicals
- Marketing
- Regulatory compliances: License applications, site monitoring, Sludge analyses, and regulatory record keeping and reporting.

8.3.6 Advantages and disadvantages
(USEPA, 2000a)

Advantages:

1 Easy to construct and reliable technology
2 Simple and flexible operation: easy start and stop
3 Less manpower requirement
4 Small area requirement

Disadvantages:

1 End product (high in pH) is not suitable for alkaline soils
2 Voluminous end product resulting in higher transportation cost.
3 Odor and dust generation potential
4 Pathogens regrowth possible in case of pH drops down to 9.5 during storage period
5 Low nitrogen and phosphorus content in end product.

8.4   CHLORINE OXIDATION PROCESS
(Campbell & Crescuolo, 1989; Metcalf & Eddy, 2003;
Negulescu, 1985; Roberts & Olsson, 1975)

In chlorine-oxidation process, the sludge is oxidise chemically with high doses of chlorine gas, which is generally applied directly to the sludge in an enclosed chamber for a short period. The oxidation process is followed by dewatering (Sand bed drying is an effective means. Belt-filter-press dewatering following conditioning with the addition of polyelectrolytes is also used). Most chlorine-oxidation units are of a prefabricated modular design, completely self-contained and skid-mounted. Application of chlorine oxidation as an exclusive means of sludge stabilisation has been limited to small plants on the order of 720 m³/h and less. The process may be used for treating any biological sludge, for treating septage, and as an auxiliary means of stabilisation to supplement existing overtaxed facilities. The sludge should be ground to ensure proper contact.

Because of the reaction of chlorine gas with the sludge, significant quantity of hydrochloric acid is formed. The acid can also dissolve heavy metals. Consequently, supematants and filtrates from chlorine-oxidised sludges may contain a high concentration of heavy metals. It has been reported that the release of heavy metals is dependent on pH, sludge metal content and the species of metal found in the sludge. Supematant and filtrate from the process may also contain high concentrations of chloramines. Implementation of chlorine oxidation requires the installation of chlorinators to feed chlorine to the process. Other chemical requirements may include sodium hydroxide and polyelectrolytes to condition the sludge prior to dewatering.

8.5   ADVANCEMENT IN PHYSICO-CHEMICAL METHODS

The new developments in non-biological methods of sludge stabilisation are discussed in this section.

8.5.1   Chemical fixation

Chemical fixation is a process of sludge stabilisation that converts the sludge into a product that can be used for landfill cover or land application. During chemical fixation, a series of chemical reactions take place by combining the dewatered sludge with chemical reagents so that a chemically, biologically, and physically stable solid is obtained. The end product has low odour potential, contains low levels of pathogens and fixes metals, which may be present in the sludge.

Most popular chemical fixation processes are: Chemfix and N-viro soil. The Chemfix process uses Portland cement and a sodium silicate setting agent to produce a sludge-derived synthetic soil (SDSS). The N-viro soil process uses lime and cement kiln dust as the additives. N-viro soil has also been produced using fly ash and limekiln dust as ingredients. Chemical fixation of dewatered sludge has a reasonable cost and is technically and environmentally feasible. The chemically fixed sludge product can be beneficially used as daily and intermediate soil cover supplement for landfills.

The treatment system consists of a central control unit, storage facilities for liquid and dry reagents, a weigh feeder and continuous-flow pug mill mixer. The dewatered sludge is fed to the weigh feeder, and then to the pug mill mixer. Dry reagent is introduced to the process mixer followed by the liquid reagent. Based on the signal received from the weigh feeder, the system operator controls the reagent ratio.

Immediately after processing, the material has the appearance of a gelatinous semi-solid. Within a few hours the processed residuals exhibit the characteristics of a moist soil due to the hydrolysis reactions that occur with the Portland cement. After curing for a period of several days, the material is similar to loamy soil. The material is feasible and may be handled with conventional earth moving equipment.

The Portland cement acts as a setting agent and provides the texture and friability of the finished product. The silicates chemically react with the materials in the sludge and bind the compounds within the matrix. The high pH results in the reduction of pathogens in the finished product.

During processing, ammonia is released due to the high pH resulting from the addition of the dry reagents. This high pH results. in conversion of the soluble ammonium ion (NH_4^+) to gaseous ammonia (NH_3). Fugitive ammonia emissions can be controlled by scrubbing the off-gases generated at the mixers and by mechanically aerating the SDSS during curing. The N-viro soil process uses similar equipment and the final product exhibits similar characteristics. The use of unhydrated lime in the N-viro process results in an exotherrnic hydration reaction, releasing heat and increasing the temperature of the product. This aids in pathogen reduction of the finished product.

Remarks

The moisture content is dependent on the amount and type of additives, curing method and climatic conditions. Greater quantities of additives result in a drier product. The addition of unhydrated lime resulting in hydrolysis and heat generation also promotes drying. Exposure to sun and wind will also accelerate curing and drying. Adverse climatic conditions should be avoided. The finished product after curing is expected to be 50–70% solid.

8.5.2 Cementitious stabilisation

Process description

As per Suprenant *et al.* (1990), simply add cement or flyash, that's the appeal of cementitious stabilisation, which turns hazardous materials into stationary and inert 'waste crete'. The same technique can now be used to produce 'oil crete'. This method was developed primarily to deal with the most dangerous types of hazardous waste, such as the radioactive variety. Now the same method is being used successfully to treat the oil field wastes/oil sludges. Cleaning up oil wastes involves mixing cementitious materials into the waste. This limits the solubility of the hazardous constituents in the waste, decreases the waste surface area exposed to the environment and improves its handling characteristics and physical properties.

Although the terms stabilisation and solidification are sometimes used interchangeably, they are actually two separate processes. The primary benefit of stabilisation is that of limiting the solubility or mobility of the hazardous contaminants, while solidification produces a strong, durable solid-waste block. Mixing cementitious materials into the oil waste and creating a solid mass of oil crete achieves the benefits of both stabilisation and solidification.

Cement and fly ash maintain the waste at a high pH in the range of 9–11, immobilizing most multivalent cations (toxic heavy metals) as insoluble hydroxides. The hydrated cementitious products formed will also chemically and physically bind metal ions. The organics present in oil can interfere with hydration (the reaction between cementitious materials and water that creates the hardened mass). Certain salts (zinc, copper and lead) may also prevent or retard hardening of the wastecrete. Mixes that do not solidify allow leaching of the waste. Oil sludges with high concentrations of particular cations can be pretreated with additives specifically chosen to immobilise those contaminants. Anions, although less toxic than cations, are also much more difficult to bind into an insoluble product.

Economics

Cementitious stabilisation is an inexpensive method of waste clean-up that does not necessarily require a remote dump site. The ideal cementitious stabilisation/ solidification treatment renders the waste chemically nonreactive and gives it physical properties that allow the land over the disposal site to be used for building sites or crops. However, wastes with high concentrations of toxic metals, organics or salts are generally not suitable for agricultural use even after cementitious stabilisation treatment.

QUESTIONS

1   What are the main physical and chemical methods for sludge stabilisation?
2   Compare the mechanisms of physical and chemical methods of sludge stabilisation.
3   Define the advantages and disadvantages of alkaline sludge stabilisation.
4   Describe the emerging non-biological sludge stabilisation methods in brief.
5   What is chlorine oxidation process of sludge stabilisation?

Chapter 9

Sludge stabilisation at small works

9.1 INTRODUCTION

The main processes used to stabilise the sludge at small water treatment plants are tabulated in Table 9.1.

9.2 GOVERNING FACTORS

At the much smaller work sites found in the countryside, supervision and maintenance is usually carried out by mobile operators and the number of visits per week is kept to a minimum. These sites therefore require sludge treatment processes which will operate reliably with minimal supervision. Other factors which should be considered when designing sludge treatment processes for small works (Ekama & Marais, 1986; Murray *et al.*, 1990; Paulsrud, 1990) are as follows:

1   The contributing population and, hence, the level of sludge production is often not accurately known because census data and drainage areas do not coincide and a large proportion of the population often commute to work or school outside

Table 9.1 Effective sludge stabilisation processes.

S. No.	Process	Process Description
1	Sludge Pasteurization	Minimum mean retention (MMR) of 30 min at 70°C or 4 h at 55°C followed by mesophilic anaerobic digestion.
2	Mesophilic anaerobic digestion	Minimum mean retention (MMR) of 12 d at temperature of 35 ± 3°C followed by MMR of 14 d secondary storage.
3	Thermophilic aerobic digestion	Minimum mean retention of 7 d with a minimum of 4 h at 55°C or higher.
4	Composting	Minimum mean retention (MMR) of 40°C for 5 d including 4 h at minimum of 55°C followed by minimum of 2 months maturation.
5	Lime stabilisation	pH not less than 12 for minimum of 2 h.
6	Lime conditioning dewatering and storage	Minimum storage period of 3 months
7	Dewatering storage	Minimum storage period of 6 months
8	Liquid storage	Minimum storage period of 3 months

the area. In addition, some rural treatment is carried out using private septic tanks.

2   The actual sludge volumes imported from other sites are difficult to assess accurately.

3   Many small sites are isolated. In the winter, bad weather conditions may reduce accessibility and the frequency of maintenance visits.

4   Screening and grit removal are often limited, if they exist at all, causing problems for downstream sludge processing plants. Horizontal-flow settling tanks are common. These have to be manually desludged, resulting in intermittent batches of sludge rather than a continuous feed.

5   Small sites are often located close to housing, so odours must be closely controlled.

9.3   PROCESS TYPES

The following criteria must be kept in mind during the selection of a sludge treatment process (Murray *et al.*, 1990). Most importantly, the process must:

1   Satisfy legislative directions or guidelines
2   Not cause odour problems
3   Produce a sludge suitable for spreading on land.

In practical terms, only anaerobic and aerobic digestion systems fulfil all three requirements.

In order to operate reliably with a digester temperature of $35 \pm 3°C$, anaerobic digestion requires a regular feed of raw sludge of greater than about 4% total solids. If the sludge supply is not regular, gas production will fall and not satisfy the heating requirements, meaning that an auxiliary fuel supply is required. If the sludge supply falls to such an extent that digestion fails, restarting the process requires the addition of an actively digesting seed sludge and considerable operator time. The difficulty of correctly sizing an anaerobic digester because of the problem of estimating sludge loading, and the added problem of intermittent sludge supply at small sewage treatment works, means that anaerobic digestion may not be sufficiently reliable for this application.

9.3.1   Thermophilic Aerobic Digestion (TAD)

Thermophilic aerobic digestion (TAD) (Figure 9.1) is considered to be much more flexible than anaerobic digestion. It requires no seeding for start-up; raw sludge is simply charged to the digester and aeration initiated. The heat generated from the aerobic degradation of organic matter raises the temperature of the digesting sludge, and this may continue to rise until it limits the rate of biological reaction. Where biodegradable matter and oxygen are in excess, temperatures of over $60°C$ have been consistently achieved, and sludge disinfection occurs at the high temperatures reached. TAD can be operated in batches or semi-continuously. The degree of stabilisation achieved is dependent on the total time the sludge resides in the digester. A liquid sludge with a uniform consistency and inoffensive odour is produced.

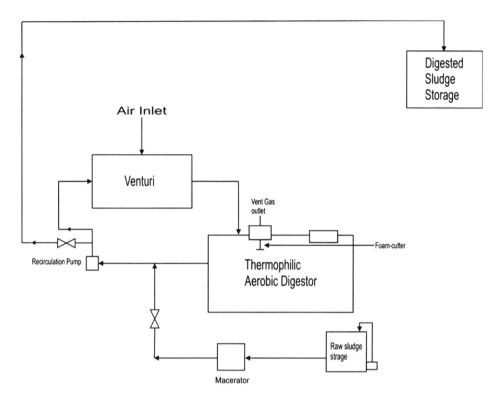

Figure 9.1 Schematic diagram of a thermophilic aerobic digestion plant.

Operational problems in TAD systems

Three main areas of operational difficulty have been encountered in TAD systems:

1 There is no mixing of sludge in the raw sludge storage tank. Sludge settles out into bands of thick sludge and water. If care is not taken during digester feeding and water is fed to the digester, the temperature rise of the digester contents is not as high or as rapid as usual because the amount of biodegradable material available is limited. TAD systems can cope with an occasional batch of thin sludge, but for consistent operation at a temperature above 55°C, feed sludges of 4–6% total solids need to be used.

2 There is often no grit removal at the works and sludge entering the digester can have a high grit content. This causes problems of excessive wear on the pump impeller and excessive grit deposition in the base of the digester. However, if an uprated pump impeller and wear plate are used, wear is minimised. All future digesters are expected to incorporate features designed to facilitate grit removal.

3 Rag causes the pump to block and the foam cutter to fail. Raw sludge screening and/or maceration are, therefore, required prior to digestion.

Performance of TAD plants

Plant performance may be monitored by the following:

1 Digester temperature
2 Degree of sludge stabilisation:

- Volatile solids (VS) destruction
- Chemical oxygen demand (COD) reduction

3 Degree of disinfection
4 Reduction of odour.

The characteristics of the treated sludge are also important, so the following should be determined:

- Consolidation characteristics
- Supernatant quality
- Fertiliser properties.

TAD produces a sludge with good consolidation properties, better than those of anaerobically digested sludge. Digestion does not seem to have any adverse effect on the fertiliser properties of sludge. However, phosphorus and potassium levels show a slight decrease after digestion. This is probably due to grit deposition in the base of the digester tank. Ammoniac nitrogen levels rise as a result of digestion and the amount of available nitrogen is, therefore, increased.

As a result of their studies, Murray *et al.* (1990) reached the following conclusions about TAD plant performance:

1 TAD is a reliable, simple-to-operate sludge treatment method suitable for small isolated works.
2 TAD is flexible in operation. It can be operated in batches or semi-continuously according to the sludge load.
3 Digester temperatures above 55°C can be obtained even during the winter months, provided the digester is well-insulated.
4 TAD plants have low maintenance requirements, the only moving parts being the recirculation pump and the foam cutter.
5 Screening or comminution of raw sludge is necessary prior to digestion. Grit removal is advantageous.
6 Treated sludge is suitable for disposal to land, because the treated sludge is:

- Stabilised
- Disinfected
- Less odorous than raw sludge
- Easily spread on land
- Of high available nitrogen content
- Easily consolidated up to 7% total solids, reducing disposal costs.

9.3.2 Low-temperature aerobic digestion

Aerobic sludge digestion in small waste water treatment works has been studied to optimise the operational conditions at lower temperatures ($<20°C$). The findings revealed that the solids retention time (SRT) should be increased as the operating temperature decreases to maintain satisfactory removal of volatile solids (Koers & Mavinic, 1977; Mavinic & Koers, 1979). At a temperature range of $5–20°C$, the system should operate at 250 to 300 degree-days (the product of SRT in days and operating temperature in °C) to achieve a satisfactory reduction of volatile solids. Hot air or waste water can be used to decrease detention time and to prevent freezing of digesters (Turovskiy & Mathai, 2006).

9.3.3 Sludge treatment wetlands

With low energy requirements and operational costs, sludge treatment wetlands offer a promising and sustainable technology for sludge management in small communities (<2000 population equivalent). Sludge treatment wetlands, also known as sludge-drying reed beds, are comparatively new sludge treatment systems (Uggetti *et al.*, 2010, 2011). Sludge treatment wetlands have been used in Europe for sludge dewatering and stabilisation since the late 1980s.

Sludge is spread directly into the basins from the aeration tanks or is homogenised beforehand in a buffer tank prior to its discharge into the wetlands. From this tank, the sludge is diverted into one of the beds (wetlands), following a semi-continuous regime. The number of beds and the bed surface may vary, according to the treatment capacity of the facility. The beds may be constructed in rectangular concrete basins or in soil basins. The bottom of the basin is covered with a waterproof sheath to close off the bed and avoid leaching. A minimum slope of 1% is desirable to ease leachate collection through a number of perforated pipes, which are placed along the bottom of the bed. These pipes also enhance aeration through the gravel filter and sludge layers. Sludge is fed in by means of pipes, which may be located in a corner of the bed, along one of the sides of the bed, or in the middle of the basin (upflow vertical pipes). In the wetlands, drying is undertaken as a batch process in such a way that sludge is fed each time to one of the beds during a feeding period that may last 1–2 days or even 1–2 weeks. Sludge is fed onto the beds and the sludge is dewatered, mainly as a result of water percolation through the sludge residue and the granular medium. Residual water content is further reduced by plant evapotranspiration (Uggetti *et al.*, 2010). In addition, stored sludge forms a dry surface film that is cracked due to the movement of the plants. The cracking of the sludge layer enhances water evaporation and oxygen transfer, which promotes a more uniform porosity along the bed and sludge mineralisation at the bottom level. Certainly, transfer of oxygen by the plants from the air to their roots, as well as transfer through the cracked surface and via filter aeration, creates aerobic conditions in some zones of the sludge layer, promoting the existence of aerobic microorganisms and ultimately improving sludge mineralisation (Nielsen, 2003, 2005a, 2005b).

Uggetti *et al.* (2011) studied the performance of full-scale sludge treatment wetlands over a period of two years. They reported that sludge dewatering increased total solids concentration by 25%, while sludge biodegradation led to volatile solids around

45% TS. Sludge treatment wetlands that employed direct land application provided the most cost-effective approach, and were also characterised by the lowest environmental impact, since greenhouse gas emissions from such wetlands were insignificant. Thus the use of wetlands for sludge treatment represents the most appropriate alternative for decentralised sludge management in small communities.

QUESTIONS

1   What are the major factors to be considered during the design of a sludge treatment facility for a small works?
2   What are the main criteria for selecting a sludge treatment method for a small works?
3   Define the major operational problems in thermophilic aerobic digestion systems.
4   Why is anaerobic digestion not a good option for a small works?
5   How does a sludge treatment wetland work? Describe in brief.

Chapter 10

Sludge minimisation technologies

10.1 INTRODUCTION

The main aim of sludge minimisation technologies is to remove organic material and water, hence reducing volume and mass, remove degradable material, which prevents subsequent odors and pathogen vectors, and remove pathogens (USEPA, 1993). This chapter will present the state-of-the-art of current minimisation techniques for reducing sludge production in biological wastewater treatment processes.

An overview of the main technologies is given considering three different strategies: in the wastewater line, in the sludge line, or in the final waste line. Any existing processes for sludge minimisation can be placed in one of these strategies (Perez-Elvira *et al.*, 2006) (Figure 10.1):

- *Treatment in the wastewater line (T1 and T2):* The first route is to decrease the production of sludge by introducing the additional stages in the wastewater treatment stage with a lower cellular yield coefficient compared to the one corresponding to the activated sludge process (lysis-cryptic growth, metabolic uncoupling and maintenance metabolism, predation on bacteria, anaerobic treatment). The idea is to reduce sludge production in the wastewater treatment rather than the post-treatment of sludge after generation.
- *Treatment in the sludge line (T3, T4, T5, T6):* The second route is to act on the sludge stage. As anaerobic digestion is the key process in sludge treatment for decreasing and stabilizing the organic solids. However, its application has often been limited by very long retention time (20–50 days) and low overall degradation efficiency (20–50%), which are generally associated with the hydrolysis stage (one of the three stages of anaerobic digestion process: hydrolysis, acetogenesis and methanogenesis) of waste activated sludge. Two options can be considered to enhance the anaerobic digestibility of waste sludge: introducing a pre-treatment process before the anaerobic digestion (physical, chemical or biological pre-treatments), or modifying the digestion configuration (two-stage and temperature-phased anaerobic digestion, anoxic gas flotation).
- *Treatment in the final waste line (T7):* The last minimisation strategy is the removal of the sludge generated in the activated sludge plant (incineration, gasification, pyrolysis, wet air oxidation, supercritical water oxidation) in order to get a final stable, dewatered and pathogen free residue. This way does not represent a

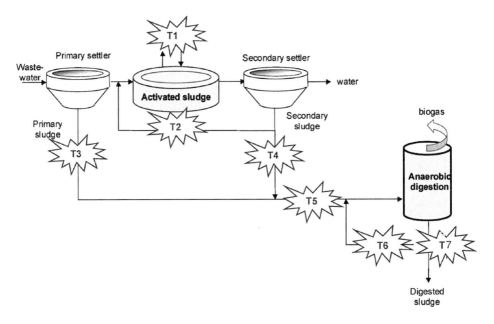

Figure 10.1   Potential location for sludge treatments in a conventional wastewater treatment plant. T1:
Treatment in activated sludge process. T2: Treatment on the activated sludge recirculation
loop. T3: Pretreatment of primary sludge before anaerobic digestion. T4: Pretreatment
of waste activated sludge before anaerobic digestion. T5: Pretreatment of mixed sludge
before anaerobic digestion. T6: Treatment on the anaerobic digester recirculation loop. T7:
Treatment in the final waste line (Reprinted from "Journal of Hazardous Materials, Vol 183
(1–3), H. Carrère, C. Dumas, A. Battimelli, D.J. Batstone, J.P. Delgenès, J.P. Steyer, I. Ferrer,
Pretreatment methods to improve sludge anaerobic degradability: A review, pp. 1–15, 2010
with permission from Elsevier).

minimisation strategy, but a post-treatment to dispose of the sewage solids. However, this chapter focused mainly on treatment in wastewater line and treatment
in the sludge line.

10.2   TREATMENT IN WASTEWATER LINE

10.2.1   Biological lysis-cryptic growth

Lysis refers to the death of a cell by breaking of the cellular membrane, through different
mechanisms. Every cell has a plasma membrane, a protein-lipid bilayer that forms a
barrier separating cell contents from the extracellular environment. When microbial
cells undergo lysis or death, the cell contents (lysate) are released into the medium.
The lysate is rich in soluble COD. The organic autochthonous substrate is reused in
microbial metabolism and a portion of the carbon is liberated as respiration products.
This results in a reduction in the overall biomass production (Low & Chase, 1999).
The biomass grew on organic lysate is different from growth on original substrate, and
is therefore termed as cryptic growth. It consists of lysis and biodegradation, where

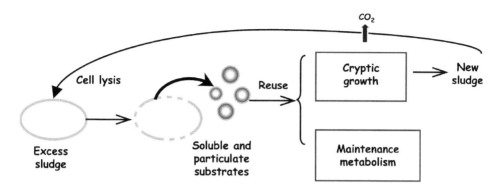

Figure 10.2 Schematic of lysis and cryptic growth (Reprinted from "Water Research, Vol 43 (7), Libing Chu, Sangtian Yan, Xin-Hui Xing, Xulin Sun, Benjamin Jurcik, Progress and perspectives of sludge ozonation as a powerful pretreatment method for minimisation of excess sludge production, pp. 1811–1822, 2009, with permission from Elsevier).

former does not occur under normal conditions, however once lysed, it becomes easy for the living cells to biodegrade the lysed cells, therefore lysis is the rate-limiting step of lysis-cryptic growth, and an increase of the lysis efficiency can therefore lead to an overall reduction of sludge production (Khursheed & Kazmi, 2011) (Figure 10.2).

Several treatment methods based on lysis-cryptic growth mechanism were studied globally including: ozonation, chlorination, combined thermal and alkaline treatment, increase of oxygen concentration and enzymatic reactions. When the treated sludge is returned to the biological reactor, degradation of the secondary substrate generated form the sludge pre-treatment takes place, hence resulting in a reduction in the sludge production (Perez-Elvira *et al.*, 2006).

High purity oxygenation

Several studies have been carried out on the excess sludge reduction by aerobic digestion, and inconsistently reported a sludge reduction from zero to 66%. McWhirter (1978) reported that the growth yield in purified oxygenation activated sludge process could be reduced by 54% compared to conventional process, even at high sludge loading rates. Boon and Burges (1974) reported that, for a similar sludge retention time, the sludge yield in the pure oxygen system was only 60% of the yield obtained in the process utilizing non-purified air. Wunderlich *et al.* (1985) revealed a significant decrease in sludge generation from 0.38 to 0.28 mg VSS/mgCOD removed as the SRT increased from 3.7 to 8.7 days in high-purity oxygen activated sludge system. It shows that the pure oxygen aeration process operated at a relatively longer SRT is more efficient in reducing of excessive sludge production. Abbassi *et al.* (2000) reported that an increase in dissolved oxygen (DO) concentration from 2 to 6 mg/L led to almost 25% sludge reduction at the sludge loading of 1.7 mgBOD$_5$/(mgMLSS.d). The increase of the DO in the bulk liquid led to a deep diffusion of oxygen, which subsequently caused an enlargement of the aerobic volume inside the flocs. As a result,

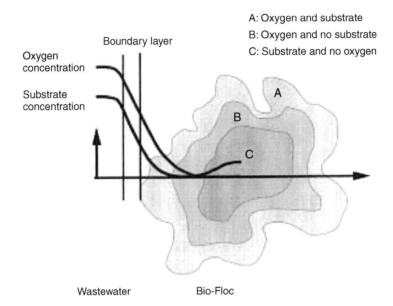

Figure 10.3 Oxygen and substrate concentration profiles within a biological floc (Reprinted from "Water Research, Vol 34 (1), B. Abbassi, S. Dullstein, N. Räbiger, Minimisation of excess sludge production by increase of oxygen concentration in activated sludge flocs; experimental and theoretical approach, pp. 139–146, 2000, with permission from Elsevier).

the hydrolyzed microorganisms in the floc matrix could be degraded and thus sludge quantity was reduced (Figure 10.3).

Several studies has been published in past on the effect of high oxygenation on sludge minimisation, however, the mechanism of lower sludge synthesis with high operating DO is not clearly known till yet. Therefore, the substantial work is required on this process. The major advantages of the high purity oxygen process in comparison of convention process are: (a) the ability to maintain a higher MLVSS concentration in the aeration tank; (b) high DO-activated sludge processes can repress the development of filamentous organisms; (c) better sludge settling and thickening can be achieved; lower net sludge production; higher oxygen transfer efficiency and more stable operation. However, the major drawbacks are (a). The efficacy of the process is not clear. (b). The mechanism is not fully known and (c). High aeration cost (Perez-Elvira *et al.*, 2006).

Consequently, high oxygen process shows great industrial potential for minimisation of excess sludge production as well as in improvement of system operation. However, economic-efficiency and energy-balance calculations as important tools for performing the cost-benefit analysis of a disintegration process shall be taken in to consideration (Khursheed & Kazmi, 2011).

Ozonation

The coupling of activated sludge system with intermittent ozonation has been well established. A portion of recycled sludge passes through the ozonation chamber, and

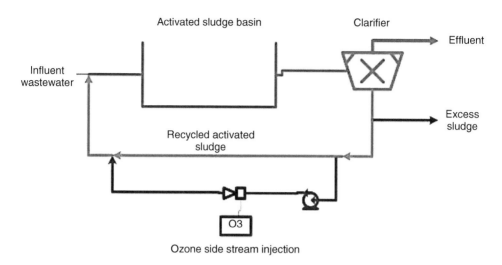

Figure 10.4 Sludge ozonation process (Source: Spartan Environmental Technologies).

the ozonised sludge is decomposed in the biological treatment (Figure 10.4). The recycling of ozone-pretreated sludge into the aeration tank will stimulate the cryptic growth (Perez-Elvira *et al.*, 2006).

The ozone attacks the cell walls of the bacteria that make up the activated sludge causing the cells to undergo lysis. Lysis is a process where the cellular wall is breached and the internal materials leak out of the cell. In the activated sludge process this leaked material, or cellular COD, goes back to the activated sludge tank where other bacteria consume the COD. As it turns out this material is very biodegradable. Essentially, the ozone damages certain sludge bacteria, so they can be consume by rest of the bacteria (biomass or sludge). This process has been studied extensively and found to work with up to a 60% reduction in excess sludge. Other benefits of the process that have been observed include: elimination of foaming problems, reduction in bulking, improvement in dewatering, improvements in settling and improvement in effluent quality (Spartan Environmental Technologies).

Kamiya and Hirotsuki (1998) reported that the excess sludge production was reduced by 50% at an ozone dose of 10 mg/g MLSS in the aeration tank per day. When the ozone dose was kept as high as 20 mg/g, no excess sludge generation was observed. Moreover, Yasui and Shibata (1994) achieved 100% sludge reduction at an ozone dose of 50 mg/g MLSS in the aeration tank per day. Most activated sludge microorganisms are lysed and oxidised to organic substances in the ozonation chamber and these organic substances can be decomposed in the subsequent biological treatment (Kamiya & Hirotsuji, 1998). The effluent quality with respect to dissolved organic carbon concentration is not influenced significantly in the ozonation mediated activated sludge processes, moreover, the sludge settleability in terms of SVI was remarkably improved in comparison of control test without ozonation. This technology with commercial name 'Biolysis O' (developed by Ondeo-Degremont) is successfully established at full-scale plants (Yasui *et al.*, 1996). In this process, mixed liquor extracted from the activated sludge basin is contacted with ozone in a reactor and returned to the activated sludge tank. A demonstration of Biolysis O in France achieved a sludge reduction of between 30 and 80%. Future research is required regarding the optimization of ozone dosage, dosing mode (continuous or intermittent), and reactor configuration (bubble or airlift reactor) (Perez-Elvira et al., 2006).

The major advantages of this technique are (a). No significant accumulation of inorganic solids in the aeration tank at optimal ozone dose rates (b). A remarkable improvement in sludge settleability (SVI) (c). Successful full-scale experience. However, the disadvantages of this technique are (a). Slight increase in TOC concentration in the effluent (although mainly composed of proteins and sugars, which should be harmless for the environment) (b). High costs operating cost (Perez-Elvira et al., 2006).

Chlorination

Chlorine is a low cost alternative of ozonation. A significant sludge reduction of 60% was observed with chlorination mediated activated sludge process (Saby et al., 2002), at the chlorine dose of $0.066\,g\ Cl_2/g$ MLSS and then recycling of chlorinated sludge with a duration of 20 h to activated sludge system. The major advantage is the lower cost of treatment than that of ozone. However, the drawbacks are: (a). Formation of trihalomethanes (THMs) (b). Significant increase of soluble chemical oxygen demand in the effluent (c). Poor sludge settleability (Perez-Elvira et al., 2006).

Thermo-chemical treatment

Approximately 60% of sludge reduction was achieved when the returned sludge was passed through a thermal reactor operating at 90°C with reaction time of 3 h (Canales et al., 1994). The chemically (acid/alkali) assisted thermal treatment was also applied to reduce the excess sludge generation. Rocher et al. (2001) showed that hybrid themo-alkaline treatment (pH 10, 60°C for 20 min) was the most efficient process to enhance the cell disintegration and achieved a 37% reduction in the excess sludge production. The major disadvantages of combined chemical and heat treatment are corrosion and odor generation (Perez-Elvira et al., 2006).

Enzymatic reactions

Enzymatic reactions are the basis of a novel wastewater treatment process, designed by combining the conventional activated sludge process with thermophilic aerobic sludge digester in which the excess sludge is solubilised by thermophilic enzyme (Sakai et al., 2000). The process includes of two different stages, one for a biological wastewater treatment and the other for a thermophilic aerobic digestion of sludge. A portion of recycled sludge from the aeration tank is passed through a thermophilic aerobic sludge digester, in which the sludge is solubilised by the thermophilic aerobic bacteria (e.g. Bacillus sp.) and mineralised by mesophilic bacteria. The solubilised sludge is returned to the aeration tank for its further degradation. Pilot scale studies revealed a 93% reduction in the overall excess sludge generation, and an efficient removal of organic pollutant (i.e. BOD). This technology with commercial name Biolysis E (developed by Ondeo-Degremont) is successfully applied at a full-scale sewage treatment plant operated for three years and showing a 75% reduction in overall excess sludge generation.

Table 10.1 Scientific efforts to minimise the excess sludge generation using lysis-cryptic growth process (Wei *et al.*, 2003a).

Process	Sludge reduction (%)	Source
Ozonation		
Full scale; 550 kg BOD/d of industrial wastewater, continuous ozonation at 0.05 g O_3/g SS	100	Yasui *et al.* (1996)
Lab scale, synthetic sewage, intermittent ozonation at 11 mg O_3/mg SS (aeration tank) d	50	Kamiya & Hirotsuji (1998)
Chlorination		
Bench scale, 20°C, synthetic wastewater, 0.066 gCl_2/gMLSS	65	Saby *et al.* (2002)
Thermal or thermo-chemical treatment		
Lab scale (90°C for 3 h); membrane bioreactor; synthetic wastewater	60	Canales *et al.* (1994)
Lab scale (60°C for 20 min, pH 10); synthetic wastewater	37	Rocher *et al.* (2001)
Increasing DO		
Lab scale; synthetic wastewater; increasing DO from 2 to 6 mg/l	25	Abbassi *et al.* (2000)

This technique consists of the thickening of mixed liquor and then passing it thorough a thermophilic, enzymatic reactor operating at about 50 to 60°C. These operating conditions enhance the development of a particular type of microbe. When activated, the microbes produce enzymes that attack the outer membrane of the bacteria present in the activated sludge, and reducing their ability to reproduce. The enzymes are released by the bacteria in such a way that they are unable to reproduce and grow. The heated, degraded sludge then passes through a heat exchanger to recover some of its energy before flowing back to the activated sludge basin. No external enzymatic source is required. A remarkable sludge reductions ranging from 30 to 80% (depending on the quantity of the sludge sent daily to the reactor) has been reported (Perez-Elvira *et al.*, 2006). The main advantages of enzymatic reaction process are: (a) Successful operation at full-scale (b). Suppress filamentous growth (c). Capital and operating cost is similar to or lower than that of conventional treatment systems. However, the drawbacks including the small increase of the SS and COD concentrations in effluent (Perez-Elvira *et al.*, 2006).

Table 10.1 summarises sludge reduction efforts under lysis-cryptic growth condition. Among these techniques on the basis of lysis-cryptic growth sludge ozonation for reducing sludge production has been successfully applied in practice.

10.2.2 Maintenance and endogenous metabolism

The cells for their maintenance use the energy obtained and captured in the form of ATP during biological oxidation, which is so-called maintenance metabolism. The maintenance energy includes energy for turnover of cell materials, active transport, motility, etc. The substrate utilization correlated with maintenance of the living functions of microorganisms is not synthesised of new cellular mass. Thus, the sludge production should be inversely related to the activity of maintenance metabolism (Chang *et al.*, 1993). Therefore, more sludge age results in increased energy consumption for maintenance, leaving less energy for synthesis. Longer SRT

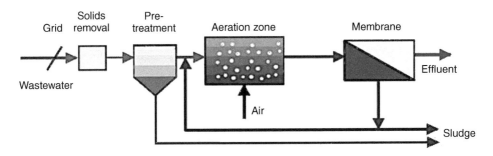

Figure 10.5 A schematic of membrane bioreactor (Source: Anja Drews, Wikipedia).

causes reduction in sludge loading rate or low food to microorganisms (F/M) ratio and reduces the sludge production (Van Loosdrecht & Henze, 1999). In case of higher concentration of reactor biomass, if the substrate is limiting to the extent that it is just sufficient to satisfy the energy requirement for maintenance than in a dynamic situation no energy would remain available to sustain the growth of microorganisms anymore. The major advantage of the endogenous metabolism is that the incoming substrate could be finally oxidised to CO_2 and H_2O, while results in a lower biomass production (Khursheed & Kazmi, 2011).

Membrane bioreactor

Membrane bioreactor (MBR) is the application of concept of increased energy consumption in cell maintenance, leaving little for growth, ideally to attain no sludge from wastewater bio-oxidation (Khursheed & Kazmi, 2011) (Figure 10.5).

In a membrane reactor, solids retention time (SRT) can be controlled independently from hydraulic retention time (HRT), which will result in a higher sludge concentration (usually 15–20 g/l), and subsequently in a lower sludge-loading rate. When this sludge-loading rate becomes low enough, little or no excess sludge is produced (Yamamoto *et al.*, 1989), however, this route is relatively expensive in terms of energy requirements. The 90% of the influent COD is oxidised to CO_2 in membrane bioreactor, and mixed liquor concentration is almost constant in the reactor, without sludge wastage (Yamamoto *et al.*, 1989). A significant reduction of 44% in biomass yield was reported by Low and Chase (1999). Rosenberger *et al.* (1999) reported a significantly low sludge production (0.002–0.032 kg/d) without sludge discharge for one year at 15–23 g SS/L and low F/M ratio of 0.07 kg COD/kg MLSS d. Thus, MBR can produce high-quality effluents with low sludge generation, which ultimately will reduce the sludge handling and disposal costs if compare with conventional activated sludge processes. Moreover, it has been successfully applied at full-scale level. However, membrane fouling and high cost of membranes are main drawbacks for wider application of MBRs (Figure 10.6). Over the past few years, considerable investigations have been performed to understand MBR fouling in detail and to develop high-flux or low-cost membranes. The essential prerequisite of maintenance of higher MLSS at high SRT in the MBR increases fouling and energy demand and decreases sludge

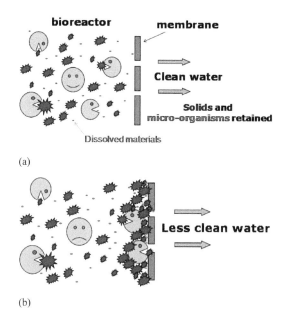

Figure 10.6 Schematic describing the (a) MBR process and (b) membrane fouling (Source: Pierre Le Clech, Wikipedia).

production. While less SRT decreases MLSS and associated energy consumption but increases sludge production, which further require energy for its treatment. Therefore, MBR operation could be optimised between these two parameters on account of energy consumption and membrane fouling (Khursheed & Kazmi, 2011). The extracellular polymeric substances (EPS) or soluble microbial products (SMP) were reported as the key foulants apart from the suspended and colloidal fractions of the waste. The carbohydrate and poly-saccharide colloidal matter of the SMP have been correlated to promote membrane fouling under certain conditions (Rosenberger *et al.*, 2006).

Apart from the high cost of membrane (which requires frequent cleaning and replacement due to fouling), the MBR is an energy intensive process. Use of ceramic membranes and anaerobic operation are two ways to make MBRs more energy efficient technique. Ceramic membranes are more fouling-resistant but are currently high in cost, which may expectedly be reduced with time owing to advances in fabrication techniques (Bishop, 2004). Moreover, anaerobic MBRs can be advantages over aerobic system since anaerobic treatment require less energy as well as generate energy rich methane gas.

10.2.3 Metabolic uncoupling

Metabolism is the sum of biochemical transformations, including interrelated catabolic and anabolic reactions. The cells synthesis is directly proportional to the amount of energy (ATP) produced via catabolism (oxidative phosphorylation). The uncoupling approach is to increase the discrepancy of energy level between catabolism and anabolism, so that the energy supply to anabolism is limited. This result in dropped growth yield of biomass without reducing the removal rates of organic pollutants in biological wastewater treatment and may therefore provide a direct mechanism for reducing sludge production (Khursheed & Kazmi, 2011; Perez-Elvira *et al.*, 2006). Uncoupled metabolism is observed under some conditions, such as in the presence of inhibitory compounds, heavy metals, abnormal temperatures, limitation of nutrients, excess energy source, and exposure of sludge to cyclic change in ATP content (Liu & Tay, 2001). In the energy uncoupling process, the rate of substrate feasting is higher than that necessary for growth and maintenance of biomass. Thus, the observed growth yield of activated sludge would be reduced markedly. It can be considered as a promising way to reduce excessive sludge production by controlling metabolic state of microorganisms in order to maximise dissociation of catabolism from anabolism.

Chemical uncoupler

In aerobic bacteria, ATP is produced by oxidative phosphorylation. This chemiosmotic mechanisms of oxidative phosphorylation can be efficiently uncoupled by the supplement of organic protonophores, which transport protons through cells' intracellular cytoplasm membrane, such as 2,4-dinitrophenol (dNP), para-nitrophenol (pNP), pentachlorophenol (PCP) and 3,3′,4′,5-tetrachlorosalicylanilide (TCS). In the presence of these compounds, the majority of organic substrate is oxidised to carbon dioxide rather than used for biosynthesis. Consequently, the growth efficiency is lowered in uncoupler containing activated sludge process (Perez-Elvira *et al.*, 2006).

Several researchers have examined the effect of uncoupler containing activated sludge process on sludge minimisation. A remarkable reduction of 50% in biomass generation was achieved with the addition of PCP and pNp, respectively, if compared with control reactor (no uncoupler) (Low *et al.*, 2000). Chen *et al.* (2002) reported a growth yield reduction of 78% at a TCS concentration of 0.8 mg/l. Chemical uncouplers may provide a promising way for sludge reduction. However, the real use of organic protonophores at full scale is impractical for several reasons, which include the inherent toxicity of protonophores. Thus, the removal of the additives is required prior to discharge the effluent in water bodies. However, it is expected that the combination of pure oxygen aeration process with the metabolic uncoupling technique would generate a novel and efficient biotechnology for minimisation of the excess sludge production (Perez-Elvira *et al.*, 2006).

The main advantage of this method is that it only needs to add a defined uncoupler dosing. However, the main disadvantages are: (a). Lack of substantial research on chemical uncoupling process (b). Most of the organic protonophores are xenobiotic and potentially harmful to the environment (c). Unexpected increase in the O_2 requirement (obtained in full-scale application) (d). Acclimation problems for the microorganism (Perez-Elvira *et al.*, 2006).

Oxic-Settling-Anaerobic process (OSA)

The principal mechanism of OSA process is alternate anaerobic–aerobic cycling of activated sludge in order to stimulate catabolic activity and suppress anabolism by uncoupling the two reactions, resulting in a reduced sludge synthesis. In OSA process, thickened sludge from a final settling tank is returned to aeration tank via a sludge

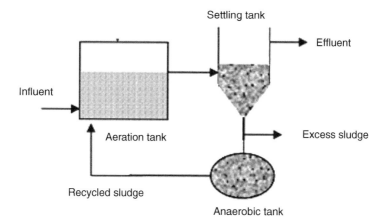

Figure 10.7 Schematic of Oxic settling anaerobic process (Reprinted from "Biochemical Engineering Journal, Vol 49 (2), Fenxia Ye, Ying Li, Oxic-settling-anoxic (OSA) process combined with 3,3′,4′,5-tetrachlorosalicylanilide (TCS) to reduce excess sludge production in the activated sludge system, pp. 229–234, 2010, with permission from Elsevier).

holding tank. In the holding tank, anaerobic conditions are maintained without addition of surplus influent substrate. A higher biomass concentration of the re-circulated sludge is maintained in the sludge holding tank at a longer retention time (Perez-Elvira *et al.*, 2006) (Figure 10.7).

Aerobic microorganisms capture energy in the form of ATP released from oxidation of organic content. The same microorganisms are unable to produce required energy when exposed to anaerobic conditions under severe food limiting condition and as a result consume their conserved ATP. Again on return to aerobic condition, augmentation of their depleted ATP reserves become first priority as against synthesis of new cell mass or anabolism. This in other words promotes catabolism and demotes anabolism by uncoupling the two reactions, thereby inducing sludge reduction (Chudoba *et al.*, 1992). Chudoba *et al.* (1992) reported that sludge yield was significantly reduced by about 38–54% in comparison with conventional activated sludge process. It appears OSA process can successfully handle problem of high excess sludge production that is particularly important when handling high strength waste streams while maintaining the economical feasibility (Khursheed & Kazmi, 2011).

The main advantages of this process are: (a) Easy to incorporate the anaerobic reactor to the conventional activated sludge process. (b) Filamentous growth can be reduced. (c) Improvement in sludge settleability and COD removal efficiency. (d) Capable to handle the surge of high organic loading

10.2.4 Predation on bacteria

Considering a biological wastewater treatment process as an artificial ecosystem (habitat for bacteria and other organisms), sludge production could be reduced by bacteriovory. Both living and death bacteria can be utilised in trophic reactions (as a

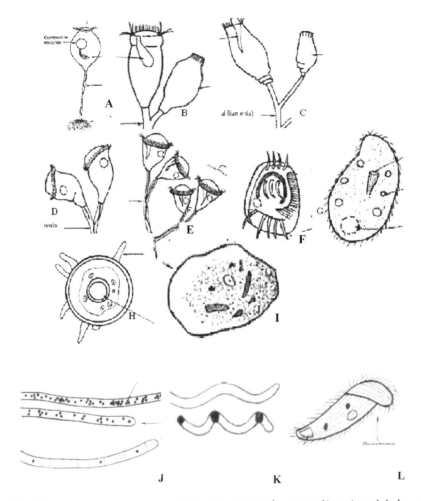

Figure 10.8 Higher bacteriovory microorganisms most commonly occurred in activated sludge system. A. *Vorticella*, B. *Epistylis*, C. *Opercularia*, D. *Carchesium*, E. *Zoothamnium*, F. *Aspidisca*, G. *Chilodonella*, H. *Arcella*, I. *Amoeba (Vahlkamphia limicola)*, J. *Beggiatoa*, K. *Spirillus*, L. *Metopus*.

food source) by higher bacteriovoric microorganisms, such as protozoa (ciliates, flag-ellates, amoeba and heliozoa) and metazoa (rotifera and nematoda), that predate on the bacteria (Figure 10.8). Protozoa are considered to be the most common predators of bacteria, making up around 5% of the total dry weight of a wastewater biomass (70% of these are ciliates) (Perez-Elvira *et al.*, 2006). The mixed liquor fauna consists of bacteria, protozoa, metazoa, where bacteria are the primary consumers, which are consumed by protozoa and metazoa. The protozoa can be classified as ciliates (free swimming, crawling and sessile), flagellates and amoeba. The metazoa consist nor-mally of rotifer, nematode and Naididae. Protozoa normally dominate the predator population in activated sludge system and metazoa in trickling filters. The presence of predators suppresses the growth of dispersed bacteria and results in favor of floc formation. As a result major portions of the sludge remain unaffected by predation activity (Khursheed & Kazmi, 2011).

Several researchers have shown reduction in sludge production due to predation of bacteria by exploiting higher organisms such as protozoa and metazoa in the different activated sludge processes modifications. Ratsak *et al.* (1993) reported a 12–43% reduction in the overall biomass production by employing the ciliated Tetrahymena pyriformis to graze on Pseudomonas fluorescens. Similarly, Lee and Welander (1996) observed a 60–80% decrease in the overall biomass production by employing the protozoa and metazoa in a mixed microbial culture.

Later, a shift on research related to sludge reduction by predators was observed towards oligochaeta apart from protozoa and metazoa mainly due to process uncertainties. With advantages such as low cost and no secondary pollution, worm (Oligochaete) technology, based on using microfauna's predation to reduce excess sludge has recently begun to receive increasing attention by the researchers. The main types of worms present in activated sludge system and trickling filters are Naididae, Aeolosomatidae and Tubificidae (Khursheed & Kazmi, 2011). Nevertheless, maintaining of high densities of worms for a long-time and no well-defined relationship between operational conditions (retention times, temperature, sludge loading rates) and worm growth are the major challenge to scaleup this process. Sludge reductions of 10–50% in the trickling filter were achieved with worms compared with 10–15% without worms, respectively. Ratsak (1994) reported that a major worm bloom (Nais elinguis, Pristina sp., Aeolosoma hemprichicii) resulted in a low SVI, lower energy consumption for oxygen supply and less sludge disposal (25–50% sludge reduction). Wei *et al.* (2003b) observed a low sludge yield (0.10–0.15 kg SS/kg COD removed) in a two-stage gravitational submerged MBR system predominated by a high worm density of 2600–3800 Aeolosoma/mL mixed liquor. A combination of a Tubificidae (17600 g/m^3) reactor with an integrated oxidation ditch (HRT:15.4 h; DO: 0.5–3.0 mg/L; 20°C) with vertical circle (HRT: 13.2–15.4 h; DO: 1.0–2.5 mg/L), was investigated for excess sludge reduction (first stage) and returned sludge stabilisation (second stage) as a new integrated system. The findings revealed that the excess sludge reduction rate was 46.4% in the first stage, and the average sludge yield of the integrated system was 6.19×10^{-5} kg SS/kg COD in the second stage (Song *et al.*, 2007).

The research outcomes regarding the sludge reduction using worms are promising, however, the control parameters of the system have not been established till yet, which is a prerequisite to scale up the process. Thus, integration of different parameters of wastewater treatment (F/M ration, SRT, MLSS concentration, treatment temperature, dissolved oxygen concentration etc.) with the population growth and density must be taken into account in future studies. The main advantage of this process is that it is using in large operations today. However, the drawbacks are: (a) The worm's growth is still uncontrollable, specially in the full-scale application. (b) High capital and operation costs (Perez-Elvira *et al.*, 2006).

10.2.5 Anaerobic treatment of sewage (ANANOX process)

The ANANOX process is an example of process integration considered for obtaining good effluent characteristics while minimizing sludge production and energy demand. The first stage of this process uses an ABR (Anaerobic baffled reactor) comprising

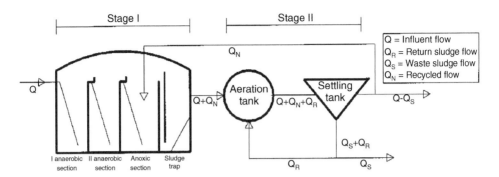

Figure 10.9 Flow diagram of ANANOX process (Source:Act Clean).

two floc sludge blanket sections, one anoxic sludge blanket (for denitrification) and a sludge trap (designed to avoid massive sludge escape from the reactor). The second stage is fed with the effluent of the first stage and is made up of an activated sludge aeration tank and a settling tank (Figure 10.9). The final effluent is partially recycled to the anoxic stage for denitrification (Perez-Elvira *et al.*, 2006).

Investigations on a pilot scale system resulted in achieving higher removal efficiencies of 90% for total COD, 90% for total suspended solids (TSS) and 81% for the total nitrogen (TN). In addition, there was a remarkably low production of sludge i.e. limited to 0.2 kgTSS/kgCOD removed (Perez-Elvira *et al.*, 2006). During a full scale study (Garuti *et al.*, 2001) with 30 m^3 ABR, 15 m^3 aeration tank and 32 m^3 secondary clarifier, a total COD removal efficiency of 95% and TSS removal of 92% was achieved at best operating conditions. The major advantages of ANANOX process are: (a) Able to achieve 30–50% reduction of sludge production; (b) Capable of achieving stringent effluent standards; (c) Robustness and versatility, and compact configuration. However, the major drawbacks are: (a) Must be subjected to numerous experiments that help to optimise the efficiency in the anaerobic phase and define the admissible values for organic load and upflow velocity in the ABR. (b) The ANANOX process is not recommended for very low sewage temperatures. Research on the anaerobic treatment of very cold effluents has pointed out the potential of new reactor concepts, like the EGSB (Expanded Granular Sludge Bed), for which research on its application to sewage is in course (Perez-Elvira *et al.*, 2006).

10.2.6 Comparative evaluation of merits and demerits of treatment in wastewater line processes

The strategies for sludge minimisation should be evaluated and chosen for full-scale application using (1) cost analysis including the capital and, operation and maintenance costs, and paybacks achieved by reduced sludge treatment and disposal, (2) possible environmental impact of the proposed technology i.e. odor problems, nutrient release and toxicity of trace chemicals in effluent, (3) demerits of each technology must be evaluated along with the possible benefits arise with the implication of techniques. Table 10.2 summarises relative merits and demerits of the treatment in wastewater line processes of sludge minimisation. Now a days, both ozonation and MBR has

Table 10.2 Relative merits and demerits of treatment in wastewater line processes (Wei *et al.*, 2003a; Khursheed & Kazmi, 2011).

Process	Merits	Demerits	Source
Ozonation	• Successfully applied at full scale	• High costs involved in ozonation • Ozone wastage	Yasui *et al.* (1996)
Chlorination	• Cheaper than ozonation	• Decrease of COD removal rate • Poor sludge settling • Formation of THM	Saby *et al.* (2002)
Thermal or thermo-chemical treatment	• Relatively simple	• Corrosion • Neutralization required • Odor problem	Rocher *et al.* (2001)
High operating DO and Pure oxygenation	• Simple stable and reliable operation, easy to implement • Better sludge settling and thickening • Higher oxygen transfer efficiency	• High aeration cost • Process efficacy is not clear • Mechanism is not fully known.	Abbassi *et al.* (2000); McWhirter (1978)
Maintenance metabolism MBR	• Flexible operation, high effluent quality, small footprint • Very high MLSS can be maintained • Applied at full-scale • Upto zero sludge growth achieved	• Membrane fouling is responsible for high cost • Sludge settling and dewatering is more difficult • Poor oxygenation: increased aeration cost • Not feasible to operate with complete sludge retention, minimum wasting is desired. • Energy requirements	Rosenberger *et al.* (1999)
OSA	• Easy retrofit of anaerobic zone to the conventional activated sludge process • Control of filamentous growth • Improvement in COD removal and settleability of sludge • Capable of handling high strength organic pollutants without serious sludge associated problems	• Sometimes high sludge production • Further research is needed to understand the process and to establish the optimum operational conditions to improve the process	Chudoba *et al.* (1992), Chen *et al.* (2003)
Predation on bacteria	• Stable operation • Relatively simple • Environmental friendly	• Unstable worm growth • Nutrients release • Kinetics is still undefined	Lee & Welander (1996), Wei *et al.* (2003a)

been functional successfully at full scale, however, other techniques based on uncoupling metabolism and predation on bacteria are still lab scale curiosity. Due to high operation costs caused from ozone production i.e. over 50% of the total operation costs (Yasui *et al.*, 1996), it is essential to reduce the amounts of ozone required for sludge treatment. The amounts of ozone required for reducing excess sludge generation are determined by the gas phase ozone concentration, the ozonation mode (continuous or intermittent ozone dose), ozone reactor configuration (bubble or airlift reactor), and the concentration of sludge treated by ozone (Wei *et al.*, 2003a). The OSA process may present a robust cost-effective solution to cope up with excess sludge problem, since addition of an anaerobic tank is only required. Nevertheless, the main disadvantage of the OSA system is higher sludge production caused sometimes by an ORP disturbance and costs raised by the addition of an anaerobic tank. High rate oxygenation has limited application and efficiency, but it does not require any additional unit, however, process economy and energy-balance required further assessment (Khursheed & Kazmi, 2011). More research is required on membrane materials, design of membrane module, impact of membrane on microbial community, membrane fouling and its countermeasures, which will lead to reduce the capital and operational cost of MBR systems. Sludge minimisation by oligochaetes can provide a promising and environmental friendly method, however, unstable worm growth is a major drawback, which must be resolved before scale up this method (Wei *et al.*, 2003a).

10.3 PROCESS IN THE SLUDGE LINE

Anaerobic digestion (AD) is the most common sludge treatment method mainly due to the generation of (1) low sludge; (2) energy-rich methane gas and (3) nutrient rich end product. However, long retention time (20–50 days) and low digestion efficiency (20–50%) limits the application of AD, which is due to slow rate of hydrolysis of waste sludge. The rate of sludge hydrolysis can be enhanced by sludge disintegration process using the sludge pretreatment processes like thermal, chemical and mechanical. These pretreatment processes are easier to manage, stable in performances and flexible during operation (Liu, 2003).

The application of sludge disintegration methods causes the microbial cell lysis by rupturing the cell wall and subsequent release of cell material (lysate) into the medium, which is rich in soluble COD. The organic autochthonous substrate is reused in microbial metabolism and a portion of the carbon is liberated as respiration products. This ends up in the reduction of overall biomass production (Low & Chase, 1999). Thus sludge pretreatment helps to (i) enhance the rate of hydrolysis rate and the entire process rate; (ii) reduce the hydraulic retention time in the anaerobic digesters, thus a large amount of sludge could be treat in the same volume; (iii) improve the biogas production; and (iv) dewatering properties of sludge.

10.3.1 Chemical pretreatment techniques

Chemical pretreatment of sludge can be achieved by the use of strong acids or alkali and ozone.

Acidic/alkaline pretreatment

Extremely alkaline conditions or extremely acidic conditions cause the disruption of sludge cells, which ultimately leads to the leakage of intracellular material into the bulk liquid. Generally, hydrochloric acid and sulphuric acid are used for acidic pretreatment of sludge. A high sludge solubilisation of 50 to 60% was achieved by the researchers at a sulfuric acid dose of 0.5–1.0 g acid/gTSS and 30s reaction time (Woodard and Wukasch, 1994). However, the pH adjustment is require before feeding the acidified sludge to anaerobic digester.

The most common chemicals used for alkaline sludge pretreatment are lime, potassium hydroxide and sodium hydroxide. The cell membrane is solubilised due to saponification of lipids. Several studies reported that the significantly higher sludge solubilisation in terms of COD solubilization (28%–63%) were achieved at different sodium hydroxide dosage and reaction time. Which ultimately enhanced the COD and VS removal, gas production, sludge dewaterability and pathogens reduction too (Lin *et al.*, 1997; Penaud *et al.*, 1999; Heo *et al.*, 2003; Neyens *et al.*, 2003). Few studies reported that alkaline pretreatment was more effective than the acidic one (Chen *et al.*, 2007). The alkaline and acidic sludge pretreatment are considered easy in operation, energy efficient and highly effective in pathogen removal. However, the odor generation, corrosion of equipments and pH adjustment of pretreated sludge are the major drawbacks (Tyagi & Lo, 2011).

Ozonation

Ozone pretreatment invlolves the sludge solubilization and subsequent mineralization of soluble organics into carbon dioxide (Lee *et al.*, 2005). A broad range of ozone dosage from 0.05 g O_3/g TS to 20 mg O_3/g TSS were studied by the researchers. A significant improvement in COD solubilisation, VS removal and biogas production was reported in earlier studies (Liu *et al.*, 2001; Braguglia *et al.*, 2012; Bougrier *et al.*, 2006; Chu *et al.*, 2009). However, the best ozone dose was recommended between 0.03 to 0.05 g O_3/g TSS, by keeping in view the treatment cost and sludge reduction potential. However, the ozone pretreatment were reported not to be capable to effectively elimintate the pathogens (Chu *et al.*, 2008; Carballa *et al.*, 2009). The reaction time and ozone demand exerted by organics present were also considered important factors to govern the rate of pathogens removal potential. Ozone pretreatment is a well-studied technique and successfully scale-up into the field. The other beneficial effects of ozone are improvement in sludge settleability and effective control of sludge bulking and foaming. However, the ozone pretreatment is an energy intensive technology.

Fenton peroxidation

Fenton peroxidation solubilised the EPS and ruptures the bacterial cell walls, which leads to release of interacellular material into the bulk liquid. The peroxidation process utilises the activation of H_2O_2 by iron salts (Fe^{2+}). Fenton peroxidation was found effective in sludge solubilisation and to achieve a high improvement in biogas generation (Dewil *et al.*, 2007). However, the process in highly corrosive in nature and expensive due to high cost of operation and maintenance (Tyagi & Lo, 2011).

10.3.2 Thermal treatment

In thermal treatment, high temperature is applied to disintegrate the sludge, which ultimately enhances the sludge biodegradability, improves the biogas generation and

Table 10.3 Summary of thermal pretreatment studies.

Treatment conditions	Anaerobic dihgestion	Findings	Source
175°C, 60 min	CSTR, 5 days HRT, 35°C	• Increase of 3% VSS degradation and of 100% methane production	Li & Noike (1992)
180°C, 60 min	Batch, 8 days, 37°C	• 90% increase of methane production • VSS solubilisation of 30%	Tanaka *et al.* (1997)
121°C, 30 min	Batch, 7 days, 37°C	• Increase of VS reduction by 30% • 32% increase in biogas production	Kim *et al.* (2003)
170°C, 60 min	Batch, 24 days, 35°C	• 45% increase in biogas production	Valo *et al.* (2004)
170°C, 30 min	CSTR, 20 days HRT, 35°C	• 61% increase in biogas production	
170°C, 30 min	Batch, 24 days, 35°C	• 76% increase in biogas production	Bougrier *et al.* (2006)
200°C, 30 min, 20 MPa	Two-stage UASB, 3.8 days HRT, 35°C	• 15% increase in methane production	Yang *et al.* (2010)

offset the requirement of digester heating. Thermal treatment can be achieved by any of two methods i.e. conventional and microwave heating.

Conventional heating

Thermal treatment disrupts the chemical bonds of the cell wall and membrane, thus solubilise the cell materials (Apples *et al.*, 2008). In comparison with carbohydrates and lipids, proteins are hardly degradable, since they are protected from the enzymatic hydrolysis by the cell wall. Thermal pretreatment destroys the cell walls in the temperature range of 60 to 180°C and makes the proteins available for bio-degradation. Earlier studies reported that thermal pretreatment significantly enhances the COD solubilisation (upto 80%), VS reduction (upto 55%) and biogas generation (upto 92%) at optimal temperature range of 160 to 180°C and treatment period from 30 to 60 min (Stuckey & McCarty, 1978; Tanaka *et al.*, 1997; Valo *et al.*, 2004; Perez-Elvira *et al.*, 2008). Thermal pretreatment is also reported effective in terms of improvement in sludge settleability (Bougrier *et al.*, 2008) and pathogens removal (Carballa *et al.*, 2009) (Table 10.3).

Earlier studies on thermal pretreatment of sludge at full-scale installation named "Cambi process", reported the significant improvement in biogas production and higher VS removal and improved dewatering characteristics. This process used live steam for sludge preheating to 100°C, minimise operational and corrosion problems, and minimizing the formation of refractory compounds (Kepp *et al.*, 2000) (Figure 10.10).

However, the thermal pretreatment is an energy intensive method having a high operation and maintenance cost with high odor potential. Moreover, fouling of the heat exchangers is another major issue (Perez-Elvira *et al.*, 2006).

Microwave thermal pretreatment

Microwave (MW) working principal based upon direct heat rapidly, which reduces the energy losses while transmitting the energy. Microwaves are the electromagnetic

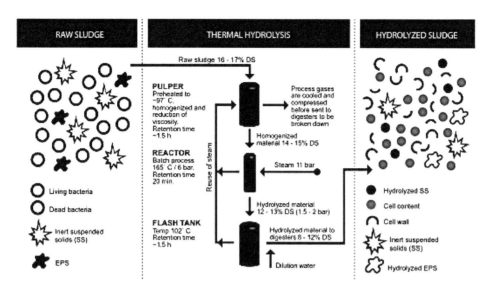

Figure 10.10 Mechanism of sludge disintegration by Cambi process (Source: Cambi recycling energy).

radiation with the frequency range of 300 MHz–300 GHz corresponding to a wavelength range of 1 m down to 1 mm (Banik *et al.*, 2003). The microwaves (domestic and industrial) are work at a frequency of 2.45 GHz. The difference in working principal of conventional and microwave heating is showing in Figure 10.11. A uniform MW field produces energy through the dipoles realignment with oscillating electric fields to produce heat both within and at the surface of the exposed material mainly due to molecular friction (Tyagi & Lo, 2011).

The sludge absorbs microwave irradiation due to high content of water (Qiao *et al.*, 2008). The microwave orientation effect is mainly caused by polarised parts of macromolecules lining up with the poles of the electromagnetic field resulting in the possible breakage of hydrogen bonds (Loupy, 2002). The microwave thermal pretreatment break the polymeric network of sludge resulting in the release of extracellular and intracellular materials into the liquid phase (Eskicioglu *et al.*, 2006). In recent years, the microwave thermal pretreatment has been studied widely in order to enhance the anaerobic digestibility of sludges mainly due to the fact that microwave heating is more rapid than conventional heating with good energy efficiency (Table 10.4).

The microwave thermal pretreatment was found to be superior over the conventional heating mainly due to faster heating, low energy consumption and less space requirement. MW pretreatment was able to significantly enhance the COD solubilisation (upto 35%), VS reduction (upto 36%), methane generation (upto 80%) at the studied temperature ranges from 91°C to 175°C (Park *et al.*, 2004; Eskicioglu *et al.*, 2008; Eskicioglu *et al.*, 2009). The MW pretreated sludge shows good dewaterability characteristics in comparison with conventionally heated sludge (Pino-Jelcic *et al.*, 2006). Moreover, MW pretreatment was found efficient in pathogens removal. It was reported that Class A type sludge can be produce by feeding the MW pretreated sludge

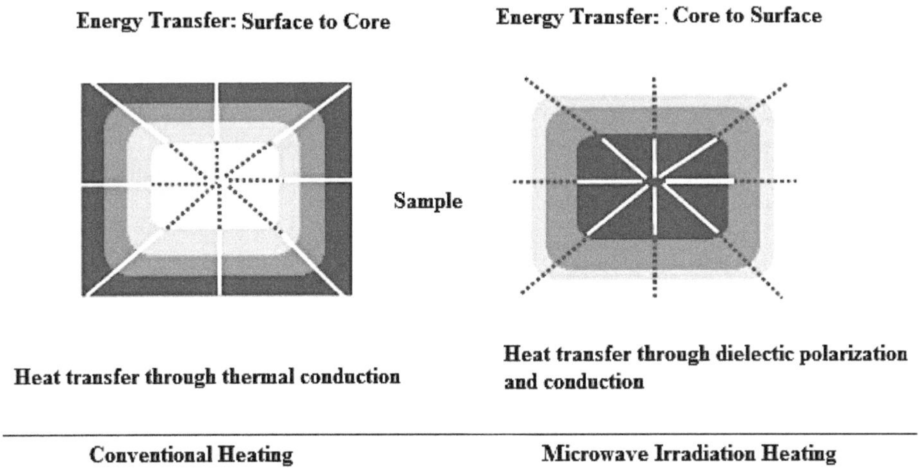

Energy Transfer: Surface to Core   **Energy Transfer: Core to Surface**

Sample

Heat transfer through thermal conduction   **Heat transfer through dielectic polarization and conduction**

Conventional Heating   **Microwave Irradiation Heating**

Figure 10.11 Difference between conventional heating and MW heating.

Table 10.4 Effect of MW treatment on the performance of anaerobic digestion.

Treatment conditions	Anaerobic Digestion	Findings	Source
91.2°C, 7 min	35°C, SRT 15 days	• 25.9% VS reduction • 23.6% TCOD removal • 64% and 79% improvement in TCOD removal and in methane production. Anaerobic digestion of MW pretreated sludge reduced the reactor SRT from 15 day to 8 day	Park *et al.* (2004)
96°C, 3% TS	Batch, 33°C 5 day SRT	30% higher biogas production over control reactor and 26% higher VS removal	Eskicioglu *et al.* (2007)
175°C, 3% TS	mesophilic batch, 35 ± 1°C	31% higher biogas production than the control	Eskicioglu *et al.* (2009)
170°C, 30 min	Batch, 35°C 30 d SRT	25.9% higher biogas production, 12% higher VS removal over control	Qiao *et al.* (2010)

(pretreatment temperature: 65°C) into mesophilic anaerobic digesters (Hong *et al.*, 2004).

10.3.3 Mechanical

In mechanical sludge disintegration, the require energy is delivered in the form of pressure, translational or rotational energy (Neyens & Beyens, 2003). Sludge disintegration

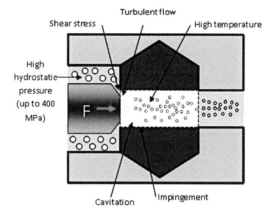

(a)

(b)

Figure 10.12 (a) Various physical phenomena simultaneously affecting sludge during high pressure homogenization (Source: University of Tennessee); (b) Commercial HPH equipment (IKA®).

by mechanical means considered very effective. However, most of the mechanical methods are energy intensive (Tyagi & Lo, 2011).

High-Pressure Homogeniser (HPH)

In HPH, sludge is pumped under high pressure and low velocity into the homogeniser, and bacterial cell are ruptured as a result of severe turbulence and large pressure differences and strong shearing force. When the sludge stream flows out, the intracellular cytoplasm and EPS are released into the bulk solution (Kleinig & Middelberg, 1998) (Figure 10.12). The cell disintegration upto 85% can be achieved by sludge compression at 60 MPa and at relatively low energy levels (30–50 MJ per m^3) (Harrison, 1991).

Table 10.5 Effect of high-pressure homogeniser pretreatment on sludge solubilisation and subsequent anaerobic digestion.

Treatment conditions	Anaerobic digestion	Findings	Source
Pressure 400 bar	Fixed bimass reactor, 2.5 days HRT, 35°C, SRT 3 days	28% improvement in VS removal	Muller *et al.* (1998)
Pressure 400 bar	Fixed bimass reactor, 2.5 days HRT, 35°C, SRT 13 days	87% improvement in VS removal	
300 bar pressure 750 kJ/kgTS	CSTR, 10–15 days HRT, 35°C	60% improvement in biogas production	Englehart *et al.* (1999)
600 bar pressure	CSTR, 20 days HRT, 36°C	18% increase in biogas production	
600 bar pressure	CSTR, 20 days HRT, 35°C	18% increase in biogas production	Barjenbruch & Kopplow (2003)

HPH is the most often used technology for full-scale operation with patent name of BIOGEST® crown disintegration system through CSO Technik, UK.

Table 10.5 summarised the results of various studies carried out to investigate the effect of high-pressure homogeniser pretreatment on sludge solubilisation and subsequent bio-digestion.

Earlier studies reported the 23% reduction in excess sludge generation and 30% improvement in biogas production during the anaerobic digestion of HPH pretreated sludge. The HPH technique has low odor potential, easy to install and HPH treated sludge shows good dewaterability characteristics. However, the demerits of the HPH is poor pathogens removal and energy intensive technology (Tyagi & Lo, 2011).

Ultrasonication

In ultrasonication, the formation, growth and subsequent collapse of micro-bubbles takes place in short times (milliseconds), which is called cavitation (Figure 10.13). Which ultimately leads to high-localised temperature (over 1000°C) and high-pressure gradients (500 bar) in the liquid medium, which disrupts the cell membrane and EPS complex and released the intercellular material in bulk liquid (Pilli *et al.*, 2011).

The application of sonication for sludge treatment at laboratory and full-scale systems was studied by various researchers, which is summarised in Table 10.6.

Earlier studies reported the US pretreatment of sludge under different intensity and specific energy input leads to the significant improvement in COD solubilisation (upto 47%), VS reduction (upto 25%) and methane generation (upto 76%) (Wang *et al.*, 1999; Bougrier *et al.*, 2006; Apul & Sanin, 2010; Braguglia *et al.*, 2012; Salsabil *et al.*, 2010). The US pretreatment was found effective in pathogens removal too at an intensity of 0.33 W/mL (Jean *et al.*, 2000; Chu *et al.*, 2001). The full-scale studies of ultrasonic sludge pretreatment revealed that three times more sludge can be assimilated in anaerobic digester with upto 50% increase in biogas generation and significant improvement in sludge dewaterability (Hogan *et al.*, 2004).

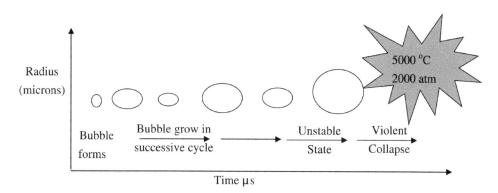

Figure 10.13 Development and collapse of the cavitation bubble (Reprinted from "Ultrasonics Sono-chemistry, Vol 18 (1), Sridhar Pilli, Puspendu Bhunia, Song Yan, R.J. LeBlanc, R.D. Tyagi, R.Y. Surampalli, Ultrasonic pretreatment of sludge: A review, pp. 1–18, 2011, with permission from Elsevier).

Table 10.6 Effect of ultrasonic pretreatment on sludge solubilisation and subsequent anaerobic digestion.

Treatment conditions	Anaerobic digestion	Findings	Source
42 kHz, 120 min		• 38.9% VS reduction • 4413 l/m^3 WAS biogas production	Kim *et al.*, 2003
20 kHz, 180 W, 60 s	Batch, 28 days, 35°C	24% increase in biogas production	Bien *et al.*, 2004
20 kHz, 7000 and 15000 kJ/kgTS	Batch, 16 days, 35°C	40% increase in biogas production	Bougrier *et al.*, 2005
5000 kJ/kgTS	Semi-continuous, 20 days HRT	36% increase in biogas production	Braguglia *et al.*, 2008
20 kHz, 108000 kJ/kgTS	Batch, 50 days, 37°C	• 84% increase in biogas production • 22.3% TSS and 29.7% VSS reduction	Salsabil *et al.*, 2009
30 kWh/m^3 sludge	• Batch • Continuous, 20 days HRT	• 42% increase in biogas production • 37% increase in biogas production • 25% increase in VS removal	Perez-Elvira *et al.*, 2009
1.5 W/mL, 30 min 1.5 W/mL, 60 min		• 15.11% COD solubilisation • 20.76% COD solubilisation	Xu *et al.*, 2010

The main advantages of US pretreatment are no odor generation, reliable operation, good sludge dewaterability, high degree of sludge solubilisation and subsequent biogas production, high level of pathogens removal and, compact design and easy retrofit in existing system. However, the sonication is an energy intensive method with high capital and operating costs (Tyagi & Lo, 2011).

Stirred ball mills

In stirred ball mills, two main methods used are the Kady mill, where two counter rotating plates generate shear, and the wet milling, which comprises of a cylindrical grinding chamber almost packed with grinding beads. The beads are forced into a rotational movement by an agitator, which generates shear and pressure forces and disintegrate the bacterial cells (Tyagi & Lo, 2011).

Earlier studies reported the significant COD solubilisation (by 47%) and VS removal (42–47%) at different energy inputs of 1–1.25 kW/m^3 TS/days and 2000 kJ/kg TS, respectively (Baier & Schmidheiny, 1997; Kopp et al., 1997). High sludge disintegration efficiency, Low odor potential and reliable operation due to high level of research and development) are the merits of this method. Nevertheless, stirred ball mill is an energy intensive technique, moreover, huge erosion in the grinding chamber and clogging issues are other demerits (Perez-Elvira et al., 2006).

Lysate-thickening centrifugal technique

The centrifugal forces produced in this thickening centrifuge are deliberately applied to cell disintegration. It permits for the continual cellular rupturing after the sludge has been thickened (Tyagi & Lo, 2011). Previous studies reported that the biogas production was increased as much as 86% by using this technique. The best results were achieved with shorter sludge retention time and a high anaerobic activity in the digesting sludge (Dohanyos et al., 1997). This technique was scaled up to full-scale at a wastewater treatment facility in Prague, Czech Republic, which resulted in a 26% improvement in biogas production (Dohanyos et al., 2004). Elliot and Mahmood (2007) observed 15–26% improvement in biogas production and 50% removal of organics in the mesophilic digester. The technique has no odor related issue, and relatively simple in installation. Low sludge disintegration, high energy demand, and wear on equipment are the major disadvantages of this technique (Perez-Elvira et al., 2006).

Jetting and colliding method

In jetting and colliding method, the cell wall was ruptured by mechanical jet and smash under pressurised conditions (from 5 bar to 50 bar) (Choi et al., 1997). The 5–7 times enhancement in soluble COD, 30–50% VS removal and notable biogas generation (upto 850 l/kg VS) were achieved upon pretreating the sludge at 30 bar pressure (Choi et al., 1997; Nah et al., 2000). However, the high cost of installation and infant stage of technology development are the major drawbacks of jetting and colliding method.

10.3.4 Combined pretreatment techniques

The combined pretreatment technique involves the combination of two different methods in order to accelerate the sludge solubilisation and subsequent bio-digestion process. It can be the combination of thermal-chemical, chemical-mechanical, thermal-mechanical and chemical-chemical. It is expected that the coupling of two different methods may have a synergetic effect on sludge disintegration.

Thermo-chemical pretreatment

(a) Conventional heating-alkali/acid

The addition of acid or alkali coupled with thermal treatment minimises the requirement of high temperature to achieve higher sludge solubilisation. Table 10.7 summarises the performance of thermo-chemical pretreatment in sludge solubilisation and subsequent anaerobic digestion.

Table 10.7 Effect of thermo-chemical pretreatment on sludge solubilisation and subsequent bio-digestion.

Treatment conditions	Anaerobic digestion	Findings	Source
300 meq HCl/L, 175°C, 60 min	Batch, 25 days, 35°C	56% increase in methane generation	Stuckey & McCartey (1978)
300 meq NaOH, 175°C, 60 min	Batch, 25 days, 35°C	62% increase in methane generation	
0.3 g NaOH/gVSS, 130°C, 5 min	Batch, 10 days, 37°C	• 50% sludge solubilisation • 31% improvement in methane generation	Tanaka et al. (1997)
7 g NaOH/L, 121°C, 30 min	Batch, 7 days, 37°C	• 67.8% SCOD removal and 30% VS removal • 38% increase in biogas production	Kim et al. (2003)
45 meq NaOH/L, 55°C, 240 min	Batch, 20 days, 35°C	• 38% COD solubilisation • 88% increase in methane production	Heo et al. (2003)
1.65 g KOH/L, pH 10, 130°C, 60 min	Batch, 24 days, 35°C	30% increase in methane production	Valo et al. (2004)
1.65 g KOH/L, pH 10, 130°C, 60 min	CSTR, 20 days HRT, 35°C	75% increase in methane generation	

The findings revealed that a higher COD solubilisation (upto 71%) can be achieved with combined thermo-chemical pretreatment at pH 12 (using NaOH) and temperature range of 140 to 170°C (Penaud *et al.*, 1999; Tanaka & Kamiyama, 2002; Andreottola & Foladori, 2006). The 100% pathogens removal was achieved after pretreatment of sludge at 100°C, pH 10 (with $Ca(OH)_2$) and 60 min reaction time (Neyens *et al.*, 2003). A full-scale installation of thermo-chemical method named "Kreproprocess" was applied on the thickened sludge (5–7% DS) acidified with H_2SO_4 (pH 1–2) and heated at 140°C (3.5 bars) for 30–40 min, which achieved a higher sludge solubilisation upto 40% (Odegaard, 2004).

Higher degree of sludge solubilisation, low chemical dosage requirement and use of effluent as a carbon source in nutrient removal are the merits of thermo-chemical methods. However, high operation and maintenance cost, corrosion issue and odor potential are the major drawbacks of this process (Tyagi & Lo, 2011).

(b) Microwave-alkali/acidic

In order to save the energy and time, the MW heating was preferred over the conventional heating. This method was reported to achieve a high degree of COD solubilisation (34%–80%), volatile solids solubilisation (40%–70%), TSS reduction (upto 35%) and improvement in biogas production (16%–44%) at temperature range of 120 to 170°C and pH 12 (Qiao *et al.*, 2008; Dogan & Sanin, 2009; Chang *et al.*, 2011). Moreover, the coliforms and *E.coli* were not detected after pretreatment at 70°C for 5 min. Thus the treated sludge can be reused for Class A biosolids application (Park *et al.*, 2010).

(c) Microwave-advanced oxidation process

This method was also found very effective in terms of high degree of sludge solubilisation and improvement in subsequent biogas generation. A significant improvement

Table 10.8 Comparative evaluation of various pretreatment methods (Tyagi & Lo, 2011).

Pretreatment technique		COD solubilization	Dewaterability	Pathogen removal	Application (Lab/pilot/full scale)	Capital cost	O&M cost	Energy Requirement
Chemical	Acid-Alkali	High	High	High	Lab/pilot scale	Low	Medium	Low
	Ozonation	High	High	Low	Lab/pilot/full scale	High	High	High
Thermal	Conventional heating	High	High	High	Lab/pilot/full scale	High	High or low	High
	Microwave	High	High	High	Lab scale	Medium	Medium	Medium
Mechanical	High Pressure Homogeniser	High	High	Low	Lab/pilot/full scale	Medium	Medium- high	Low
	Ultrasonic	High	High	High	Lab/pilot/full scale	High	High	High
	Electric Pulse Power Method	High	–	–	Lab	High	High	High
	Gamma Irradiation	High	–	High	Lab scale	High	High	High
	Stirred ball mills	High	High	No	Lab/pilot/full scale	High	High	High
	Lysat-thickening centrifugal technique	Low	–	–	Lab/pilot/full scale	Low	Medium	Low
	Jetting and Colliding Method	High	–	–	Lab/pilot scale	High	High	High
	High Speed Rotary Disc Process	High	–	–	Lab	High	High	High
Hybrid technique	Thermo-chemical	High	High	High	Lab/pilot/full scale	Medium	High	Medium
	Mechanical-chemical**	High	–	–	Lab/pilot/full scale	Medium	Medium	Low
	Thermomechanical*	High	–	–	Lab/Full scale	High	High	High
	Wet oxidation (Aqua-Reci)	High	–	–	Pilot/Full	High	High	High

*(thermal, explosive decompression and shear forces); **(alkali-high pressure homogeniser)

in COD solubilisation (18%–34%) was achieved at lower treatment temperature of 80°C and H_2O_2 dosage of 1 g/g TS. Moreover, the complete removal of fecal coliforms was observed at 70°C with 0.08% H_2O_2 dosage (Lo *et al.*, 2008; Eskicioglu *et al.*, 2008).

Chemical-mechanical pretreatment
(Alkaline-ultrasonication and ozonation-ultrasonication)

The alkaline pretreatment weakens the cell walls and makes them more susceptible to ultrasonication. The combined alkaline-ultrasonication pretreatment improve the COD and VS solubilisation from 51%–70% at pH 12 (using NaOH or KOH) and ultrasonic specific energy range from 7500 to 12000 kJ/kg TS (Liu *et al.*, 2008; Jin *et al.*, 2009; Cho *et al.*, 2010).

Ozone is a very powerful oxidizing agent, which can provide an effective pretreatment upon integration with ultrasonication. Earlier study reported that 40 times enhancement in COD solubilisation upon O_3 treatment (O_3 dose: 0.6 g/h) for 60 min followed by 60 min ultrasonic treatment (0.26 W/mL) (Xu *et al.*, 2010).

10.3.5   Comparative evaluation of merits and demerits of treatment in sludge line processes

The critical evaluation and detailed discussion about all the studied pretreatment methods, it was observed that each pretreatment method has advantages and drawbacks too. Table 10.8 summarises a comparative evaluation of merits and demerits of treatment in sludge line processes.

On one side, the pretreatment methods were effective in sludge solubilisation and improvement in biogas generation, pathogens removal and improvement in sludge dewaterability. However, the high capital, and operation and maintenance cost and energy intensive nature are the major concern. The combined pretreatment (thermal-chemical, thermal-mechanical, mechanical-chemical) shows some promising results, however, the most of them are still lab scale curiosity and need extensive research and development in order to scale up the techniques (Tyagi & Lo, 2011).

QUESTIONS

1   What are the three routes to achieve the sludge minimisation at source?
2   Give the comparative evaluation of merits and demerits of treatment in wastewater line processes for sludge minimisation.
3   What is biological lysis-cryptic growth, and how sludge minimisation can be achieved by using this process?
4   Describe the ANANOX and OSA processes of sludge minimisation in detail.
5   Why sludge hydrolysis is a difficult step in sludge digestion process? How sludge pretreatment helps to enhance the sludge digestion rate?
6   What are the chemical methods of sludge pretreatment? Describe the mechanism of pretreatment of alkaline and ozonation processes.

7   Define the thermal methods of sludge pretreatment. What is the difference between conventional and microwave heating? Give the merits and demerits of both methods.

8   How ultrasonication method pretreats the sludge and upto what degree of sludge solubilisation can be achieved? Describe the major factors affect the sludge pretreatment by ultrasonication.

9   What are the major factors affects the sludge pretreatment?

10 What are the hybrid pretreatment methods and how they are advantageous if compare with other pretreatment methods?

11 Describe the merits and demerits of treatment in sludge line processes.

12 What are the processes in the final waste line for sludge treatment?

13 What are the major advantages and disadvantages of sludge pretreatment methods?

14 How many sludge pretreatment methods have been implemented at full scale?

Chapter 11

Sludge disinfection and thermal drying processes

11.1 DISINFECTION OF SLUDGE

(Metcalf & Eddy, 2003; Mitsdorffer *et al.*, 1990)

11.1.1 Introduction

Sludge disinfection is necessary before its application to land mainly because of the presence of pathogenic microorganisms. Sludge disinfection minimises the risk of pathgenic contamination and ensure the safety of public health.

There are several ways to destroy pathogens in liquid and dewatered sludges. The following methods have been used to achieve pathogen reduction beyond that achieved by stabilisation:

1   Pasteurisation for 3 min at 70°C
2   High pH treatment, typically with lime at a pH above 12.0 for 3 h
3   Long-term storage of liquid digested sludge (60 d at 20°C or 120 d at 4°C)
4   Complete composting at temperatures above 55°C and curing in a stockpile for at least 30 d
5   Addition of chlorine to stabilise and disinfect the sludge
6   Disinfection with other chemicals
7   Disinfection by high-energy radiation.

Some stabilisation processes also provide disinfection. These processes include chlorine oxidation, lime stabilisation, heat treatment and thermophilic aerobic digestion (TAD).

Anaerobic and aerobic digestion (excepting TAD) does not disinfect the sludge, but greatly reduces the number of pathogenic organisms. Disinfection of liquid aerobic and anaerobic digested sludges is best accomplished by pasteurisation or long-term storage. Long-term storage and composting are probably the most effective means of disinfecting dewatered aerobic and anaerobic digested sludges.

11.1.2 Sludge pasteurisation

The two methods that are used for pasteurising liquid sludges involve (i) the direct injection of steam and (ii) indirect heat exchange.

Because heat exchangers tend to scale up or become fouled with organic matter, it appears that direct steam injection is the more feasible method. However, the equipment presently used for sludge pasteurisation may not be cost-effective for plants with capacities of less than $0.2 \, m^3/s$ (5 Mgal/d) because of its high capital costs. Pasteurisation for small plants may be achieved by direct steam injection into the tank trucks that transport the sludge to the disposal site.

11.1.3 Sludge storage

Liquid digested sludge (i.e. without dewatering) is normally stored in earthen lagoons. Storage requires that sufficient land should be available. In land application systems, storage is often necessary to retain sludge during periods when it cannot be applied because of weather or crop considerations. In this case, the storage facilities can perform a dual function by providing disinfection as well. Because of the potential contamination effects of the stored sludge, special attention must be devoted to the design of these lagoons with respect to limiting percolation and the development of odours (see also Section 10.3.2).

11.2 THERMAL DRYING OF SLUDGE
(Keey, 1972; Marklund, 1990; Metcalf & Eddy, 2003; Turobskiy & Mathai, 2006)

The main purpose of thermal drying of sludge is to evaporate water and reduce the moisture content of the sludge below that achieved by conventional dewatering processes, so that it can be incinerated efficiently or processed into good fertiliser.

11.2.1 Introduction

An improved heat balance and an increased throughput rate are possible only by reducing the water volume to be evaporated, that is, by increasing the percentage of dry solids. Sludge drying is a unit operation that involves reducing water content by vapourization of water to the air. In sludge-drying beds (refer to Section 6.3.1), vapour pressure differences account for evaporation to the atmosphere. In mechanical drying apparatus, auxiliary heat is provided to increase the vapour-holding capacity of the ambient air and to provide the latent heat of evaporation.

In terms of the economic aspects, mechanical dewatering by means of centrifugation or otherwise cannot be sufficiently improved, and additional sludge drying in the form of a separate process step is required. Depending on the final usage – for example, direct incineration, storage for incineration fuel or storage for fertiliser – the sludge can be dried to between 40 and 60% solid content for direct incineration, and up to 90% solid content for storage.

When it is to be sold as a commercial fertiliser, waste water sludge is commonly heat dried to less than 10% moisture to permit grinding of the sludge to reduce its weight, and to prevent continued biological action. Heat drying also precedes sludge incineration (described in Section 12.3) to make it more efficient. Sludge incineration may then itself provide the heat required for heat drying as well as for other plant operations.

11.2.2 Process theory

Under equilibrium conditions of constant-rate drying, mass transfer is proportional to (i) the area of wetted-surface exposed, (ii) the difference between the water content of the drying air and the saturation humidity at the wet-bulb temperature of the sludge–air interface, and (iii) other factors, such as the velocity and turbulence of the drying air expressed as a mass transfer coefficient. The pertinent equation is:

$$\omega = k_y(H_s - H_a)A \qquad (11.1)$$

where,

ω = evaporation rate (kg/h)

k_y = mass-transfer coefficient of gas phase [kg/m^2 · h per unit of humidity difference; $\Delta H = (H_s - H_a)$]

H_s = saturation humidity of air at sludge–air interface (kg water vapour/kg dry air)

H_a = humidity of drying air (kg water vapour/kg dry air)

It is evident that drying may be accomplished more rapidly on a finely divided sludge by exposing new areas to the airstream. Furthermore, maximum contact between dry air and wet sludge must be achieved to ensure a maximum value of ΔH. These factors must be considered in the selection of drying apparatus for sludge in a waste water treatment plant.

11.2.3 Process mechanisms

Thermal drying

The two options for thermal sludge drying are convection drying and contact drying (Figure 11.1).

In the convection drying system, heated air, flue gas or vapour is passed over a sludge layer where it absorbs water from the wet sludge. In closed systems, which work with heated vapour, the surplus vapours are condensed in a condenser, and in open systems the carrier gas (e.g. vapour or flue gas) leaves the dryer.

In the case of contact drying, the system is closed and the necessary heat is transferred through the vessel wall. Saturated steam or thermal oil can be used as the heating medium. By using rotators, sludge is interchanged very rapidly at the heated wall. The water present is evaporated and thus the use of a carrier gas with all its condensation and odour problems is not encountered.

Continuous thin-film sludge dryer

The principle of thin-film processing is well known and practiced in evaporation and distillation technology. The same principle of a mechanically agitated product film on a heated surface has also been applied to the continuous drying of products from a

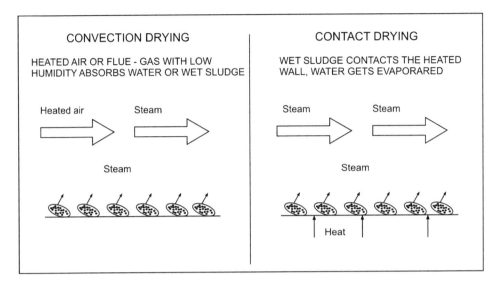

Figure 11.1 Thermal drying of residential sludge.

solution, slurry, paste or cake. For liquid feed, a vertical-type thin-film dryer is used. For the drying of slurry or cake-like feedstock, a horizontal-type thin-film dryer is more suitable.

11.2.4 Thermal dryers

The heat required for sludge drying is generally obtained from a combustion process. In some drying processes the exhaust gases transfer the heat to the sludge by direct contact. Examples of such direct dryers are as follows:

- Belt dryers
- Multiple-hearth dryers
- Drum dryers
- Fluidised-bed dryers
- Flow dryers
- Grinding dryers.

At the end of these drying processes a mixture of flue gas and water vapour is always obtained, which typically contains some quantity of the dried product in the form of particles and vapours, mostly organic. Cleaning of this contaminated gas vapour mixture poses significant difficulties.

However, indirect drying systems transfer the thermal energy of the exhaust gases to the sludge via heated surfaces and an intermediate heat transfer medium (e.g. steam,

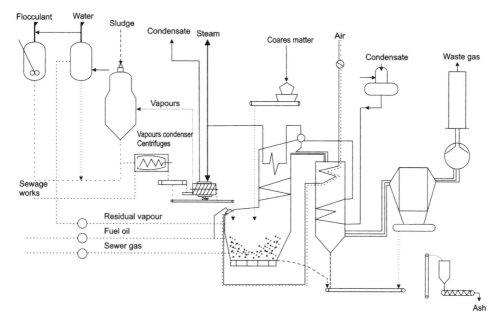

Figure 11.2 Flow sheet of a mechanical sludge dryer.

hot water or thermal oil). The exhaust gases and the vapour from the dryer are evacuated separately. The exhaust gases are cleaned via conventional processes. The vapours incorporate only a minor proportion of solid particles and can undergo a special cleaning treatment. Examples of indirect dryers are as follows:

• Screw dryers
• Disk dryers
• Thin-film dryers
• Kneading dryers.

This apparatus is, without exception, compact. With a view to saving cost and energy, it is aimed at a self-supporting combustion. Here, the calorific value of the fuels, including the heat of the combustion air, should meet the heat requirements of the process for (i) the evaporation and superheating of water, (ii) the flue gas superheating to at least 800°C, and (iii) the heat losses.

11.2.5 Mechanical dryers

The flow diagram in Figure 11.2 illustrates the general procedure for drying sludge mechanically. The vapours that are evacuated from the dryer condense through intimate contact with the fresh sludge, which absorbs the heat of condensation. The dewaterability of the sludge is considerably improved by heating to 58°C. The residual vapours are constantly exhausted and blown into the combustion chamber. The entire vapour system, including the dryer, is maintained under a slight vacuum, which avoids malodours. The dryers and the entire vapour system are of corrosion-resistant material, and the vapour pipework is heated and insulated.

The dried sludge is transported on drag conveyors to two incinerators and falls through shafts that are provided at the top of the combustion chamber, directly into a hot fluidised bed. The flue gases leave the combustion chamber at a temperature ranging between 800 and 855°C. Finally, the flue gases, at approximately 200°C, pass through the electrostatic precipitator, where the fine ash particles precipitate. The ash is stored in silos and is moistened prior to transportation.

11.2.6 Thermal-drying alternatives

Sludge-drying beds, if used for a sufficiently long time, may serve to fulfil the function of a heat drying process. However, there are also five mechanical processes that may be used for drying sludge: (i) flash dryers, (ii) spray dryers, (iii) rotary dryers, (iv) multiple-hearth dryers, and (v) the Carver-Greenfield process or oil-immersion dehydration. Flash dryers are the type most commonly in use at waste water treatment plants. The use of sludge dryers is normally preceded by dewatering.

Flash dryers

Flash drying is the rapid removal of moisture by spraying or injecting the sludge into a hot gas stream. Flash drying involves pulverising the sludge in a cage mill or using an atomised suspension technique in the presence of hot gases. One operation involves a cage mill that receives a mixture of wet sludge or sludge cake and recycled dried sludge, which is why it is also known as the fluid–solid heat drying system (Figure 11.3). The mixture contains approximately 50% moisture. The hot gases and sludge are forced up to a duct in which most of the drying takes place and then to a cyclone, which separates the vapour and solids. It is possible to reduce the moisture content to 8–10% with this operation. The temperature of the dried sludge is about 71°C and that of the exhaust gas is 104–149°C. The dried sludge may be used as fertiliser or it may be incinerated in the furnace in any proportion up to 100% of production. The exhaust gas treatment facility consists of a deodorising preheater, a combustion air heater, an induced-draft fan and a gas scrubber. Odours are destroyed when the temperature of the gas from the cyclone is elevated in the deodorising preheater. Part of the heat absorbed is recovered in the combustion air preheater. The gas then passes through a scrubber to remove dust before discharge to the atmosphere. Flash drying is an energy-intensive process because of its high energy requirement and significant operational and maintenance costs.

Spray dryers

A spray dryer uses a high-speed centrifugal bowl into which liquid sludge is fed. Centrifugal force serves to atomise the sludge into fine particles and then spray them

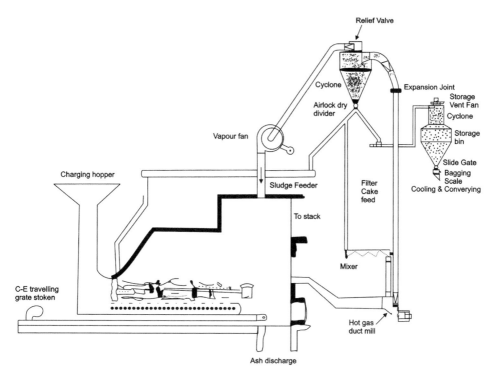

Figure 11.3 Flash-drying system for sludge.

into the top of the drying chamber, where a steady transfer of moisture to the hot gases takes place, as shown in Figure 11.4. A nozzle may be used in place of the bowl, as long as the design prevents clogging of the nozzle.

Rotary dryers

Rotary kiln dryers have been used in several plants for the drying of sludge and for the drying and burning of municipal solid wastes and industrial wastes (Figure 11.5). The main component of a rotary drying system is a rotary drum dryer installed with an inclination of 3–4° to the horizontal. Under gravity, the sludge moves along the drum from its raised (charging) end to the lower (discharge) end. The rotational speed of the drum is 5–8 rpm. Auxiliary components of the drum dryer system include a mixer for blending raw dewatered sludge and recycled dry sludge, a drum feeding screw, a furnace with a fuel burner for heating the air, a cyclone and scrubber for separating particulates from the hot exhaust gases, a heat exchanger, a dried-sludge extraction screw and a storage silo. Both direct and indirect rotary dryers have been developed for industrial processes. In direct-heating dryers, the hot gases are separated from the drying material by steel shells. In indirect-heating dryers, the hottest gases surround a central shell containing the material, but return through it at reduced temperatures. Thus, for example, one

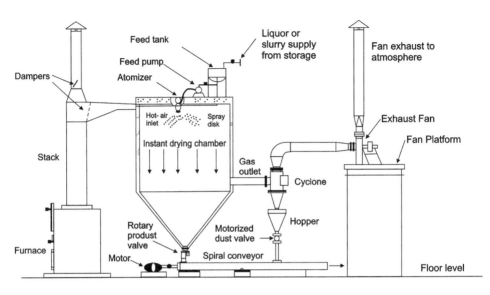

Figure 11.4 Spray dryer with parallel flow.

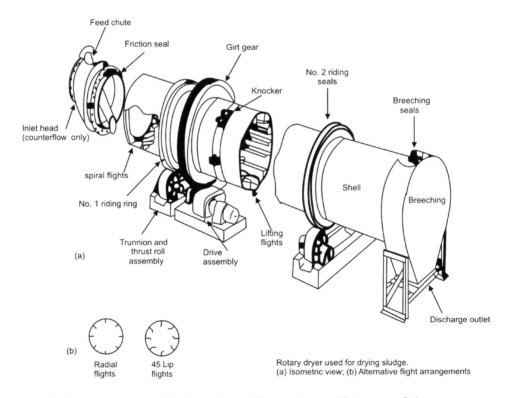

Rotary dryer used for drying sludge.
(a) Isometric view; (b) Alternative flight arrangements

Figure 11.5 Rotary dryer used for drying sludge: (a) isometric view; (b) alternative flight arrangements.

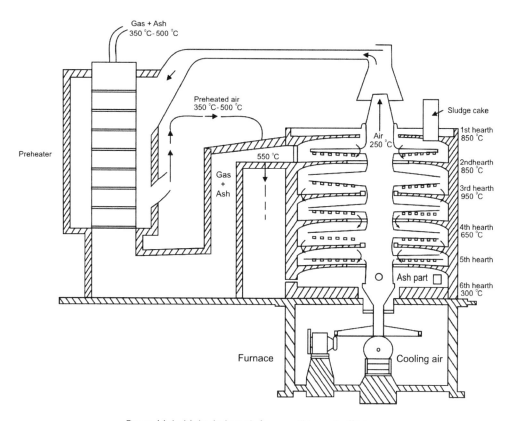

Figure 11.6 Multiple-hearth furnace, "Herreshoff"-type.

manufacturer produces a variety of models, with evaporation capacities ranging from 1000 to 10,000 kg/h for direct dryers, and from 350 to 2000 kg/h for indirect dryers. Coal, oil, gas, municipal solid wastes or the dried sludge itself may be used as fuel. Fins (flights) or louvres may be installed for lifting and agitating the material as the drum revolves.

Multiple-hearth dryer

As shown in Figure 11.6, the multiple-hearth dryer consists of a cylindrical, refractory-lined steel shell containing a series of horizontal refractory hearths located one above the other (there are six hearths in Figure 11.6). The feed system consists of belt conveyors, ribbon screws and bulk flow conveyors. There are three combustion zones in a hearth: drying at the top, burning in the middle and cooling at the bottom. The temperatures in the hearth, in each zone, vary approximately within the limits shown in Figure 11.6.

A multiple-hearth incinerator is frequently used to dry and burn (i.e. inciner-ate) sludges that have been partially dried (dewatered) by vacuum filtration. This is a countercurrent operation in which heated air and the products of combustion pass through the finely pulverised sludge that is continuously raked to expose fresh surfaces.

In multiple-hearth incinerators, sludge cake is fed on to the topmost hearth and moved from hearth to hearth by fins or teeth attached to horizontal, hollow (for air cooling) arms branching from a vertical, central, hollow shaft. The sludge cake loses moisture, ignites, burns and cools. Hearth temperatures rise to a maximum in the centre of the incinerator. The exhaust gases are passed through a preheater or recuperator and heat the air blown into the furnace to support combustion. Cooling air is taken in through the central shaft.

Supplemental fuel is injected into the furnace on a few of the upper hearths. Digested sludge always requires supplemental fuels. Gases and fly ash, after leaving the preheater where they have heated the air for combustion, are passed through a scrubber where the ash is entrapped by water and is removed after settling, while the gases are released into the atmosphere (McCarty *et al.*, 1969).

The heat requirements are dictated by the temperature of the sludge, its moisture content and by the efficiency of the furnace. Furnace efficiency usually varies between 45 and 70% in terms of total heat recovery (including credits on stack gas and latent heat of evaporation) and total heat input. Multiple-hearth furnaces have a combined efficiency of evaporation and incineration of about 55%. Sludge cake produced by the vacuum filtration of chemically conditioned, raw, plain-sedimentation sludge and its mixture with fresh, trickling-filter or activated sludge will ordinarily supply sufficient heat for self-incineration. However, most digested sludges will not, and auxiliary fuel, possibly sludge-digester gas, is required for these. Incinerators of solid municipal wastes are another possible source of heat.

Oil immersion dehydrators

The drying of sludge can also be accomplished using a patented oil-immersion process known as the Carver-Greenfield process. Operationally, a light oil is mixed with dewatered sludge. The oil–sludge mixture, which can be pumped easily and minimises corrosion and scaling, is then passed through a four-stage falling-film evaporator in which water is removed because it has a lower boiling point than the oil carrier. After evaporation, what remains is essentially a mixture of oil and dry sludge. The solids are removed from the oil with a centrifuge, and the remaining oil can be separated into a light oil and a heavy oil residue by exposing it to superheated steam.

The dry solids are suitable for further processing (e.g. pelletising as a fuel source) or disposal. The recovery of energy and heat from the dried sludge, using an incinerator, pyrolysis reactor or gasifier, is an option that should be investigated when the Carver-Greenfield process is being evaluated. Heat recovered from the dried sludge could be used to supply the energy requirements of the process. If the solids are to be disposed of in landfill, they can be mixed with the solids from a centrifuge or vacuum filter to reduce dewatering costs.

General considerations

The two most important control measures associated with the heat drying of sludge are fly ash collection and odour control. Cyclone separators with efficiencies of 75–80%

are suitable for vent-gas temperatures up to 340 or 370°C. Wet scrubbers have higher efficiencies and will condense some of the organic/volatile matter in the vent gas, but may carry over water droplets. Sludge drying occurs at temperatures of approximately 370°C, whereas 650–760°C is required for complete incineration (the incineration process is described in Section 12.2.1). To achieve the destruction of odours, the exhaust gases must reach approximately 730°C. Thus, if the gases evolved in the drying process are reheated in an incinerator to a minimum of 730°C, odours will be eliminated. At lower temperatures, partial oxidation of odour-producing compounds may occur, resulting in an increase in the intensity or disagreeable character of odour produced (Metcalf & Eddy, 2003; Shen, 1979).

11.2.7   Design considerations

The most significant features of the design of a heat-based drying technique are outlined below.

Moisture content of feed sludge. Effective dewatering is necessary to reduce the moisture content of feed sludge. A low moisture content in sludge will ensure effective and energy-efficient sludge drying, as thermal dehydration of sludge requires a large amount of energy, in comparison to mechanical dewatering methods.

Storage. Storage is required for dewatered sludge as well as dried biosolids. For dewatered sludge, a minimum of three days? storage capacity is recommended. However, storage for the dried biosolids will depend on the final disposal plan. The dried biosolids are usually sold seasonally and so a 90-day storage capacity may be sufficient.

Emission and odour control. The sludge drying and storage facilities must be equipped with proper emission- and odour-control equipment. Wet scrubbers, cyclone separators and baghouses, or a combination of these, remove particulate matter from ambient air. Afterburners and chemical scrubbers are normally used to control odours.

Fire and explosion hazards. Biosolids are comprised of protein, carbohydrate and lipid, which in solid form can burn readily if enough oxygen and a sufficiently warm temperature are available. Baghouses, wet scrubbers and cyclone separators can be used to remove combustible particulate matter from the ambient air. In addition, a well-trained workforce, proper safety controls and effective monitoring can reduce the chances of fire or explosion-related accidents.

Side streams. Water vapour condensation from dryers produces odorous liquid side streams, which contain both organics and ammonia. This liquid must be recycled regularly to the influent stage of the waste water treatment plant.

Heat recovery. Sludge dryers should be designed to recover the heat generated and reuse it to make the process more energy-efficient. The recovered heat can be used to preheat the feed sludge and combustion air as well as to support other plant heating necessities. In addition, because of its good fuel value, the dried sludge can be used as a heat source itself (Turobskiy & Mathai, 2006).

QUESTIONS

1   Why is sludge disinfection necessary before disposal?
2   How does sludge drying help in sludge management?
3   What do you understand by thermal drying of sludge? Describe the mechanism.
4   What are the dryers? Describe the types. How do they help to achieve sludge drying?
5   What are the design considerations for heat-based drying methods?

Chapter 12

Thermal treatment and sludge disposal

12.1 INTRODUCTION
(Metcalf & Eddy, 2003; Stathis, 1980)

Sludge volume reduction using thermal treatment can be achieved in following ways:

1   Partial oxidation and volatilization of organic solids by pyrolysis to end products with calorific value, or
2   Partial or total conversion of organic solids to oxidised end products, primarily carbon dioxide and water, by incineration or wet air oxidation.

Sludges processed by thermal reduction are usually dewatered untreated sludges. It is normally unnecessary to stabilise sludge before incineration. In fact, such practice may be detrimental, because stabilisation, specifically aerobic and anaerobic digestion, decreases the volatile content of the sludge and consequently increases the requirement for an auxiliary fuel. An exception is the implementation of heat treatment ahead of incineration. Heat-treated sludges dewater extremely well. Therefore, the sludge is normally rendered auto combustible, i.e. no auxiliary fuel is required to sustain the burning process. Sludges may be subjected to thermal reduction separately or in combination with municipal solid wastes.

The thermal reduction processes discussed in section 12.3 include multiple hearth incineration, fluidised-bed incineration, flash combustion, incineration, pyrolysis, wet air oxidation, and re-calcination. Before discussing these processes, it will be helpful to review some fundamental aspects of thermal reduction.

12.2 PROCESS TYPES

This section describes the various sludge reduction processes by means of thermal treatment viz. complete combustion, incomplete combustion and pyrolysis processes.

12.2.1 Incineration

Incineration of sludge implies complete combustion of all the organic substances present in the sludge. The predominant elements in the carbohydrates, fats and proteins composing the volatile matter of sludge are carbon, oxygen, hydrogen and nitrogen

(C-O-H-N). The approximate percentages of these may be determined in the laboratory by a technique known as ultimate analysis.

Oxygen requirements for complete combustion may be determined from knowledge of the constituents, assuming that carbon and hydrogen are oxidised to the ultimate end products CO_2 and H_2O. The formula becomes

$$C_aO_bH_cN_d + (a + 0.25c - 0.5b)O_2 = aCO_2 + 0.5cH_2O + 0.5dN_2 \qquad (12.1)$$

The theoretical quantity of air will be 4.35 times the calculated quantity of oxygen, because air is composed of 23% oxygen on a weight basis. To ensure complete combustion, excess air amounting to about 50 per cent of the theoretical amount will be required.

Heat requirements will include the sensible heat Q_s in the ash, plus the sensible heat required to raise the temperature of the flue gases to 760°C (i.e. 1400°F) or whatever higher temperature of operation is selected for complete oxidation and elimination of odours, less the heat recovered in pre-heaters or recuperators. Latent heat Q_1, must also be furnished to evaporate all of the moisture in the sludge. Total heat required Q may be expressed as:

$$Q = \varepsilon Q_s + Q_1 = \varepsilon C_p W_s(T_2 - T_1) + W_w\lambda \qquad (12.2)$$

where,
C_p = specific heat for each category of substance in ash and flue gases,
W_s = mass of each substance,
T_1 and T_2 = initial and final temperatures, and
λ = latent heat of evaporation/kg.

It should be obvious that reduction of moisture content of sludge by dewatering and/or drying the sludge, is the principal way to lower heat requirements and may determine whether additional fuel will be needed to support combustion.

The main merits and demerits of the incineration process are as follows (Turobskiy & Mathai, 2006):

Merits

- Reduces the volume and weight of wet sludge cake by approximately 95%, thereby reducing disposal requirements.
- Complete destruction of pathogens.
- Potentially recovers energy through the combustion of waste products, thereby reducing the overall expenditure of energy.

Demerits

- High capital and operating costs.
- Reduces the potential beneficial use of biosolids.
- Highly skilled and experienced operating and maintenance staffs are required.
- If residuals (ash) exceeds the prescribed maximum pollutant concentrations, they may be classified as hazardous waste, which requires special disposal.

- Gaseous discharges to atmosphere (particulates and other toxic or noxious emissions) require extensive treatment to ensure the safety of the environment.

12.2.2 Incomplete combustion

Organic substances may be oxidised under high pressures at elevated temperatures with the sludge in a liquid state by feeding compressed air into the pressure vessel. The process, known as wet combustion, was developed in Norway for pulp-mill wastes, but has been revised for the oxidation of untreated wastewater sludges pumped directly from the primary settling tank or thickener. Combustion is not complete; the average is 80–90% completion. Thus, some organic matter, plus ammonia, will be observed in the end products. This incomplete combustion reaction can be expressed by the following formula:

$$C_aH_bO_cN_d + 0.5(ny + 2s + r - c)O_2$$
$$= nC_wH_xO_yN_z + sCO_2 + rH_2O + (d - nz)NH_3 \tag{12.3}$$

where
$r = 0.5[b - nx - 3(d - nz)]$
$s = a - nw$

The results obtained from this equation can also be approximated by the COD of the sludge, which is approximately equal to the oxygen required in combustion. The range of heat released per 1 kg of air required has been found to be from 2.6 to 3.0 MJ. Maximum operating temperatures for the system vary from 175 to 315°C with design operating pressures ranging from 1 to 20 MN/m².

12.2.3 Pyrolysis

Because most organic substances are thermally unstable, they can, upon heating in an oxygen-free atmosphere, be split through a combination of thermal cracking and condensation reactions into gaseous, liquid, and solid fractions. Pyrolysis is the term used to describe the process. In contrast to the combustion process, which is highly exothermic, the pyrolytic process is highly endothermic. For this reason, the term destructive distillation is often used as an alternative term for pyrolysis.

The characteristics of the three major component fractions resulting from the pyrolysis are:

1 A gas stream containing primarily hydrogen, methane, carbon monoxide, carbon dioxide and various other gases, depending on the organic characteristics of the material being pyrolyzed.
2 A fraction that consists of a tar and! or oil stream that is liquid at room temperatures and has been found to contain chemicals such as acetic acid, acetone, and methanol.
3 A char consisting of almost pure carbon plus any inert material that may have entered the process.

For cellulose ($C_6H_{10}O_5$), the following expression has been suggested as being representative of the pyrolysis reaction:

$$3(C_6H_{10}O_5) = 8H_2O + C_6H_8O + 2CO + 2CO_2 + CH_4 + H_2 + 7C \qquad (12.4)$$

In Eqn. (12.4), the liquid tar and/or oil compounds normally obtained are represented by the expression C_6H_8O. It has been found that distribution of the product fractions varies dramatically with the temperature at which the pyrolysis is carried out.

12.3 THERMAL PROCESSES
(Brechtel & Eipper, 1990; Liao, 1974; Metcalf & Eddy, 2003; Negulescu, 1985; Shen, 1979; Turobskiy & Mathai, 2006)

Usage of several type of thermal processes, viz. multiple hearth incineration, fluidised bed incineration, rotary kiln incineration and emerging incineration technologies, flash combustion, co-incineration, wet air oxidation and co-pyrolysis, have been described in this section.

12.3.1 Multiple hearth incineration process

The multiple hearth process converts dewatered sludge cake to an inert ash. Because the process is complex and requires specially trained operators, multiple hearth furnaces are normally used only in plants larger than 720 m³/h, i.e. 0.2 m³/s (5 Mgal/d). They have been used at facilities with lower flows, where land for the disposal of sludges is limited. They are also used in chemical treatment plants for the recalcining of lime sludges.

As shown in Figure 11.6 [and explained in section 11.2.6 (iv)], sludge cake is fed on to the top hearth and is slowly raked to the centre. From the centre, it drops to the second hearth where the rakes move it to the periphery. Here it drops to the third hearth, and is again raked to the centre. The hottest temperatures are on the middle hearths where the sludge burns and where auxiliary fuel is also burned as necessary to warm up the furnace and to sustain combustion. Preheated air is admitted to the lowest hearth and is further heated by the sludge as it rises past the middle hearths where combustion is occurring. The air then cools as it gives up its heat to dry the incoming sludge on the top hearths.

The highest moisture content of the air is found on the top hearths, where sludge with the highest moisture is heated and some water is vaporised. This air passes twice through the furnace. Cooling air is initially blown into the central column and hollow rabble arms to keep them from burning up. A large portion of this air, after passing out the central column at the top, is recirculated to the lowest hearth. This furnace may also be designed as a drier only [see section 11.2.6 (iv)]. In this case, a furnace is needed to provide hot gases, and the sludge and gases both proceed downward through the furnace in parallel flow. Parallel flow of product and hot gases is frequently used in drying operations to prevent burning or scorching a heat sensitive material.

Here it may be noted that the feed sludge must contain more than 15% solids, because of limitations on the maximum evaporating capacity of the furnace. Auxiliary fuel is usually required when the feed sludge contains between 15 and 30% solids. Feed sludge containing more than 50% solids may create temperatures in excess of the refractory and metallurgical limits of standard furnaces. Average loading rates of wet cake are approximately $40 \, kg/m^2 h$ of effective hearth area, but may range from 25 to $75 \, kg/m^2 h$.

In addition to dewatering, required ancillary processes include ash-handling systems and some type of wet scrubber to meet air pollution requirements. Recycle flows consist of scrubber water. Scrubber water comes in contact with, and removes, most of the particulate matters in the exhaust gases. The recycle BOD and COD is nil, and the suspended solids content is a function of the particulates captured in the scrubber.

Ash handling may be either wet or dry. In the wet system, the ash falls into an ash hopper located beneath the furnace, where it is slurried with water from the exhaust gas scrubber. After agitation, the ash slurry is pumped to a lagoon or is dewatered mechanically. In the dry system, the ash is conveyed mechanically by a bucket elevator to a storage hopper for discharge into a truck for eventual disposal as fill material. The ash is usually conditioned with water.

The merits of multiple-hearth furnaces include combusting both primary and secondary sludge, as well as trash from screens, scum from settling tanks and oil separators, dirty grit from grit chambers, and industrial wastes. They are characterised by their simplicity of service and by the reliability and stability of operation during significant variations in the quantity and quality of sludge treated. The furnaces can be installed in the open air. The demertis of multiple-hearth furnaces include high capital cost, large area required, presence of rotating mechanism in the high-temperature zone, and frequent failure of the rake devices (Turobskiy & Mathai, 2006).

12.3.2 Fluidised bed incineration process

Developed by Dorr-Oliver, the method by which sludge is incinerated on a fluid bed is known as "Fluo-solids" also. It has been applied for quite some time in the oil refining industry, chemical industry, etc. but only recently in sludge incineration. The incinerator consists of a metal cylinder with a fluid bed made inside of inert solids such as sand, through which, a fluidizing gas-air is permanently blown.

During burning, an intense and violent mixing between the inert solids and gases takes place which makes the inside temperature constant. An extremely rapid transfer between gases and solids takes place, because of the large surface area available (Shen, 1979). The schematic diagram of fluidised bed incinerator is shown in Figure 12.1.

When combustible solids are introduced into the fluid bed, due to high heat transfer rates the particles heat up rapidly to the ignition temperature. At the same time, the oxygen from the fluidizing air, blown permanently through the bottom of the incinerator, reacts quickly with the combustible solids. The heat released from the reaction is taken up by the bed. The sludge introduced into the incinerator, usually primary sludge, is previously passed through a hydro-cyclone to have the sand removed. After that, sludge is sent into a thickener and then to a centrifuge (or a vacuum filter) for good dewatering, after which it is introduced into the incinerator. Ash and gases from the incinerator pass either directly through a scrubber, the ash being retained in a cyclone and the gas being released into the air, or through a preheater, when it heats the fluidizing and combustion air, and enters the scrubber. The incoming air from

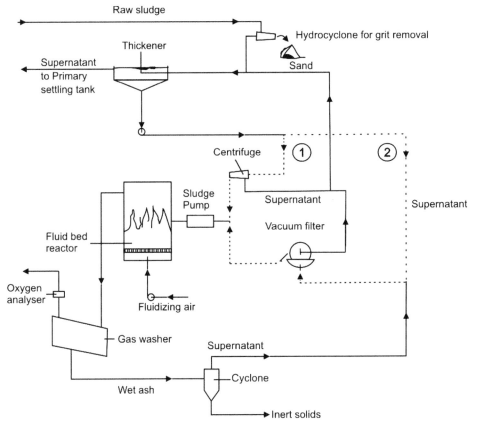

① or ② different circuits (flows) function of dewatering method

Figure 12.1 Fluid bed incinerator "Fluo-solids" type.

fluidization and combustion is heated to 540°C by the hot exhaust gases from the incinerator. Similar to the multiple hearth furnace, the fluid bed incinerator also has an outside source of supplemental fuel. Even though with pre-heating the fuel cost is considerably reduced, the capital cost of the pre-heat system is approximately 10–15% of the fluid bed plant investment. Further, maintenance costs of the preheating system are considerable compared to other auxiliary equipment.

Fluid bed system capacities are a function of superficial space velocity or the rate at which the fluidizing gas is forced through the fluid bed. Space velocities are chosen to meet a given set of feed conditions and will generally be in the range of 0.6–1.1 m/s. Units of up to 10 m in diameter and 15 m in height are usual for the incineration of sludge from municipal wastewater treatment plants (Liao, 1974).

The combustion process of fluidised-bed incineration is controlled by varying the sludge feed rate and the airflow to the reactor to oxidise all the organic material completely and to eliminate the need for sludge digestion. If the process is operated continuously or with shutdowns of short duration on untreated sludge, there is no need for auxiliary fuel after start up.

Like the multiple hearth, the fluidised bed, though very reliable, is complex and requires the use of trained personnel. For this reason, fluidised bed incinerators are normally used in large plants, but they may be used in plants with lower flow ranges where land for the disposal of sludges is limited.

The major advantages of the fluid-bed incinerator are:

(i)   Good mixing of the sludge and combustion air;
(ii)   Drying and combustion take place concurrently within the bed and freeboard at high temperatures, thus reducing the potential air pollution problems;
(iii) The incinerator has no moving parts;
(iv)   The unit can be operated 4–8 h/day with little reheating when restarting, since the sand bed serves as a heat reservoir;
(v)   A mechanical system for ash removal is no longer necessary, because the ash is separated from incinerator by up-flowing combustion gases.

Dorr-Oliver has also developed a fuel-fired horizontal spiral-flow incinerator especially adapted to the needs of small wastewater treatment plants. In this process, thickened sludge is atomised as it enters the combustion chambers by a strong blast of compressed air. The sludge particles are oxidised rapidly and exit as ash with the combustion gases. The ash is then separated and disposed off in the same manner as with the Fluo-Solids process.

12.3.3 Rotary kiln incineration process
(Turobskiy & Mathai, 2006)

Rotary kilns (or drum kilns) are used most often in the calcination of cement clinker and claydite and for incineration of sludge mixed with municipal solid waste (co-incineration). The drum is installed with an inclination of 2 to 4° in the direction of the external furnace at the lower end. The furnace is cylindrical in shape, lined with refractory bricks, and equipped with gas–oil burners. The dewatered sludge (mixed with municipal solids waste for co-incineration) is charged at the upper end of the drum. The sludge dries as it moves through the drying zone and burns in the incineration zone with the liberation of heat. The hot ash falls through an opening in the external furnace chamber and enters the air cooler, from which it goes by pneumatic transport into the receiving hopper and is then hauled to the ash dump. After cooling the ash to 100°C, the hot air goes to the furnace for use in combustion. In the drying zone, the temperature of the exhaust gases is 200 to 220°C, and moisture of the sludge decreases from 65 to 85% to 30 to 40%. In the combustion zone, the length of which usually does not exceed 8 to 12 m, the temperature reaches 900 to 1000°C.

The advantages of rotary kiln incinerators include low emission of heat and low emission of particulates with the exhaust gases, the possibility of treating sludge with high ash and high moisture, and the possibility of installing the rotating section of the kiln in the open air (the furnace section and the charging chamber are usually in buildings). The drawbacks are their cumbersome size, substantial weight, high capital cost, and relative complexity of operation.

12.3.4 Flash combustion process

In flash combustion, a variation of flash drying part or all of the dried sludge is incinerated. Part of the sludge may be incinerated to decrease fuel requirements when drying
the sludge for fertilizer. All the sludge is incinerated when there is no market for the
fertilizer. Flash combustion for total sludge incineration is not competitive with multiple hearth or fluidised bed incineration, and future use for that application is not
expected.

12.3.5 Co-incineration process

Co-incineration is the process of incinerating wastewater sludges with municipal solid
wastes. The major objective is to reduce the combined costs of incinerating sludge and
solid wastes. At present, co-incineration is not widely practiced. The process has the
advantages of producing the heat energy necessary to evaporate water from sludges,
supporting combustion of solid wastes and sludges, and providing an excess of heat
for steam generation, if desired and required to use for some plant operations such as
to generate in-plant electric power, without the use of auxiliary fossil fuels. In properly
designed systems, the hot gases from the process can be used to remove moisture
from sludges to a content of 10–15 per cent. An electrostatic precipitator is used to
clean the exhaust gases. It has been found that the direct feeding of sludge filter cake
containing 70–80 per cent moisture over solid wastes on travelling or reciprocating
grates is ineffective. For operations without heat recovery, a disposal ratio of 0.5 kg
of dry wastewater solids to 2.25 kg of solid wastes is fired in normal operation. In the
case of the water walled boiler with heat recovery, the ratio is approximately 0.5 kg of
dry (industrial plant) solids to 4.0 kg of solid wastes.

12.3.6 Wet-air oxidation process

The Zimmerman process (refer Figure 12.2) involves wet oxidation of untreated sludge
at an elevated temperature and pressure. The process is the same as that discussed under
heat treatment (see sections 6.2, 11.2 and Figure 11.1), except that higher pressures
and temperatures are required to oxidise the volatile solids more completely. Untreated
sludge is ground and mixed with a specified quantity of compressed air. The mixture
is pumped through a series of heat exchangers and then enters a reactor, which is
pressurised to keep the water in the liquid phase at the reactor operating temperature of 175–315°C. High-pressure units can be designed to operate at pressures up to
20 MN/m^2. A mixture of gases, liquid, and ash leave the reactor.

The liquid and ash are returned through heat exchangers to heat the incoming
sludge and then pass out of the system through a pressure-reducing valve. Gases
released by the pressure drop are separated in a cyclone and released in the atmosphere.
In large installations, it may be economical to expand the gases through a turbine to
recover power. The liquid and stabilised solids are cooled by passing through a heat
exchanger and are then separated in a lagoon or settling tank or on sand beds. The
liquid is returned to the primary settling tank and the solids are disposed off by landfill.
The process can be designed to be thermally self-sufficient when untreated sludge is
used. When additional heat is needed, steam is injected into the reactor vessel.

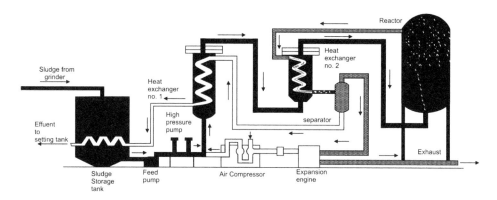

Figure 12.2 Schematic of Zimmerman wet oxidation process.

A major disadvantage associated with this process is the production of high-strength recycle liquor. The liquor represents a considerable organic load on the treatment system. The BOD content of the liquor may be as high as 40–50% of that of the unprocessed sludge, the COD typically ranges from 7000 to 10,000 mg/L. It is reported that a fraction of the COD is refractory and has been found difficult to remove by physical or chemical processes such as coagulation and activated carbon absorption.

Wet air oxidation has been implemented in only a limited number of installations since its introduction in the early 1960s. Some of these units have subsequently been taken out of service. Recent innovations may render this process more acceptable. In one system, the feed sludge is preconditioned by the addition of acid to lower the pH to 3. As a result, the process may be operated at lower pressures of $4\,MN/m^2$ and temperatures of $230°C$. In general, the wet-air oxidation system will not be applicable to treatment plants smaller than $0.2\,m^3/s$ (i.e. 5 Mgal/d).

12.3.7 Co-pyrolysis process

Pyrolysis is the destructive distillation and decomposition of organic solids at temperatures ranging from 370 to $870°C$ in the absence of air or other gases, which support combustion. Like incineration, pyrolysis reduces the volume of solid wastes and produces a sterile end product. Unlike incineration, it offers the potential advantages of eliminating air pollution and producing useful by-products.

At present, considerably more data are available for pyrolysis of municipal solid wastes than for the eo-pyrolysis process with wastewater sludge. The development of operating parameters for the optimum production of pyrolysis products, such as fuel gases, oils, and char, is necessary to make this process economically viable. Operation of pyrolysis equipment at the commercial scale is required to determine such parameters as operating temperature, operating pressure, and residence time in the pyrolyzer. A typical flow sheet for the eo-pyrolysis of solid wastes and sludge is shown in Figure 12.3.

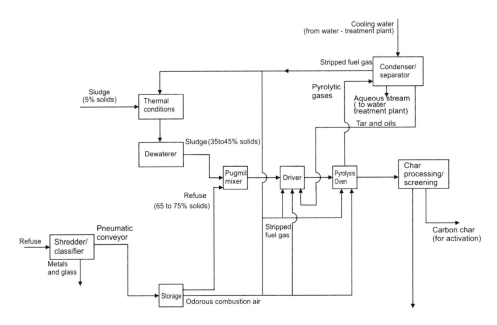

Figure 12.3 Flow chart for the co-pyrolysis of municipal solid wastes and sludge.

Aside from the residual chars and wastewater streams, treatment of effluent com-bustion gases from the processes is required. Gas scrubbers using water as the scrubbing medium are proposed for these systems. If the residual chars or solid material from the pyrolysis process are not used as fuels or as aggregate materials in road construction, land filling is the recommended disposal method. Because of the inert nature of the chars or ash from pyrolysis reactors, and because of the relatively small quantity of ash from these processes, effluent gases comprise the major pollutant to be treated.

12.4   PROCESS ADVANCEMENTS

12.4.1   Melting treatment of sewage-sludge
(Perry, 1973)

Introduction

With the spread of sewage systems and the increase of the sewage volume to be treated, sewage sludge production increases every year. However, places suitable for sludge disposal are limited and final disposal has already become a serious problem.

One solution to this problem, melting treatment of sewage sludge, has been attract-ing wide attention, particularly in Japan. Melting treatment reduces sludge volume, stabilises chromium and other harmful substances, and enables effective utilization of generated slag for construction material.

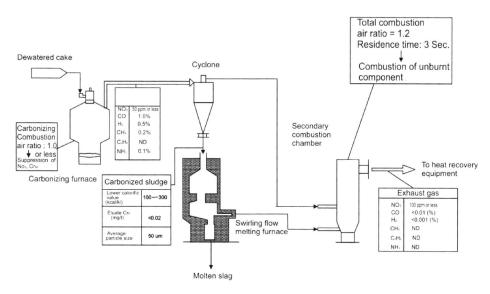

Figure 12.4 Basic configuration of carbonizing-melting system.

Basic system construction

Sewage sludge swirling flow melting furnace systems generally fall into two broad categories:

(a)   carbonizing melting systems (CMS), and
(b)   drying-melting systems (DMS)

Of these systems, carbonizing melting system, as shown in Figure 12.4, is more popular.

Basic configuration of CMS

The basic configurations of CMS are:

1   Dewatered cake is fed into a fluidised bed furnace and carbonised in a reducing atmosphere at a combustion air ratio of 1.0 or less.
2   By use of a cyclone in a reducing atmosphere, decomposed residue is separated from the exhaust gas and collected. In the residue, unburnt components primarily comprise of carbon remains and the calorific value of the residue ranges from 100 to 300 kcal/kg. This calorific energy can be used for melting the sludge.
3   The residue separated and collected by the cyclone is discharged into the swirling flow melting furnace without lowering the temperature (hot charge).
4   The residue is melted to make slag in the swirling flow melting furnace.
5   Exhaust gases from the carbonizing furnace and the melting furnace are mixed in the secondary combustion chamber to bring the total combustion air ratio to about 1.3 to achieve complete combustion.

System characteristics

The characteristics of the system with the above-mentioned configuration include the following:

1   Because fluctuations of cake properties are absorbed in the carbonizing furnace, the properties of the carbonised sludge are stabilised, enabling stable melting treatment.
2   The volume of carbonised sludge is reduced and the melting furnace can be made compact (about 1/3 the size of the dry cake melting furnace).
3   Properties of the carbonised sludge become best suited for the swirling flow melting furnace, resulting in extremely easy hanuling and maintenance.
4   NO_x and Cr^{6+} generation can be suppressed. Setting the combustion air ratio to 1.0 or less in the carbonising furnace can suppress generation of Cr^{6+} and NO.
5   Selection can be made between incineration and melting.

The carbonizing process can be individually operated, allowing for incineration and/or melting.

12.4.2   Sustained Shockwave Plasma (SSP) destruction method
(Mathiesen, 1990)

Introduction

When landfill space becomes scarce, an ideal sludge processing method may be that which can destroy the sludge where it is produced, at the rate it is produced and leave behind only clean air and non-hazardous useable residue.

Incineration, be it multiple-hearth or fluidised bed, is too slow and too limited to process parameters dictated by the need to sustain a combustion reaction under certain conditions and with only sludge combustibles and additional fossil fuel as the energy sources. This inflexibility, coupled with inevitable process disturbances, contributes to gaseous effluents forming, which are difficult or impossible to control. Moreover, the combustion residue, the ash, is a hazardous waste, and has to be handled and disposed of as such. Yet a method of destruction by rapid and controlled oxidation can be very attractive. However, the demands on sludge combustion are many and contradictory, such as:

1   Destruction of dioxins requires sufficiently high temperature for a certain period of time, followed by quenching to restrict reforming.
2   Retention of volatile heavy metals requires low temperatures.
3   Sludge contains up to 5% w/w of nitrogen on a dry basis, hence the oxidation requires lower temperature and low oxygen partial pressure to suppress NO_x.
4   Destruction of all organic species requires high temperature and high oxygen potential.

Thus the solution cannot be found in combustion systems where the only possible process control parameter is the practice of flooding the oxidation zone with excess air in order to keep temperatures down while attempting to achieve full oxidation.

All this can be achieved in a multi-stage SSP powered rapid throughput oxidation system, where optimum conditions can be established in each stage. Its attractions are cost-efficiency, flexibility to meet changes in sludge composition, flow rate and environmental safety.

Process description

The process comprises three sections: drying, oxidation and gas cleaning. The oxidation step is the most important part.

(a) Drying: Dewatered sludge at 80% moisture (w/w) enters the plant. It is further dried to a moisture content of 8–10% (w/w) in two dryers operating in series, a rotary drier followed by a flash drier. The heat required for drying is derived from the air-cooling system of the rapid oxidation part of the plant.

   The off-gases from the dryers reach a temperature of about 90°C and the drying sludge reaches 60°C. The drying process also serves to break up the drying sludge to a particle size required for the oxidation step.

   The reasons for the drying step are the need to limit the gas flow through the oxidation section and to minimise the power input in the plasma treatment stage. This allows rapid thermal energy release in the oxidation process given by the exothermic oxidation of sludge and this heat is used for drying but not electrical energy.

(b) Oxidation: The oxidation chamber comprises three sections:

   (i)   Plasma zone
   (ii) Primary oxidation zone, and
   (iii) Secondary oxidation zone

The three-oxidation zones are arranged vertically with the sludge fed in from the top and falling through the zones with all reactions being completed by the time the inert material reaches the bottom. The treatment achieves complete oxidation and thus destruction of all carbonaceous and organic matter.

The oxidation chambers have been designed to meet all requirements for emission of NO_x and CO. In addition, the process provides acid gas suppression by means of alkali or alkaline earth metal salt addition to the feedstock, which react with sulphur in the reducing plasma section.

From the oxidation step, the products are as follows:

(a) Glassy solids, containing heavy metals. Up to 90% of the non-oxidisable material exits the process in this phase.
(b) Fine, carbon-free particles, less than 10 μm in diameter.
(c) Oxidation reaction gases, containing quantities of SO_2, HCl, HF and extremely small quantities of volatile metals or metal compounds.

The back-end gas cleaning system is designed to handle these impurities carried by the off-gas releasing a clean gas to the atmosphere.

(i) The plasma zone: It is the sustained shockwave plasma environment through which all sludge particles are passed by means of free fall and initiate the reactions. The retention time is very short (about 0.9 s), but sufficient. In the plasma environment, the

particles of sludge not only experience the thermal energy transfer from the plasma, but also interact with the acoustic and electrical energy forms present.

The SSP plasma can be seen as continuous vertical DC discharge, through which a superimposed high frequency DC pulse is fired from a central cathode to a set of anode segments mounted along the inner wall of the plasma section below the cathode. A conical plasma volume is established, around the perimeter of which the pulsating discharge travels at speeds of up to 100,000 revolutions/min. The pulses are fired at rates of up to 100,000/s.

The purpose of the plasma treatment in this wholly reducing step is to prepare the sludge for the rapid destruction to follow in the two chambers below. It is also to provide for the reduction of metals from the oxide state to a metallic state, whereby they tend to group together, thereby preventing, to a significant extent, volatilization of heavy metals and their compounds. Although this mechanism does not remove all metals from the gas stream, it does contribute to the effectiveness of the back-end gas cleaning effort.

(ii) The primary oxidation zone: The sludge particles, highly reactive and in a state of beginning oxidation, enter the sub-stoichiometric primary oxidation zone, where air is introduced. All oxidisable solids are gasified. The solid residue is present here as a molten mass and is collected at the bottom of the shaft as solidifying, carbon-free glassy substance. This coalescence of the slag further limits any metals emission to the gas.

The collected solid material is a useful material for cement manufacture or for use as aggregate. It contains, in unleachable form, significant portions of the sludge heavy metals as well as sulphur.

(iii) The secondary oxidation zone: The gas from the primary zone is treated here to fully oxidise CO and H_2 to CO_2 and H_2O. The gas exiting the secondary zone carries certain contaminants, which are removed with the gas cleaning equipment. The contaminants at this stage are limited to acid gases, particulates and metal vapours of extremely low concentration.

(c) Gas Cleaning: The exit gas from the oxidation unit is cleaned, using wet scrubbers, etc. in five stages and cooled in a step-wise fashion to meet thermal requirements for optimum removal efficiencies.

Mercury vapours may survive the system at concentrations above those stipulated, particularly if sudden increase of mercury concentration occurs in the sludge feedstock. For this reason, a selenium filter is installed, bringing the maximum mercury emission to below 10 Nano grams per cubic metre.

12.4.3 Gasification

The necessity and incentive to develop and use nonconventional sources of energy is significantly increasing world over day by day. Current enthusiasm for the use of hydrogen as an alternative transportation fuel is founded on the expectation that the hydrogen win be produced from renewable resources at a competitive price. One method of achieving this goal is the steam reforming of sewage sludge and other biomass (Xu & Anal, 1998), as represented in the following equation:

$$C_6H_{10}O_5 + 7H_2O \rightarrow 6CO_2 + 12H_2 \tag{12.5}$$

In the above idealised, stoichiometric equation, cellulose (represented as $C_6H_{10}O_5$) reacts with water to produce hydrogen and carbon dioxide, mimicking the commercial manufacture of hydrogen from methane by catalytic steam reforming chemistry. More realistically, a practical technology must convert the cellulose, hemicellulose, lignin, and extractive components of the biomass feed stock to a gas rich in hydrogen and carbon dioxide, but also including some methane and carbon dioxide. Unfortunately, biomass feedstock does not react directly with steam to produce the desired products (Bouchard et al., 1990, 1991). Instead, significant amounts of tar and char are formed, and the gas contains higher hydrocarbons in addition to the desired light gases. The recent work of Herguido et al. (1992) nicely illustrates this situation. In a fluid bed operating at atmospheric pressure, Herguido and Co-workers observed yields of char from the steam gasification of wood sawdust in the range of 20–10 wt%, and yields of tar decreasing to 4 wt%, as the temperature of the bed increased to 650–775°C. But at the highest temperature, only 80% of the carbon in the feedstock is converted to gas. By employing a secondary, fluidised bed of calcined dolomite operating at 800–875°C, Delgado and co-workers (1997) were able to convert almost all the tar to gas. Nevertheless, the char by-product is not converted and represents an effective loss of gas. Thus the total steam reforming of biomass feedstock as envisioned by the above equation, remains an elusive goal. However, Xu and Anal (1998) carried out a research work with the objective to identify conditions, which evoke the desired total steam reforming chemistry. It was found that a semi-solid gel could be made from 5 wt% (or less) corn starch in water. Sewage sludge and wood sawdust can be mixed into this gel and suspended therein, forming a thick paste. This paste is easily delivered to a supercritical flow reactor by a cement pump. Above the critical pressure of water, digested sewage sludge and wood sawdust can be steam reformed over a carbon catalyst to gas, composed of hydrogen, carbon dioxide, methane, and a trace of carbon monoxide. There are effectively no tar by-products. The liquid water effluent from the reactor has a low TOC value, a neutral pH, and no colour. This water can be recycled to the reactor.

The results obtained from this research work and earlier experiments conducted by Xu et al. (1996) and Matsumura et al. (1997) indicate that digested sewage sludge behaves similarly to wood sawdust, but is more prone to plug in the entry region of the reactor due to the high ash content of the feedstock. The amount of minerals exiting from the reactor was found to be less than that in the tap water.

12.4.4 Sludge-ash solidification
(Nishioka et al., 1990)

Where enough land disposal sites are not available, the sludge is incinerated after drying and sludge ash is treated to reduce the volume significantly. In particular, an ash-melting process is quite promising. The melting processes, however, have disadvantages, such as the requirement of a high temperature, above 1000°C, and the complexity of their process equipment.

On the other hand, the hydrothermal hot-pressing (HHP) technique is a solidification method to solidify the sludge ash. The equipment for the HHP technique is simple (Figures 12.5 and 12.6). It is chiefly composed of an autoclave, heater, and compressor. The temperature for solidification treatment is lower than the melting temperature

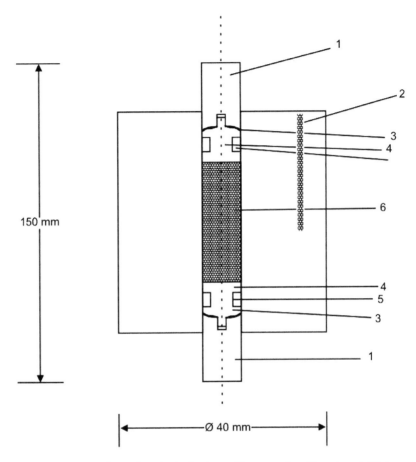

Figure 12.5 Cross-section of autoclave for hydrothermal hot-pressing: (1) push rod, (2) thermocouple well, (3) gland packing, (4) cast rod, (5) space for lost water, and (6) sample.

for ash melting. The solidification of the sludge ash by means of the HHP technique may be an effective post-incineration method. As an additive for solidification, glass powder is adopted.

When the mixture of sludge ash and glass powder is solidified by the HHP technique, the distinct advantage is that two wastes are used together to make a useful product.

12.4.5 Important patents
(Pergamon PATSEARCHER)

This section contains abstracts and, where appropriate, illustrations of a few issued United States patents and published patent applications for treating sludges filed from different countries under the Patent Cooperation Treaty. Copies of complete patents

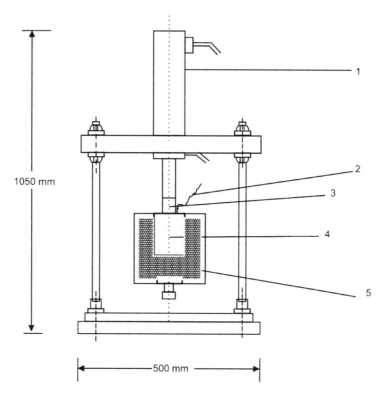

Figure 12.6 Scheme of apparatus of hydrothermal hot-pressing: (1) Pressure ram, (2) Thermocouple, (3) Push rod, (4) Autoclave, and (5) Induction coil.

announced in this section are available from the source mentioned in bibliographical reference section.

(a) 4880586: Method for the treatment of animal excrement and sewage sludge: (Wolfgang, Baader, 3300 Braunschweig, Federal Republic of Germany).

Method for the treatment of animal excrement material, particularly liquid manure and sewage sludges, in which the material containing a high percentage of liquid is mixed with a dry additive, preferably the re-circulated dry material, and after forming into briquettes which can be heaped is ripened in heaps penetrable by air, in which for the ripening process briquettes are employed having a volume predominantly between 30 and 50 cm³. The briquettes are conveniently formed in a press to have a diameter of between 30 and 40 mm, and a length of between 40–80 mm. For a larger diameter briquette, provision is made to form a hole in the briquette.

(b) 4902431: Method for treating wastewater sludge: (John P. Nicholson, Jeffrey C. Bumham assigned to N-Viro Energy Systems Ltd.)

It is a method of decontaminating wastewater sludge to a level that meets or exceeds USEPA process to further reduce pathogens standards, wherein lime or kiln dust and/or other alkaline materials are mixed with wastewater sludge in sufficient quantity to raise the pH of the mixture to 12 and above for a predetermined time and drying the resulting mixture.

(c) 4881473: Method and apparatus for treating oil water solids sludges and refinery waste streams: (James L. Skinner assigned to Atlantic Richfield Company)

Heavy hydrocarbon containing sludges such as refinery waste streams, oil storage tank sludges and marine oil tanker ballast are treated by passing the sludges in a flow stream through an indirect drier to vaporise liquids having a boiling point at atmospheric pressure of less than about 700°C and to provide substantially dried solid particles discharged from the drier. The dried particles, containing heavy hydrocarbons as a coating or as a part of the solids, are conducted to a combuster/oxidiser and exposed to a high velocity flow stream of oxygen containing gas, typically low pressure forced air, to bum the residual hydrocarbons in the solids and to reduce heavy metals to oxides. Sludge handling system includes a rotary disk type indirect drier connected to a lift pipe type combuster/oxidiser for thorough exposure of the dried solids to an oxygen-containing atmosphere. A centrifugal or cyclone type gas-solids separator receives the flow stream discharged from the lift pipe and a portion of the dried solids may be diverted after discharge from the drier back to the drier inlet to reduce the moisture content of the sludge flow stream introduced to the drier to minimise caking and clogging of the drier itself.

(d) 4919775: Method and apparatus for electrolytic treatment of sludge: (Eiichi Ishigaki, Sakaide, Japan assigned to Ishigaki Kiko Co. Ltd.)

A multiplicity of electrode plates, anodic and cathodic, are alternately arranged in an electrolytic tank side by side at short intervals in a row. A circulating current of the sludge being treated is caused to be produced within said electrolytic tank. Said circulating current includes an upward current flowing upward between individual electrode plates and a surface current flowing above the row of the electrode plates and near the surface level of the sludge in a specified direction. Gas bubbles produced between the individual electrode plates and flock present between the electrode plates are carried to the surface of the tank by said upward current. The surfaced bubbles and floes being then carried by said surface current to a location, which is generally outside the surface area above the electrode plates. Air bubbles thus collected at one side of the electrolytic tank are beaten out by a shower of sludge falling thereon.

(e) 4925571: Method for the simultaneous transport and pasteurization of sewage sludge: (A. Kirk Jacob, M. Joseph Willis, A. Pierc Hardman assigned to Organic Technology Inc.)

A process for treating sewage sludge having a predetermined water content to produce a soil conditioner, comprising: (a) transporting the sludge aboard a vehicle, and (b) heating the sludge during the transporting step (a) to a temperature and for a time sufficient to pasteurise the sludge without substantially reducing the water content of the sludge. The treatment is conducted aboard any suitable vehicle, such as a ship, rail road car or truck. The heat required to pasteurise the sludge may be obtained from the vehicle's engine waste heat and/or an auxiliary heat source. The sludge is pasteurised during the vehicle's transit from one destination to another and thereby eliminates the requirements of performing a separate pasteurization process before the sludge has been unloaded from the vehicle. The present invention protects the public health and saves both time and money as compared with conventional process of the prior art.

(f) 4926764: Sewage sludge treatment system: (Broek Ios Van den, 3972 KC Driebergen, Netherlands)

Gaseous discharge from a pelletizing drier used in the treatment of sewage sludge is partially directed back to a combustion chamber that generates an effluent, which is fed to the drier. Volumetric requirements of a gas scrubber and an after burner are reduced to the volume of gaseous discharge not recycled back to the combustion chamber. A concentrated stream of sewage sludge is mixed with a quantity of dehydrated particulate matter and supplied to a rotary pelletizing drier. Fuel and air undergo a combustion process and are mixed with additional air and part of the gaseous discharge in the combustion chamber, which generates a hot gaseous effluent that is directed through the drier. The effluent removes moisture from the mixture of concentrated sludge and dehydrated particulate sludge and the gaseous discharge. Entrained materials are initially separated from the gaseous discharge by cyclone separators. A gas flow-proportioning valve is disposed in a duct system interconnecting the cyclone separators, gas scrubber and combustion chamber for directing a portion of the gaseous discharge back to the combustion chamber.

(g) 4936983: Sewage sludge treatment with gas injection: (Charles Long, Philip M. Grover assigned to Long Enterprises Inc.)

This invention relates to an apparatus for treating sewage sludge in a hyperbaric vessel, in which the sludge is oxygenated by injecting an oxygen-rich gas into the sewage sludge and then dispersing the mixture of sludge and oxygen-rich gas into the upper portion of a hyperbaric vessel for further interaction with an oxygen rich atmosphere. The oxygen rich gas is injected into the sewage sludge by delivering the gas to a combination gas and sludge mixing and dispersing assembly. The gas and sludge are mixed within a plurality of channels formed in the assembly before the mixture is dispersed from the channels.

(h) EP 0348707 A1: Process for treating oil sludge

A process for treating petroleum refinery sludge to produce a coke-like residue product wherein an oily petroleum refinery sludge containing organic solid material boiling above 550°C and water is heated to a temperature above the boiling point of water and below the thermal cracking temperature of hydrocarbons and water in the sludge forms steam used to steam strip any light hydrocarbons from the organic solid material which is recovered as a solid coke-like residue product (Lane, 1990).

(i) US 5868942 A: Process for treating a waste sludge of biological solids

A process for treating a pathogens (bacteria, viruses, parasites) containing sludge including the steps of mixing the sludge with calcium oxide, ammonia and carbon dioxide so as to elevate a temperature of the mixed sludge to between 50°C and 140°C and to elevate a pH of the mixed sludge to greater than 9.8, pressurizing the mixed sludge to a pressure of greater than 14.7 p.s.i.a., and discharging the pressurised mixed sludge. The sludge has a water content of between 65% and 94% by weight. Ammonia is added to the sludge in the form of either ammonia gas, ammonium hydroxide, ammonium bicarbonate or as a byproduct of the reaction of the calcium oxide with the water in the sludge. Carbon dioxide is added to the sludge in the form of carbon dioxide gas or a reactant of ammonium bicarbonate. The pressurised mixed sludge is discharged by flashing the sludge across a restricting orifice and by evaporating a liquid component of the flashed sludge (Boss & Shepherd, 1999).

(j) US 20020148780 A1: Method of enhancing biological activated sludge treatment of waste water, and a fuel product resulting therefrom

Cellulose-based catalytic media is introduced into the waste water treatment system with a very simple in-line eductor injection system to enhance biological treatment, to improve settability of the biomass, and to produce a biomass fuel. The cellulose-based catalytic media particles create a feeding site for the microbes that provides a rich food consisting of the organic load that has been absorbed and the naturally occurring glucose and protein. In addition, the cellulose-based catalytic media naturally contains a carbohydrate known as glycocalyx which functions as a flocculent by causing the smaller suspended solids in the final clarifier to "stick" together and form larger, heavier particles. The larger, heavier solid particles produce a biomass sludge that settles better and faster. The settled biosolids, which now contain fractions of the cellulose-based catalytic media, are then dewatered and dried to create a biomass fuel (Tiemeyer, 2002).

12.4.6 Advancements in sludge utilisation

Utilization of wastewater sludge to produce L-lactic acid

A variety of biodegradable plastics have recently been developed (Doi & Fukuda, 1994). However, one of the important issues need to be tackled by the industry to promote the use of biodegradable plastics is their high production cost. Peimin *et al.* (1997a, b) has estimated the production costs of L-Iactic acid, which is precursor of biodegradable plastic, and has found that the cost of raw materials is about 40 per cent of the total manufacturing costs. Therefore, achieving a low production cost for L-Iactic acid is dependent primarily on what we choose for raw materials. Thus in case the wastewater sludge can be efficiently used as a raw material for the production of polylactic acid, then a considerable reduction in costs would be possible.

From the above point of view, many studies have been published regarding the production of biofuel from cellulosic material (Lynd *et al.*, 1991; Wymann & Goodmann, 1993). Some studies are concerned with production of lactic acid from cellulosic materials, that is usable in various industries (Xavier & Lonsane, 1994). Schmidt and Padukone (1997) conducted a study to use lactic acid, produced from waste paper, as a raw material for biodegradable plastic. However, Nagasaki *et al.* (1999) have succeeded to produce lactic acid from wastewater sludge. In their study the high concentrations of cellulose contained in the sludge, derived from a paper manufacturing facility, were found to be convertible to L-Iactic acid at a rate as high as 6–91 g/L. To achieve such a high conversion rate, the sludge must be pretreated with cellulose. This pretreatment includes inoculation of the sludge with lactic acid bacterial, strain LA1, after the sludge has been subjected to enzymatic hydrolysis.

Recycling the wastewater sludge nutrients via land application

Application of organic residuals such as municipal wastewater sludge to land is being widely promoted as a cost-effective disposal alternative, with agricultural, forest, and range lands as well as land reclamation sites increasingly used for land application. However, mass balances in a number of long-term studies on land application of sludge have been unable to account for up to half of the sludge-applied metals, challenging the concept of long-term metal immobilization in soil (Bell *et al.*, 1991; Dowdy *et al.*, 1991; McBride *et al.*, 1997; Richards *et al.*, 1998). Further, the loss of metals from fields that received sludge may possibly represent an environmental risk, and a better understanding of the factors governing metal mobility is necessary. Hence, the recycling of nutrients from wastewater sludge (biosolids) via land application is a desirable goal, but potential movement of sludge-applied trace metals is of concern and an area of ongoing research (Steenhuis *et al.*, 1996, 1999).

Turning sewage and sludge into power

British water company Thames Water is turning sewage into electricity to provide power for 38,000 homes throughout southern England. In collaboration with the Renewable Energy Company, Thames is operating two methods: methane gas and steam-to make use of effluent usually disposed off at sea, to provide renewable energy at 22 of its treatment plants. At Becton and Bexley, east of central London, sewage sludge is incinerated to provide steam to generate power. Becton, one of the largest sludge treatment plants in Europe, will eventually dispose off three million tonnes of sewage annually, providing 60 MW of power (IEI, 1999).

Converting sludge into construction board

According to an IEI-document (1999) the Bridgewater Paper Company's plant-sludge is being converted into construction board and used for brick making in a recycling system under trial by Salford University in north-west England. The process involves collecting the waste material in a single tank with a flocculating agent to assist de-watering. When shredded, pressed into flat board and then dried, it has no bending strength. However, the sludge contains small discrete lumps of clay distributed in a randomly orientated mass of fibres. It has been observed that if the sludge is removed prior to flocculation and subjected to the same drying procedure, it does have good bending strength. The Salford researchers have found that when the fire-rich streams are collected separately and used for brick making, the sludge left behind can be used for board making. In brick making, the white china clay's fine fibres modify the pore structure, giving bricks improved heat resistance, although it is necessary to moni-tor firing emissions for increased toxicity. Both pre-flocculated and post-flocculated sludge can be used to advantage as additives in plasterboard. Anchorage, indentation and impact resistance are improved, but there is a slight reduction in fire-resistance. Manufacturing boards wholly from negative waste material appears attractive, but the material is slow to de-water and difficult to keep flat when drying. The use of a heated plate press is likely to solve the latter problem, but it is not a viable solution for board containing over 40% of water. This had led the researchers to examine other methods of operation. At present, the factory produces 45,000 tonne of cellulose fibre and 55,000 of clay filler bound in 100,000 tonne of water annually (IEI, 1999).

Use of biochar generated from sludge

The carbon in biochar applied in soil for agriculture resists degradation and can sequester carbon in soil away from the atmosphere for thousands of years. As biochar is porous it also absorbs gas. Char from sewage sludge has an excellent absorption function of any type of gas. Hence char from sewage sludge can be used for Air Pollution Control in a large scale.

QUESTIONS

1   How thermal treatment helps to reduce the volume of sludge?
2   Name the different thermal treatment methods. Describe each in brief.
3   Define the incineration process. What are the merits and demerits of incineration?
4   Describe the advancements in sludge utilization.
5   Name the three patents of sludge treatment. Explain each in brief.
6   How you will differentiate the incineration, pyrolysis, wet air oxidation process, co-pyrolysis and gasification on the basis of operational conditions?

Chapter 13

Sludge disposal methods, problems and solutions

13.1 INTRODUCTION

Ever since mankind has mobilised into large towns and urban areas they have been concerned with the effective and safe disposal of liquid and solid wastes generated by domestic and/or commercial activities. Normally, the untreated liquid wastes were discharged into a stream, river or the ocean, but as population densities have increased this option has become increasingly unviable. Consequently, treatment systems were introduced to remove soluble as well as suspended organic matter from the sewage resulting in the production of sludge containing about 1 to 2% by weight dry solids. Conventionally this material was then disposed off either to agricultural land or to holding basins where it could be dried to a solid cake. This happy situation could not be continued mainly for two reasons; firstly, as treatment processes became more widely adopted, the sheer volume of sludge production increased dramatically while, at the same time, the availability of traditional disposal routes became limited. Secondly modem industrial civilization has resulted in a number of hazardous and highly toxic materials finding their way into sewage sludge, thus making it unsuitable for conventional methods of disposal.

Sludge disposal is presently a worldwide problem and a wide variety of disposal routes have been adopted as per the local conditions. The technologies utilised for sludge disposal as well as the environmental effects incurred are dictated largely by the choice of final resting place which can either be land, air or water.

Disposal of sewage sludge to land is widely practiced. Both landfill operations and application to agricultural soil are used, but the main limitations arise from such factors as: pathogens, toxic organics, heavy metals, and transport and application difficulties. Sometimes at considerable cost, however, a variety of technologies is available to circumvent these difficulties. Current research findings indicate that while organochlorines at low concentrations in the soil do not transfer to crops and are eventually degraded by soil microorganisms, heavy metals do accumulate in the soil and have a significant transfer to the food chain. Thus, a cost effective means for removing heavy metals from sludge is required before disposal of sludge through this route. Nevertheless, this practice is coming under increasing pressure as suitable sites become less available and controls on hazardous and toxic materials become more stringent.

Disposal of sludge to the air largely employs high temperature incineration or pyrolysis. Incineration converts all the organic matter present in the sludge to carbon dioxide (CO_2) and leaves behind most of the inorganic material as a dry ash. Pyrolysis, on the other hand, produces a liquid hydrocarbon as well as CO_2 and ash. In some circumstances the liquid hydrocarbon can be used directly as a diesel fuel. However, all high temperature operations/processes require considerable add on equipment, viz. stack gas scrubbers, to prevent gross air pollution, and are very capital intensive.

Disposal of sewage sludge to the ocean has also been widely practiced, but is now coming under increasing pressure to be banned because of its perceived adverse effects to the oceanic environment. However, the results of sludge dumping appear to be very site specific and are heavily influenced by the range of the sludge constituents. Materials of concern are organic matter, bacteria and virus, organochlorines, heavy metals, oil and grease, and nutrients such as nitrate and phosphate. In well flushed and dispersive ocean environments the major concern is that of organochlorines, which accumulate up the food chain and can cause severe health effects in higher species. On the other hand heavy metals, when present at low concentrations, do not appear to accumulate through the ocean food chain. Fish are able to separate and cleanse low levels of heavy metals by use of the enzyme metallothionein. Organic matter, oil and grease and pathogens generally cause problems only when discharged close to beaches or shellfisheries. In enclosed coastal seas with restricted flow and circulation the major environmental problem is eutrophication caused by the addition of nutrients.

13.2   SLUDGE DISPOSAL PROBLEMS AND SOLUTIONS

13.2.1   Land disposal

Land disposal is generally considered the ideal option for sewage sludge disposal for a number of reasons. If suitable land is available within 10–15 km of the treatment plant, then land disposal can avoid excessive processing of the sludge and benefit can be gained from the nutrient content of the sludge as well as its soil conditioning properties. However, land disposal option also has the problems which can be summarised under a number of headings viz. pathogens, heavy metals, toxic organics, and transport and application difficulties.

Pathogens
(Hashimoto et al., 1991; Strauch, 1988)

Most pathogens of the raw sewage get concentrated into the sewage sludge and present a considerable hazard when any handling of the sludge is considered. They can be separated into four categories e.g. viruses, bacteria, protozoans and larger parasites such as human roundworms, tapeworms and liver flukes. Such microorganisms can cause disease in humans, as the exposure and transmission may occur in several ways such as by inhaling sludge aerosols or dust, eating vegetables or fruit contaminated by sludge, drinking water contaminated by run-off, or by eating meat from livestock infected whilst grazing pastures fertilised with sewage sludge. Because of these dangers, sludge applied to land must be specially treated before its disposal to greatly reduce the number of pathogenic microorganisms and very careful handling and management techniques must be employed to minimise the risk of infection.

Solutions: Sludge sterilization or pasteurization can be achieved by a number of techniques as discussed in Chapter 11. These techniques may employ various combinations of high temperature, extended time and high pH. Simple temperature/time combinations can be employed, for example: 65°C for at least 30 minutes, 70°C for at least 25 minutes, 75°C for at least 20 minutes, or 80°C for at least 10 minutes. However, at still higher temperatures a reaction time of less than 10 minutes is not normally allowed. Also, the sludge must be pulverised prior to pasteurization to reduce the size of larger particles and thus to achieve an effective pasteurisation.

The sludge can be heated to the required temperature by gas fired burners or large continuous microwave ovens. However, from the viewpoint of energy input, a highly efficient way is thermophilic aerobic digestion, where the heat generated by the oxidizing bacteria is sufficient to maintain auto-thermal operation at 50–60°C. Again there is a certain temperature/time combination, which is required to achieve efficient pasteurization. Further, two-stage reactors are recommended to evade the microbiological disadvantages of hydraulic short-circuiting. For such systems a residence time of 5 days with temperatures in the range of 50 to 60°C gives the reliable pasteurization.

When properly insulated, the addition of unslaked lime (CaO) to dewatered sludge results in a heating of the sludge to temperatures between 55°C and 70°C. This occurs by virtue of exothermic reactions of the calcium oxide with the available water. However, the pH of the lime-sludge mixture must reach at least 12.6 and the temperature of the whole mixture must be maintained at equal to or more than 55°C for two hours to achieve efficient disinfection. Another way of achieving elevated temperatures is by composting the sludge, which is nothing but another variation of thermophilic aerobic digestion. However, temperature control is more difficult in this case, which results in application of more conservative conditions. This requires forced aeration or regular turning of the pile to ensure the uniform distribution of effective temperature in each part of the composting material for the necessary reaction time. The initial water content of the composting material must be in between 40–60% and the reaction temperature in the pile must be at least 55°C for three weeks.

A new technique for disinfecting sewage sludge, which is highly effective and quick, is known as electron-beam disinfection. The Japanese have developed this technology in combination with a highly efficient composting process to produce a safe material for use in intensive agriculture. The technique relies on irradiating a thin layer of sludge with an electron beam from a Cockcroft-Walton accelerator. The low penetrating capacity of the electron beam is overcome by a machine, which continuously produces a thin layer of sludge cake before feeding it to the electron-beam machine. Experiments demonstrate that a radiation dose of 5 kGy is sufficient to achieve complete disinfection and can be achieved in 0.5 seconds at a dose rate of 10 kGy/sec. The great advantage of this approach is that subsequent composting operations can be carried out at optimal conditions for conversion of organic matter without any consideration for achieving disinfection, resulting in a significant reduction of composing time from at least 10 days to around 4 days. Also, the resulting compost can be safely handled in situations of intensive agriculture. Nevertheless, the machine for producing the radiation is sophisticated and quite expensive.

Heavy metals
(Rulkens *et al.*, 1988; Swinton *et al.*, 1988)

A major limiting factor on the disposal and application of sewage sludge to agricultural land is the presence of heavy metals. Even in sludges from non-industrial regions problems can arise from zinc (phytotoxicity), lead, copper, and even cadmium. In particular, cadmium has been found to accumulate in the food chain and strict limits have been placed on its acceptable level in sewage sludge.

Solutions: One possible way to greatly reduce the severity of the treatment conditions required to remove heavy metals from sludge is the use of magnetic ion exchange resins. The sludge is treated with an oxidant, and then intimately mixed with the magnetically recoverable resin. The mixture is acidified with mineral acid. As the pH is lowered, metal ions are released into solution and almost immediately get adsorbed by the ion exchange resin, which is usually a chelating or strong acid type. The resin thus acts as a sink for the heavy metals. Consequently, quite high extraction levels of heavy metals can be achieved under moderate processing conditions e.g. pH 2.5 instead of pH 1. The resin is recovered from the sludge by magnetic means and regenerated to recover the metals in a concentrated stream. The resin can then be reused. The pH of the sludge is raised after the extraction step with lime. The opera ting cost of this process is significantly lower than more conventional techniques, provided a high recovery of the expensive magnetic resin is achieved. However, it is yet to be proven in a fully commercial application.

Toxic organics
(Bosma *et al.*, 1988; Kampe, 1988)

Many pesticides such as chlordane, dieldrin and heptachlor have been found in sewage sludge, along with a variety of other chlorinated organic compounds such as: polychlorinated biphenyls. Such compounds when ingested by animals tend to accumulate in the body fat where they can persist for many months or years. As such compounds tend to become more concentrated in each animal group up the food chain, human beings are at significant risk of having the highest concentrations of these pesticides, which have caused cancer in laboratory animals.

Some plants such as lucerne, cowpeas or oats are known to take up organochlorine pesticides into their tissues, although the route of uptake has not been clearly determined. While there is a potential problem in applying contaminated sludge to grazing land, evidence is emerging that, in general, chlorinated hydrocarbons do not accumulate in soils and have a limited transfer to crop plants. In a German study, where contaminated sludge was applied to soils at a high rate, polychlorinated biphenyl levels were found to have increased in the soil by 5 to 17 fold and polycyclic compounds 5 to 10 fold. However, no regular transfer to crop plants was detectable even in the mg/kg range. The author concluded that the organic substances investigated did not need to be limiting factors for the use of sewage sludge in agriculture. This work, however, needs confirmation by more experiments in different soil types.

Solutions: If a particular sludge is highly contaminated with organochlorine residues, there is only a limited range of technologies available to detoxify it. Thermal technologies, such as high temperature incineration, can destroy the organochlorines either by complete oxidation or by dechlorination under reducing conditions. However, such technologies are quite expensive and should only be applied in extreme circumstances. It is well known that certain species of aerobic and anaerobic bacteria have the ability to breakdown organochlorine compounds into innocuous by products. Such bacteria have been applied experimentally to the rehabilitation of contaminated soil, but have been little studied for the detoxification of sewage sludge. However, anaerobic digestion of sewage sludge is widely practised. It appears eminently possible that specific anaerobic bacteria, which are capable of degrading organochlorines, could be added to such digesters. Further research in this area is needed to determine the potential of this approach.

Transport and application difficulties
(Hall, 1988; Nihlgard, 1985)

The major problem with land disposal that has been encountered worldwide is the decreasing availability of suitable disposal sites. There are mainly two causes of this situation. Firstly, high population densities and an increase in sewage treatment facilities have resulted in a large increase in sludge production volumes. The second factor is that restrictions placed on the rate of application of sludge to agricultural soil have resulted in longer distances for transport of the sludge to a suitable disposal site. Nevertheless, there are a number of other options, which can be pursued.

Solutions: The simplest and cheapest one is to put stabilised sludge, normally of solids content 1–2% w/w, into large storage lagoons surrounded by a significant buffer area. Further, screening provided by shrubs and trees is sometimes necessary to prevent nuisances from insects and odours. During warm dry weather the ponds dry out to form a solid sludge cake, which can be removed by a mud cat dredge. The dried sludge is stored onsite until all available storage lagoons have been filled.

As and when sludge has to be transported over any significant distance, sludge disposal costs rise significantly. If transport distances are less than 10 km, then the liquid sludge of 2–3% w/w solids can be tankered and disposed off by either land spreading or soil injection. Soil injection has several advantages over other methods of land application. These include; odour minimisation, minimum contamination of surface waters, aesthetically more acceptable, reduced risk of exposure to the public, farm workers and livestock. Furthermore, liquid sludges contain more nitrogen (as nitrates), incorporated sludge is more readily degraded and nutrients become available to plants, and dewatering of sludge is not required, nor is the need to plough sludge into soil. However, there are a few disadvantages also of this option. Disadvantages of soil injection include; need for expensive and specialised equipment, compaction of soil when conditions are too wet, excessive surface disturbance and pasture die off when soils are too dry, and agricultural demands are seasonal.

Despite the potential advantages of soil injection, land spreading is still the most common method of land disposal of sludge. However, with this technique there is the potential problem of damage in soil structure through the use of heavy sludge spreading vehicles, which on the other hand, are necessary to ensure the spread of sludge evenly and at the target application rate. Nevertheless, recent research into improved equipment design and operational practice has overcome many of these problems.

Environmental and health problems of land spreading of sludge relate mainly to odour and the potential for disease transmission through aerosols. These problems can be managed by techniques previously discussed, but a potential problem, which has not yet been fully understood and addressed, is ammonia volatilization with its potential impact on global soil acidification. It has been hypothesized that increased levels of ammonium nitrogen in the soil can cause damage to the plant or tree in a number of ways. In the soil, uptake of ammonium by roots is usually accompanied by oxidation and a release of hydrogen ions or organic acids. Also, when ammonia volatilises uptake of ammonium through the leaves is, in a similar way, followed by a release of hydrogen ions or organic acids. In a complex mechanism it has been argued that such processes are involved in the spread of dieback in European forests. If such a mechanism is proved to be indeed operating then this factor may also place another limitation on the rate of application of sludge to the agricultural soils.

If transport distances become much longer than 10 km then trucking of the liquid sludge becomes very costly and extensive dewatering of the sludge becomes necessary. There are a number of ways of dewatering sewage sludge (this topic is discussed in detail under disposal to air i.e. in the section 13.2.2) all of which add significantly to total disposal costs. However, it is the only way in which sewage sludge can be economically transported over long distances to a variety of sites. Depending on distance and road traffic conditions, transport of sludge can become as expensive as some thermal disposal technologies. In such cases, on site incineration of the dewatered sludge becomes an important consideration.

Normally, the nutrient content of the sludge has a favourable effect on plant growth and is seen as a positive factor. However, as the removal of nutrients from sewage becomes more widely practised, the nutrient content of the sludge, particularly phosphorus, is likely to rise significantly. Application of such sludge to land will have to be closely monitored and controlled to avoid leachate to groundwater or runoff to surface waters causing a nutrient build-up in these receiving waters. For example, eutrophication problems are becoming more widespread in Australia with agricultural runoff being a major contributor to the water bodies. It is, therefore, important to ensure that the land disposal of sludge does not increase this problem.

13.2.2 Air disposal
(Booker et al., 1991; Worner, 1990)

Disposal of sludge to air (i.e. conversion of sludge into gaseous form), apart from conversion of organic matter to methane in anaerobic digesters, is generally avoided because of its high cost and adverse environmental impacts. In many ways, high temperature oxidation or pyrolysis of sludge is considered as a course of last resort when access to all other disposal options is denied for physical or environmental reasons. The only reason the Japanese incinerate a high proportion of their sludge is that suitable land disposal sites are minimal and, in a nation that is greatly dependent on healthy fisheries, ocean dumping is considered to be an environmentally unacceptable proposal.

Nevertheless, due to certain advantages, the use of thermal technologies for the destruction and disposal of sewage sludge to air is increasing rapidly worldwide. Advantages are mainly a drastic reduction in transport costs and the destruction of pathogens and toxic organics in the sludge. Heavy metals normally end up in the ash, but problems can arise with some of the more volatile metals such as mercury. However, modem flue gas treatment systems are available to remove most toxic materials from incineration smoke stacks. The essential disadvantages of thermal technologies are a high-energy cost (generally in dewatering the sludge), a requirement for expensive equipment and the resultant air pollution.

Given the advantages and disadvantages enumerated as above, there appear to be only two scenarios in which thermal disposal technologies may be applied. The first would be in major metropolitan areas where the long distances to suitable disposal sites and the large volume of sludge produced make other disposal options difficult and expensive to apply. The second scenario would occur when the sludge contains such a high level of toxic materials that other disposal options are again discarded. This situation could occur when a town is centred largely on chemical, mining or metallurgical industry. An interesting sewage treatment technique, which takes advantage of high temperature industrial processes, is being developed in Wollongong, Australia. It has been found that blast furnace dust can be used as an aid in the physico-chemical clarification of raw sewage in much the same way as magnetite is used in the Sirofloc process. However, instead of stripping the sewage sludge from the fine particles of blast furnace dust, the sludge is used here as a suitable source of carbon to aid the smelting of the dust to form agglomerates of pig iron. This process totally destroys the sewage sludge and toxic organic materials it may contain and, at the same time, immobilises most of the heavy metals within the pig iron. However, applicability of this kind of process would appear to be viable only close to large sources of blast furnace dust.

Incineration
(Dewling et al., 1980; Gore & Storrrie, 1977; Theis et al., 1984)

Incineration technologies are now highly developed. The main equipments used are multiple hearth and fluidised bed furnaces, as discussed in chapter 10. The prime pollution problems from incineration are those arising from gaseous emissions and ash disposal. However, air pollution can be greatly reduced by a range of systems. E.g. particulates can be removed by either wet systems (scrubber) or dry ones (electrostatic precipitator, bagfilter or cyclone). The most popular system amongst these is the venturi scrubber, because problems are created in other devices by the sticky nature of sludge fly ash. Other gaseous emissions, e.g. SO_2 and HCI, can be controlled by scrubbers, while odours and organic compounds can be reduced by after-burning. Catalytic reduction, however, is the only technique able to remove oxides of nitrogen i.e. NO_x.

Further, attention has been drawn by pesticides and polychlorinated biphenyls (PCBs) contained in sewage sludge, which have been found to be the most thermally resistant of the chlorinated hydrocarbons. Test results have shown that 94% reduction of PCBs is achieved at 430°C and 99.9% at 600°C with detention times of the order of 0.1 sec. With proper furnace operation and efficient gas cleaning systems, PCBs in sludge do not appear to represent a major hazard. Such a situation, however, does not appear to prevail for the most volatile of heavy metals i.e. mercury. A study conducted on the fate of heavy metals in sludge incineration found that even after a water scrubbing system to remove particulates, 97.6% of the mercury in the sludge reported to the exhaust gases. While, for all other metals, 99% ended up in either the ash or the wash water.

The other main product of incineration is ash, which is generally disposed off to landfill where the main concern revolves around the leachate characteristics. The ash consists mainly of insoluble sulphates, silicates, phosphates and refractory metal oxides, some of which may be soluble. An analysis of metals present in the incinerator ash has shown a range of metal concentrations e.g. cadmium 70 to 900 mg/kg, zinc 900 to 24,000 mg/kg. Only mercury was found in low level concentration range of 2–9 mg/kg, which reflects the very high percentage transfer of mercury to the exhaust gases.

Thickening and dewatering
(Gildemeister, 1988; Lipke, 1990; Wedag, 1990)

Thickening and dewatering processes of sewage sludge, which have been discussed in detail in chapter 4 and chapter 6, respectively, are essentially required to perform whenever the sludge is to be either incinerated or sent to landfill. While the basic process of the both operations is same, thickening is applied to dilute sludges of 1–2% w/w solids whereas dewatering occurs for sludges generally greater than 5% w/w solids. Incineration of the sludge, without any external energy input, is possible only if the sludge is dewatered to a cake solids content of at least 30–35%, depending on organic matter content in the solids. When the sludge is going to be disposed off in landfills, the volume of sludge should be minimised to reduce transport costs and dumping areas. In many cases the transportation cost savings alone are sufficient to justify improved dewatering, but another reason is to attain the required geotechnical stability of the landfill.

Centrifugation, which accelerates the sedimentation process by the application of centrifugal forces, can be used both for thickening and dewatering of sludges. Historically, centrifuges have been largely applied to dilute sludges with solids content in the range 1–2% w/w dry solids and have produced the sludges in the range 7–13% w/w dry solids. Their main advantages have centred around the totally enclosed continuous operation, which greatly reduces the amount of odorous air that needs to be treated. From housekeeping point of view also, centrifuges have the advantage of operating without sprays or exposed sludge and have low manpower requirements. Their main disadvantages have been high rates of wear and consequent maintenance, and the production of a sludge cake with relatively high water content. However, new developments in centrifuge technology indicate that these traditional disadvantages can easily be overcome. High wear rates can be reduced by the addition of hard surfacing materials in key locations and adopting improved servicing.

The "HI-COMPACT" process is a new development for the extensive mechanical dewatering of sludges. The process is based on the introduction of an interstitial drainage system into the substance to be dewatered by coating the surface of small pellets of it with a powdery layer of a drainage substance. Compressing a stack of these pellets results in a compact block interwoven with a network of drainage layers. Water within the pellets then has only a short distance to travel against a high flow resistance until it reaches the next drainage layer. In the drainage system, the water flows via the filter cloth without encountering any strong resistance. The process can dewater sludge cake up to the solids content between 55 and 65% at a pressure of 50 bar and with charging times between 1.5 and 2 minutes. This result compares favourably with the existing filter press technology, which generally has an upper limit of only 45% w /w dry solids. When transport distances to the landfill site are significant, the extra complexity of the "HI-COMPACT" technology appears economically justifiable. Further, more exotic technologies, which have been applied to the dewatering of sewage sludge, are electro-dewatering (Lockhart, 1986), acoustic dewatering (Ensminger, 1986) and vacuum combined with electrical and ultrasonic fields (Murlidhara *et al.*, 1986). These are, however, yet to receive wide acceptance in practice.

Wet air oxidation
(Cunningham & Duwer, 1989; de Bekker & Van den Berg, 1988)

An alternative thermal destruction technique to incineration, which does not require extensive dewatering of the sludge, is wet air oxidation. When liquid sludge with high organic solids content is brought to a high temperature (>175°C) and pressure (>10 MPa) in the presence of oxygen, decomposition reactions take place in the liquid phase. Organic matter is oxidised to carbon dioxide and water with the ash produced remaining in the water. The process can be controlled by the concentration of oxygen in the liquid sludge and autothermal operation can be maintained on a liquid feed so long as the chemical oxygen demand (COD) is greater than 15,000 mg/L. The process can be carried out either in a high-pressure reactor vessel or in a deep shaft called the Vertech reactor. Effluent from the reactor is a three-phase mixture, which must be passed through both a gas and solids separation system. The liquid effluent contains a quantity of ammonium acetate and other easily biodegradable components and soluble salts. Wet air oxidation technology, as for incineration, would only be considered where all other disposal options had been eliminated.

Sludge melting and other techniques
(Campbell, 1988; Murakami *et al.*, 1991; Yasuda, 1991)

While incineration may appear as an expensive and radical solution to sludge disposal under conditions such as that of Australia, to the Japanese it represents only an intermediate step in the minimisation process of the final solid waste volume. Such are the limitations on solid waste disposal sites in Japan that they even go so far as to smelt incinerator ash at temperatures up to 1500°C. At this point of temperature, even the inorganic solids melt to form a liquid slag, which then forms a hard, glasslike impervious aggregate when cooled under the right conditions. Such aggregates can be successfully used as building or construction materials. Another similar technology employed by the Japanese is the press burning of sewage sludge ashes to form interlocking bricks. In this technique, incinerated sludge ashes are placed in a metal mold, compressed under a pressure of 1 tonne/cm^2 and then burned in a furnace at 1050°C for 3 hours. The volume is reduced to about 25% of the original ash volume and becomes a strong impervious brick of red to black colour depending on the iron to manganese ratio in the ash.

Apart from incineration, the other thermal technique, which can be applied to dewatered sewage sludge, is pyrolysis. This technique generally results in the production of a liquid fuel and an ash, which can then be burnt to provide heat energy for drying the sludge.

Environment Canada's 'oil from sludge technology' is the one of the best-developed processes in this area. Dried sludge is heated to 300–350°C in the presence of a catalyst in an oxygen free environment for about 30 minutes. It is presumed that catalyzed vapour phase reactions convert the organics to straight chain hydrocarbons, much like those present in crude oil. Condensation of the product vapours results in the formation of two immiscible phases, oil and reaction water. In this process, oil yields have ranged from a low of 13% for an anaerobically digested sludge to a high of 46% for a mixed raw sludge. Char yields have ranged from 40 to 73% at the optimum operating temperature. If good revenue can be obtained from the oil, when it is of adequate quality to be used as diesel fuel, then the process, although expensive, compares favourably with incineration.

Anaerobic digestion

In comparison to incineration or pyrolysis technologies, biological techniques can provide conversion of sludges to a valuable resource, methane, in a cost effective manner. While only a fraction of the sludge is converted to methane, this technology can still be classified under disposal to air. Anaerobic digestion is widely utilised in sewage treatment plants, its main purpose being to stabilise organic sludges prior to disposal on land or water. During this process a plethora of anaerobic bacteria carry out a wide range of reactions to break down complex organic molecules. In 1983, Gujer and Zehnder (1983) proposed a six-step system in the anaerobic conversion of high molecular weight degradable organics to CH_4 and CO_2. These stages are comprised of (i) hydrolysis of proteins, lipids and carbohydrates, (ii) fermentation of sugars and amino acids, (iii) anaerobic oxidation of long-chain fatty acids and alcohols, (iv) anaerobic oxidation of intermediates such as the volatile fatty acids (with the exception of acetic), (v) conversion of acetate to CH_4 and, (vi) conversion of H_2 to CH_4.

As discussed in chapter 7, a wide range of microorganisms is involved in this complex series of reactions and understanding their interrelationships and dependencies is still a matter of research. Despite this lack of understanding, engineering developments have provided a variety of equipment designs in which the digestion process is carried out. A major classification in reactor design is between non-attached biomass and attached or fixed film processes. Examples of a non-attached biomass reactor include the simple continuously stirred tank reactor (i.e. the contact process where settled biomass is continually recycled to a stirred tank) and the upflow anaerobic sludge blanket (where biomass forms compact granules which are then gently fluidised in an upflow of feed). Attached film processes utilise a variety of surfaces on which the biomass is grown, either in a fixed bed mode or in an expanded or fluidised bed operation. Each design appears to have advantages in particular situations and pilot plant trials seem to be the best means for determining the optimum design.

13.2.3   Water disposal

Mankind has always found it convenient to dispose off untreated wastes to rivers and oceans. The river option quickly became untenable as the limited water flow resulted in catastrophic effects such as anoxia and fish kills. However, the ocean option has remained open and is still widely practised around the world. Sludge dumping is a major component of waste disposal to the ocean and is coming under increasingly close scrutiny because of possible adverse environmental effects. This development is perhaps best illustrated by the moves of the United States Environmental Protection Agency (US EPA) to ban ocean dumping of sludge completely (Marshall, 1988). However, it has been pointed out that sludge has to be disposed off somewhere and each disposal route has its own environmental effect. A careful assessment of the ocean disposal option is, therefore, necessary to clearly define under what circumstances, if any, it would be preferred over land or air. Such considerations would certainly have ramifications for major coastal cities.

The main advantage of sludge disposal in the ocean is its simplicity and consequent low cost, although this last factor can be upset if deep sea dumping is required. For example, when in New York, the dumping site was shifted from 12 mile to a site 106 miles offshore, extra sludge dewatering equipment was installed to reduce the volume of sludge and thus to limit the number of barge trips per day. The main disadvantage of ocean dumping is, of course, its possible adverse environmental effects discussed as hereinafter.

Environmental effects of sludge disposal in the ocean

There are a number of components of sewage sludge, which have the potential to cause environmental problems in the ocean. These components are: organic matter, oil and grease (especially hydrocarbon residues), bacteria and virus, heavy metals, organochlorines, and nutrients.

(a) Organic matter (Gunnerson, 1963): The main problem with organic matter is its tendency to cause sags in the dissolved oxygen level when it is decomposed by bacteria in the water. Marine bacteria (Organotrophic bacteria), as opposed to injected terrestrial bacteria, are the primary decomposers of organic matter in the oceans. Marine bacteria utilise particulate organic matter found in the residue of plants and animals, as well as dissolved organic matter, which partly originates during photosynthesis and excretion. With the breakdown (i.e. mineralisation) of organics by Organotrophic bacteria, carbon dioxide and nutrients are released in the form of simple soluble inorganic ions that can be utilised by plants and phytoplankton. At the same time, Chemolithotrophic bacteria oxidise reduced inorganics e.g. ammonia and hydrogen sulphide into nitrate and sulphate for energy. However, the rates of mineralization of sewage constituents can be quite rapid, exceeding those from marine sources several fold. This may cause severe oxygen sag resulting into death of oceanic creatures. To minimise this effect, the sludge-dumping site should be carefully chosen to maximise dispersion and avoid the build-up of any sediment on the floor of ocean.

(b) Oil and grease (Priestley, 1992): The oil and grease in sewage sludge, although more difficult to degrade, can still be metabolised by bacteria in the ocean. However, if not dispersed effectively, oil and grease in the sludge can accumulate to form floating grease balls. In this form the oil and grease is partly protected from degradation and can survive intact for a long period. Bacteria and virus within the grease ball are also protected and can present a significant environmental hazard if washed up on a populated beach. Nevertheless, pretreatment of the sludge, such as by anaerobic digestion, can greatly reduce the level of oil and grease in the sludge.

(c) Bacteria and virus (Cabelli, 1983; Fattal *et al.*, 1986): The main concern about bacteria and virus present in sewage sludge dumped into ocean relates directly to

public health. While anthropogenic bacteria are known to die off relatively rapidly in the ocean, there is some evidence of pathogen transmission to infants (up to 4 years age) who bathe in lightly polluted waters. It has been observed that conventional primary and secondary sewage treatment processes do not remove pathogens significantly, rather concentrate these microorganisms into the sludge largely unharmed. As a result, ocean dumping of the sludge does inject a large quantity of pathogens into the marine environment.

The main danger from pathogens in ocean-dumped sludge seems to arise from their concentration in shellfish and subsequent consumption by humans. It has been found concentration factors of seven to tenfold for various microorganisms in shellfish, although the shellfish themselves appeared not to be harmed by the micro-organisms. Thus, as a preventive measure, the correct selection of the dumping site can avoid any significant transfer of pathogens to the human population.

(d) Heavy metals (Calabrese *et al.*, 1982; NOAA, 1979): Acute toxic effects of heavy metals have been demonstrated in laboratory experiments primarily through the interaction of metals with animal tissue enzymes. As the heavy metals in sewage are largely concentrated into the sludge, concern has been expressed about their disposal in the ocean. Opinions differ as to which metals are more toxic but the following are of primary concern: Mercury (Hg), Cadmium (Cd), Lead (Pb), Silver (Ag), Copper (Cu), Zinc (Zn), Tin (Sn), Chromium (Cr), and Selenium (Se).

The main concern with heavy metal pollution in the ocean has been bioaccumulation, where metals are concentrated through the marine food web. However, extensive studies conducted on both the east and west coasts of the United States have indicated that many of the previously accepted beliefs concerning the toxicity of metals to marine life are erroneous. These studies have shown that high concentrations of toxic metals in animal tissues are not necessarily the cause of metabolic disorders in marine life. The argument used to explain this kind of results is that heavy metals have always been present in the ocean, *albeit* at very low concentrations, and the fish have responded by developing their own detoxification system. This means that the ocean disposal of sludge should not result in any significant sublethal toxic effect on marine organisms provided the heavy metal content of the sludge is strictly limited. Also, in case proper care is taken in the harvesting and handling of fish and shellfish, the consumption of toxic compounds in seafood can be controlled.

(e) Organochlorines (Keith *et al.*, 1970; MCZM, 1982; Sherwood, 1982; Young *et al.*, 1979): Organochlorine compounds, viz. dieldrin, heptachlor, hexachlorbenzene (HCB) and polychlorinated biphenyls (PCBs) find their way into the ocean via a number of routes. For example, agricultural runoff to rivers, atmospheric deposition and industrial discharges through sewerage systems have all been identified as contributors to the problem. As with heavy metals, organochlorines in sewage are also largely concentrated in the sludge and thus subsequent dumping of the sludge into the ocean significantly increase the local organochlorine content of the water. Bioaccumulation of such materials in the ocean food chain has been the main issue of concern. In contrast to the situation with heavy metals, however, the evidence accumulated to date indicates that there is indeed a problem.

Kepone, DDT and PCBs are the chlorinated hydrocarbons that have received greater attention as a result of their impact on coastal or estuarine environments. Because of low solubility in water these persistent chemicals are mainly attached to fine suspended particles, but find their way into the fatty tissues of marine organisms. Various sublethal effects, e.g. fin rot disease, have been documented in marine organisms in association with the exposure to these materials.

The above considerations do not entirely rule out ocean dumping of sewage sludge, rather the need is highlighted for strict control on the discharge to the sewer of industrial effluents containing such compounds. Moreover, it should be noted that the diffuse sources of organochlorines, such as agricultural runoff and atmospheric deposition, also need to be controlled. Further, the experiences observed with organochlorines also tend to make one question the impact of other synthetic organic compounds in the ocean. Almost nothing is known at present about the long term fate of detergent breakdown products (e.g. nonylphenol) in the ocean. However, past experience gives us reason to be cautious.

(f) Nutrients (Bell, 1990): Although most sewage treatment plants are not designed to remove the inorganic nutrients, e.g. nitrogen (N) and phosphorus (P), a considerable amount of these elements (especially N) are organically bound in sewage sludge. Further, even though nutrients are essential for primary production in the ocean, excessive amounts may ultimately reduce local diversity. It is because of the fact that the excessive build-up of nitrogen and phosphorus compounds can lead to the condition of eutrophication, which is characterised by increased production of a few species of algae and phytoplankton and subsequent decline in other types of species. This occurs because much of the disproportionate plant material (i.e. algae and phytoplankton) cannot be consumed by predators, it is instead decomposed by bacteria. This process reduces the available oxygen content in the water column, and as the oxygen supply decreases, predatory species disappear. Some examples of acute eutrophication and anoxia are found in marine waters where a restricted water circulation limits the dilution of nutrients and the resupply of dissolved oxygen. Enclosed coastal seas are the prime targets for such phenomena.

Because of the sensitivity of their fisheries both the Japanese and the Europeans are becoming particularly concerned about eutrophication problems. In Australia also, eutrophication of partially enclosed and coastal seas is a basic problem in several regions. For example, Port Phillip bay is a classic case of an enclosed coastal sea. Further, the Great Barrier Reef Lagoon, because of limited exchange with oceanic waters and the fragility of the reef ecosystem, appears now to be facing a notable eutrophication problem. In such regions ocean dumping of sludge would be terribly harmful. However, this situation does not completely rule out the ocean dumping option. In deep well flushed waters with a major exchange with the ocean, eutrophication problems most likely will not arise.

13.3 RELEVANT CASE STUDIES
(Ahmad *et al.*, 1998; Philip, 1985; Preistley, 1992; Willett *et al.*, 1984)

13.3.1 Canberra's sewage treatment plant (Australia)

For a nutrient removal plant with a real difference there is nothing in Australia quite like the Lower Molonglo Water Quality Control Centre, Canberra's sewage treatment plant. This plant was designed to protect the water quality of the Murrumbidgee River.

It utilises an integrated combination of physical, chemical and biological processes to produce a very high quality effluent (BOO <5 mg/L and total P <0.15 mg/L).

The treatment starts with lime clarification of raw sewage in primary settling tanks followed by biological nitrification in a conventional extended aeration activated sludge plant. As waste activated sludge is recycled to the head of the plant, the major sludge wastage occurs from the primary sedimentation tanks. At an average dry weather flow of 75 ML/d the Lower Molonglo plant produces on a dry weight basis about 4 tonne/day of grit, 0.15 tonne/day of scum and 45 tonne/day of sludge. The sludge from the primary sedimentation tanks usually has a solids content of 6–8% dry weight, a high figure directly attributable to the lime content. This sludge is fed to a horizontal continuous solid bowl centrifuge, where it is dewatered to about 35% solids dry weight with the aid of polymer dosing.

The original design of the plant was prepared during 1971–72, prior to the dramatic price increases in petroleum products occurred in mid 1970s. The original solids process design was based on the use of a two-stage centrifuge operation where lime rich solids were recovered in the first stage and fed to a multiple hearth furnace operating at around 1100°C. The idea was to convert calcium carbonate to calcium oxide for recycling in the process. The higher organic fraction remaining was to be processed through the second stage centrifuge for maximum dewatering prior to incineration and landfill disposal as a sterile ash. The rapid price increases of petroleum products during the 1970s rendered lime recovery uneconomic. Consequently, the sludge from the centrifuge was administered to incinerate in one of two multiple hearth furnaces to produce an ash with a weight around 20% of the original sludge. Careful operation of the furnaces reduced fuel oil consumption to around 800 l/day. Later, auto thermal operation was also achieved but in relative terms this was found to be a costly operation. Moreover, the ash still needed to be disposed off.

Further, a study by the Central Soil Investigation and Research Organization (CSIRO) has looked at the feasibility of disposing of sewage sludge from the Lower Molonglo plant direct to agricultural land. These investigations included consideration of land disposal of ash from the furnaces. The CSIRO studies found that although the ash contained less than the required amount of calcium (16% compared to 30%) to be sold as a liming agent, it was more effective than either lime or dolomite on a clover pasture. The better performance was attributed to its phosphorus and trace metal content. As Australia's normally acid soils require pH adjustment, it was found that ash could be applied at rates up to 20 tonnes/hectare without toxicity problems. It also resulted in significant improvement in crop yields with lucerne. However, some toxicity problems were noted with rye grass when application rates equalled 10 tonnes/hectare.

The main thrust of the CSIRO trials was to examine the effects of applying the sludge cake from the centrifuge to soils. This disposal route was aimed to obviate the need for incineration and could greatly reduce costs. In fact, the sludge proved effective in increasing the yield of crops and was an effective liming agent for soils, reacting more rapidly than agricultural lime. The sludge also provided nitrogen, which was not present in the ash after incineration. Application rates of up to 80 tonne/hectare dry matter were possible without toxic effects to the crops tested. The presence of micronutrients such as copper, zinc and boron were found to be an advantage, especially as boron deficiency is widespread in ACT forest areas. It was found that the major limitation on the land disposal of sludge from the Lower Molonglo plant was transportation costs. Disposal was only economic over a limited area, an area that was too small to accept the full sludge production of 140 wet tonnes per day. Nevertheless, sludge is still incinerated at Lower Molonglo and the ash is supplied to the farmers.

13.3.2   Oily sludge farming (India)

The petroleum refining process generates the oily sludge in considerable volumes. The common sources of this type of sludges are the bottoms of storage tanks, equipment cleaning operations, wastewater treatment, and contaminated soil from minor spills on refinery grounds. Technological and economical factors currently discourage the reclamation of oil from these sludges beyond steam pit treatment and they pose a disposal problem. In India, so far, the most common technique used for the disposal of oily sludges is by landfilling. However, land farming is widely practised for the disposal of oily sludges in the developed countries. A field scale two years project was undertaken in India at Gujarat Refinery premises with a view to study oily sludge application to the soil as a disposal alternative along with the cultivation of crop (millet). For this experimentation purpose, an area of about 1000 m^2 within the refinery premises was selected. It was cleared of bushes, stones, etc. and thoroughly ploughed by tractor. The site was developed into five large size plots, each containing four, 4 m × 4 m, experimental beds. Further plots were added for the second cycle of experiments.

It was observed that the soil samples from plots with the highest application rate of 100 L/m^2 oily sludge had the oil content same as that of the soil samples from plots receiving lesser application of 50 L/m^2 oily sludge. In fact the oil content of soil samples collected from various depths after 5 and 10 months showed that insignificant amounts of oil leached up to a depth of 165 cm. It was seen that leaching of oil was very limited for both the application rates. The high initial rate of disappearance of oil immediately after application of sludge, therefore, could be due to the presence of n-alkanes, n-alkyl aromatic and aromatic compounds of C$_{10}$–C$_{22}$ range which are least toxic and most biodegradable. It could have also been due to the initial spread of the oily sludge, which might have been relatively uneven, and thus the average sample might have not been exactly representative. Or, it could be due to volatilization and photochemical oxidation of the lighter fraction of the oil applied in the initial period of about 15 days. It is known that up to 20–40% of crude oils may volatilise from soil. Therefore, the sharp decline immediately after the application of oily sludge cannot be taken as the rate of biological decomposition. Later it was found that the biological degradation/decomposition rate of oily sludge is about 0.0025 kg oil/kg soil per month. In terms of volumetric aerial loading rate for sludge containing 80% oil, this amounts to be equal to 0.94 L sludge/m^2 per month. Therefore, leaving two months every year for resting of soil and preparation, the yearly permissible application rate comes out to be 9.4 L sludge/m^2 per year. Thus it was inferred that at this application rate, most of the applied oily sludge would be degraded in a year and the soil would maintain its productivity.

It should be pointed out that the above-recommended rate of 9.4 L sludge/m^2 per year is for the sludges containing 80% oil. However, for well treated sludge from steam pits where the oil content will be much less, the application rate of sludge could be correspondingly higher. For example, in case of steam pit treated sludge having 20% oil, an application rate of 37.6 L sludge/m^2 per year may be used. Nevertheless, it has been observed that with repeated applications, an accumulation of hydrocarbons occurs in the landfarm soil. This residue consists mainly of high molecular weight and relatively inert waxes and asphaltic compounds, which are eventually oxidised or humified and the completion of the process requires a long period of time.

It is concluded that in the present study because of very high (50 as against 9.4 L oily sludge/m^2 per year) oil application rates, there was negative effect on agricultural productivity. However, with the application rates of 9.4 L oily sludge/m^2 per year, it is expected that there will be minimum effect on the productivity and it would be possible to obtain a useful product from the land, but there should be caution with edible crops because of the potential risk of build-up of undesirable constituents such as heavy metals. Steam pit treated sludge at the above-recommended rate (37.6 L/m^2 per year) could also be used in a green belt of tree plantations around the refinery. This would optimise the land requirement for sludge disposal. If such a disposal method is adopted for oily sludge, then a regular monitoring programme would also be necessary.

13.3.3   Coffs harbour: A sensitive ocean environment (Australia)

At Coffs harbour, Australia, sewage from a number of local communities is treated before ocean discharge. What makes Coffs Harbour an interesting and controversial case is that the discharge is into a very sensitive ocean environment. In fact, the area has been classified as a marine national park. Its uniqueness determines from the fact that in this area warm coastal currents from the north meet colder currents brought up from the south.

The sensitivity of the ocean environment needs a high level of treatment. Coffs Harbour has the facility of not only conventional biological secondary treatment but also full nutrient removal via chemical precipitation and biological nitrification/denitrification. This treatment is followed by tertiary polishing in lagoons to render a high quality effluent (i.e. having 10 mg/L suspended solids and 10 mg/L BOD) for discharging into the ocean. However, it is to be noted that despite such a high level of treatment, controversy has still raged over the issue of ocean discharge versus land disposal. Nevertheless, unfortunately with their high seasonal rainfall, the Coffs Harbour weather patterns make ocean disposal virtually unavoidable during some months of the year.

13.3.4   Sustainable sludge management at a large Canadian wastewater treatment facility
(LeBlanc et al., 2004)

The Greater Moncton Sewerage Commission's 115,000 m^3/d wastewater treatment facility in New Brunswick, Canada, has established an integrated, long term, sustainable, cost effective plan for the management and beneficial utilization of sludge namely: composting, mine site reclamation, landfill cover, land application for agricultural use, tree farming, sod farm base as a soil enrichment, topsoil manufacturing.

Sludge as a soil additive in agriculture: The Commission actively promotes sludge as a soil additive in agriculture as a beneficial reuse. The Commission transports the material to the farm field at no cost, while, farmers carry out the sludge spreading. However, sludge must meet the metal restrictions.

SOD Farming: The use of sludge on the sod farm replaces a significant portion of nitrogen and phosphorus which acquired from commercial fertilizers as well as providing much needed organic matter, thus improving the soil structure. Moreover, lime added to the sludge also assists in reducing the soil acidity.

Landfill cover: The Commission has had successful experience in the beneficial use of sludge for cover the landfill sites at Moncton and Sackville. The landfill now integrates a setup of walking, hiking and biking trails. The use of sludge as landfill cover reduces the need of topsoil, which normally stripped from local farms.

Mine site rehabilitation: The mining site at Central New Brunswick is being regarded and conditioned with lime stabilised sludge at an application rate of 30 dry tonnes per hectare.

Golf courses: Composted sludge were used in the soil conditioning for the construction of a local, family owned golf course. The project not only saved this family business considerable money but also eliminated the need for importing topsoil.

Tree farming: Sludge compost/topsoil mixtures are being used as a planting medium in the nursery and as a transplanting medium with great success. Young trees grow faster while losses are kept to a minimum. The widespread forestry product industry offers great potential and could benefit from application of lime stabilised Sludge. Practical experience gained at a local Christmas tree farm will be used on a larger scale. Areas where trees were subject to wind damage recovered and growth rate improved.

13.4 SUMMARY

Sewage sludge, a by-product of raw water and wastewater treatment plants, can be disposed to any of the three final resting places which are: the land, air and water. Each disposal route has its own environmental impacts and a detailed knowledge of the effect of individual sludge components is needed to accurately assess such impacts. For instance, organochlorine residues cause severe problems in aquatic biota through bioaccumulation, but do not appear to be transferred to plants from agricultural soil. On the other hand, toxic heavy metals are accumulated by edible crops, but there is little evidence to suggest that they cause problems at low concentrations in the ocean. Thus the effective sludge disposal policies need to take into consideration all of this detailed scientific knowledge, but at the same time to recognise the remaining areas where knowledge is inadequate. In nutshell, the sludge disposal options are being restricted all around the world as environmental standards are tightened. This trend will increase the emphasis on development of cost effective technologies to solve environmental problems associated with each disposal route. Moreover, a variety of sludge disposal routes will still be required to handle the increasing volume of sludge.

QUESTIONS

1   What are the major problems related with sludge disposal?
2   Define the air and water disposal of sludge.
3   What are the environmental effects of ocean disposal of sludge?

4   What are organochlorines? How they are hazardous to environmental health?
5   What are the differences between Class A and Class B biosolids?
6   What are the different methods for pathogen removal during sludge treatment?
7   Describe any two case studies:

   a. Canberra's sewage treatment plant (Australia)
   b. Oily sludge farming (India)
   c. Coffs harbour: A sensitive ocean environment (Australia)
   d. Sustainable sludge management at a large Canadian wastewater treatment
       facility

Chapter 14

Energy and resource recovery from sludge

14.1 INTRODUCTION

Historically, it was common to see schematics that showed the water treatment scheme in detail and an arrow at the end that simply said "sludge to disposal" (Neyens *et al.*, 2004). However, presently the situation is altogether different as increasing sludge production and its proper management is a big challenge for waste management authorities mainly due to increasingly stringent sludge reuse and disposal regulations. The population explosion resulting in the generation of wastewater volume, which is triggering to increased wastewater treatment facilities. Activated sludge process is a generalised secondary treatment system with a fact that over 90% wastewater treatment systems are using the conventional activated sludge process. The worldwide sludge generation is about 50 million tonnes dry solids per year, which will increase upto 83 million tonnes of dry solids by 2017. The main bottlenecks in sludge management are:

- Sludge treatment and disposal accounts 40–60% cost of total plant operation cost
- Land application of sludge is restricted by pathogens, emerging contaminants, heavy metals etc.
- Incineration is an efficient method in sludge volume reduction but high capital cost, ash generation and discharges of heavy metals, CO_2 and N_2O to the atmosphere are the main issues
- Landfills are facing pressure due to high building, operational and land cost, and emissions and loss of resource

Drying and landfill disposal are the sludge management options with the most significant greenhouse gas emissions. Based on a typical urban wastewater treatment plant treating 100,000 cubic meters/day, producing 80 tons/day of dewatered sludge with 80% moisture content; carbon footprint indicated as tons of carbon dioxide equivalent/year. In Europe and the US, greenhouse gas emissions from sludge treatment are considered now as an important benchmark while assessing alternative sludge treatment technologies (ADB, 2012). At the same juncture, rising energy demand resulting in shrinking reserve of fossils fuels and fuel price hikes. Thus, following drivers prompting increased interest in extracting value added crops from sludge: (a) tightening global regulations for sludge disposal, (b) sludge handling issues, (c) fuel prices spikes and depleting fossil fuel reserves, (d) climate change concern and (e) public awareness (Tyagi & Lo, 2013).

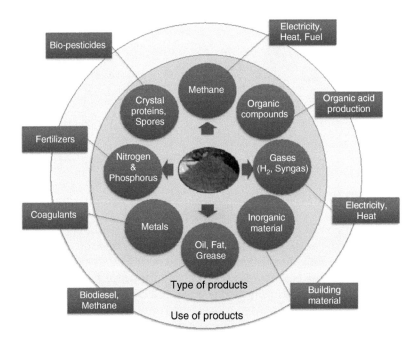

Figure 14.1 Beneficial reuse of sludge.

Nutrients (primarily nitrogen and phosphorus) and energy (carbon) are two components in sludge, which are technically and economically feasible to recycle (Campbell, 2000). Various routes to recover resources and energy from sludge are (Tyagi & Lo, 2013):

(a)   Bio-chemical: Anaerobic digestion, microbial fuel cell, anaerobic-aerobic treatment
(b)   Thermo-chemical: Incineration and co-incineration, gasification, pyrolysis, wet air oxidation, super-critical water oxidation, hydrothermal treatment
(c)   Mechanical-chemical: ultrasonication

Following products can be recover using the above-mentioned technical routes: biogas, fuel gas, syngas, biodiesel, electricity generation, nutrient (nitrogen and phosphorus), heavy metals, hydrolytic enzymes, bio-pesticides, bio-plastic, commercial fertilizers and construction material (Figure 14.1).

14.2   CHARACTERISTICS OF SLUDGE

Two types of sludges are generated in a wastewater treatment plants: (a) primary sludge usually contains 5% to 9 % total solids (b) secondary sludge or waste activated sludge, having a total solids concentration range between 0.8% and 1.2%. Primary

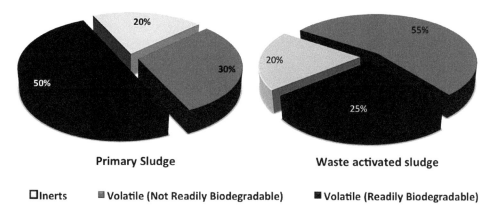

Primary Sludge   Waste activated sludge

☐ Inerts   ■ Volatile (Not Readily Biodegradable)   ■ Volatile (Readily Biodegradable)

Figure 14.2 Composition of primary sludge and waste activated sludge (Data Source: NACWA, 2010).

and secondary sludges contain 60–80% and 59–88% (%TS) of organic matter, which is decomposable and produces the offensive odors (Figure 14.2). Municipal sludge is composed of non-toxic organic carbon compounds (approximately 60% on dry basis), nitrogen (1.5–4%TS) and phosphorus (0.8–2.8%TS) containing components. The inorganic compounds found in sludge are silicate, aluminates, calcium and magnesium containing compounds. Few heavy metals are also present in the sludge (Zn, Pb, Cu, Cr, Ni, Cd, Hg). The water content varies from a few percent to more than 95% (Tchobanoglous *et al.*, 2003). Energy content of sludge is laid in volatile solids, which shared 65% of dry solids (DS). One pound of dry biosolids contains 6000–9000 BTU (british thermal units) energy content.

14.3   CONSTRUCTION MATERIAL

Sludge can be reuse for the production of construction materials such as bricks, cement, pumice, slag and artificial lightweight aggregates. The organic and inorganic compounds in sludge offer the valuable resources for production of construction material. The waste sludge can be used in different forms as dewatered sludge, dried sludge or incinerated sludge ash. In Japan, thermal solidification of inorganic compounds of sludge are attempted to produce the beneficial products. In solidification process, the incinerated sludge ash is processed at high temperature of 1000°C (Rulkens, 2008).

For Portland cement manufacturing, the most promising and rapidly adapting method is to directly introduce the dewatered sludge into the Portland cement kilns. This method is considered cost-effective mainly due to the benefits of neither requiring new incinerators nor making extra cost of operation. For artificial lightweight aggregate (ALWA), the mixture of incinerated ash, water (23% w/w) and a small amount of alcohol-distillation waste (a binding reagent) is transferred to a centrifugal pelletizer. The pellets are dried at 270°C for 7–10 minutes and then processed at 1050°C into a fluidised-bed kiln. After that, the spherical shape pellets are air-cooled and a hard film forms on their surface, however, the inside remains porous. Due to their lower specific gravity, greater sphericity, and less compressive strength, the major reuses of ALWA are water-infiltrating pavement, planter soils, thermal insulator panel, flower vase additive, substitution of anthracite media of rapid sand filters, and in walkways (Spinosa, 2004).

Slag is another option to reduce the sludge volume and the immobilization of heavy metals. Air and water cooled slags are two different kinds of slag, which are vitreous, and attain the standards of the crushed gravel used for concrete, however, their compression strength is inferior in comparison with natural gravel. Air-cooled slag is used as a concrete aggregate and back-filling material, ready mixed concrete aggregate, roadbed materials, permeable pavement, interlocking tiles (Spinosa, 2004). Sludge pumice is manufactured with the similar methodology as bricks, by the addition of sieving and crushing procedures. The main use of sludge pumice is for under-layer of athletic turfs, mainly due to its characteristics of quickly drains the excess water, however, holds the adequate moisture, thus upholding the necessary conditions of the athletic arena (Spinosa, 2004).

The bricks can be manufacture by mixing the sludge ash with clay, which is similar to conventional bricks in physical properties and appearance, however, quite heavier than conventional bricks. In Tokyo, first full scale plant for sludge brick manufacturing was started in 1991 with a production capacity of 5500 bricks per day using 15000 kg of incinerated sludge ash. There is no heavy metals leach from the sludge bricks, even under acidic environment (<3 pH) (Spinosa, 2004).

14.4 NUTRIENTS

In sludge, phosphorus and nitrogen are the key nutrients of interest, mainly present as proteinaceous material. The natural resources of phosphorus are depleting rapidly, and sooner or later no reliable source of phosphorus will remain. Sludge is one of the main source of majorly available phosphorus streams in present day, thus it is necessary to develop the bio-refinery for phosphorus recovery from sludge.

Several efforts have been made so far recently in the direction of phosphorus recovery from sludge through crystallization method. Sludge solubilisation and its subsequent transformation to ammonia and phosphates could be used to produce calcium phosphate and magnesium ammonium phosphate (struvite), which can be used as plant fertilizers (Wong et al., 2006). Several types of treatment like thermal (conventional, microwave), chemical (acidic, alkaline, advances oxidation process) alone and in combination (thermo-chemical) were applied under different sets of treatment conditions to release the phosphate and ammonia from the sludge to bulk solution. The bacterial cells and hardly degradable organic compounds could be disintegrate under harsh thermal, chemical or thermo-chemical treatment conditions, which ultimately leads to the release of stored phosphorous, polyphosphate and nitrogenous compounds into the bulk liquid. The phosphorus (as total phosphorus, orthophosphate) and nitrogen (as total nitrogen, ammonia, TKN) release of ranges from 38%–96% and 36%–53%, respectively, were reported into the literature (Liao et al., 2005a; Liao et al., 2005b; Wong et al., 2007; Danesh et al., 2008). The ash material produced during super critical water oxidation (SCWO) treatment of biosolids is an option for a quite simple process to recover phosphates from biosolids. Aqua reci process has been used at laboratory and pilot scale to recover the phosphate from biosolids ash by caustic and acids extraction method.

Several commercial technologies for phosphorus recovery from sludge are Crystalactor®, SEPHOS, P-RoC, Phostrip®, Aqua-Reci, BioCon, OSTARA, KREPO and Kemicond, which are based on the thermal-chemical, physical-chemical and thermal treatment to solubilise sludge, release phosphorus in solution and recover through precipitation (Tyagi & Lo, 2013). KREPO technology, developed in Sweden, involves the following steps: acidification, heating with steam, hydrolysis in a pressurised reactor, organic sludge separation and precipitation of iron phosphate from the centrate. The Kemicond^{TM} is a modified KREPO process included acidification with H_2SO_4 and H_2O_2 followed by two stage dewatering, found to be less complicated because of no pressurised heated reactor requirement and less land and carbon footprint (Kalogo & Monteith, 2008). Seaborne technology, developed and under trial at full scale in Germany, involves incineration, acid treatment, desulphurization, methane production, heavy metals separation, and struvite precipitation (Berg & Schaum, 2005). The main advantages of the technology are recovery of multiple nutrients with no heavy metals and organic pollutants and H_2S-free biogas (Müller *et al.*, 2007). BioCOn and SEPHOS are other emerging technologies using incinerated sludge ash for phosphorus recovery as phosphoric acid and, aluminium phosphate and calcium phosphate, respectively. However, none of them has not yet been investigated at full-scale. ARP (Ammonia Recovery Process) technology, commercialised by the ThermoEnergy Corporation, uses a proprietary resin system to extract a commercial-grade fertilizer (ammonium sulfate) from a nitrogen-rich sludge sidestream. The technology was scaled up to pilot plant level in New York in 1998 and investigated for its feasibility. However, no recent updates about the development of technology was documented (Kalogo & Monteith, 2008).

14.5 HEAVY METALS

Sludge, mainly of industrial origin, can be a good source of recovering the heavy metals such as cadmium (Cd), chromium (Cr), nickel (Ni), zinc (Zn), lead (Pb), copper (Cu) and mercury (Hg). The land application of metal laden sludge is restricted mainly due to leaching issues, which caused soil and ground water contamination, and ultimately affect the human and animal health (Tyagi & Lo, 2013). Therefore, metal-bearing sludges needs to be treated to extract the metal ions, or stabilise the metals in solid forms before disposal (Hsieh *et al.*, 2008).

Thermal (conventional, microwave), chemical (acid) and mechanical (sonication) methods were applied alone and in combination to leach out and extract the metals ions from the sludge. The significant heavy metals recovery of Cu (upto 93%), Ni (upto 98.8%), Zn (100.2%) were reported under different sets of thermal and thermal chemical treatment conditions (Perez-Cid *et al.*, 1999; Wu *et al.*, 2009). The effect of sonication-acidification leaching process on the separation and recovery of heavy metals from the printed circuit board (PCB) waste sludge was studied by Li *et al.* (2010). They reported significantly higher recovery of Cu (97.42%), Ni (98.46%),

Zn (98.63%), Cr (98.32%) and Fe (100%) (100 min contact time, and 100 W ultrasonic power, pH 4.0). Ultrasonic waves induced liquid cavitation that can blow out the surface of solids, which generates highly reactive surfaces, causes short-lived high temperatures and pressures at the surface, produces surface defects and deformations, forms fines and increases the surface area of friable solid supports, ejects material into solution (Kim *et al.*, 2006), thus increasing ion movement from surface of metal bind compounds into liquid phase and enhancing the leaching of metals into the leachate from the sludge (Xie *et al.*, 2009).

14.6 BIOFUELS

14.6.1 Biogas

Biogas production from anaerobic digestion is a common practice at wastewater treatment facilities to stabilise the organic matter, reduce the sludge volume and pathogens reduction, and recovery of energy rich methane gas. Anaerobic digestion produced the biogas with a high proportion of energy rich methane gas (60%) and remaining carbon dioxide (40%). Methane gas is used for electricity and heat generation at wastewater treatment plants. Electricity costs upto 80% of total operational cost of WWTP, however, energy recovery through methane can cover approximately half of this cost (Deublein & Steinhauser, 2008). Biogas can be use for large number of application such as production of heat and electricity, as vehicle fuel, chemical manufacturing, and electricity co-generation (Tyagi & Lo, 2013).

Biogas is the ideal fuel for combined heat and power (CHP) applications. An analysis accomplished by the CHP Partnership observed that if CHP were installed at all 544 wastewater treatment facilities in the U.S. (influent flow rates >5 mgd and that operate anaerobic digesters), then approximately 340 MW (340,000 kilowatt hours) of electricity could be generated – enough to power 261,000 homes (NACWA, 2010). The biogas can be used as fuel to generate electrical power using engine generators, turbines, or fuel cells. The gas turbines could be used (micro-turbines, 25–100 kW; large turbines, 4100 kW) with low emissions, efficiencies comparable to spark-ignition engines and low maintenance. Dual fuel engines are also very popular with good power efficiency. The heat generated during the power production process can be used for building heating or cooling, or in the treatment process itself. Fuel cells are considered to become the small-scale power plant of the future, having the potential to reach very high efficiencies and low emissions. Special interest for biogas is focused on hot fuel cells (>800°C) where CO_2 does not inhibit the electrochemical process, but rather serves as a heat carrier. Biogas can be used as vehicles fuel (SenterNovem, 2008), provided it is upgraded to natural gas quality, and application in the same vehicles that use natural gas (NGVs) becomes possible (Appels *et al.*, 2008).

Hydrolysis is a rate-limiting step in anaerobic digestion of sludge due to complex structure of sludge. Several methods have been developed to enhance the sludge solubilisation (pre-hydrolysis) in order to promote the biogas production. Several thermal, chemical, mechanical technologies alone and in combination were developed. Cambi® (160–180°C, 600 kPa, 30 minutes) and BioThelys® (150–180°C, 800–1000 kPa, 30–60 minutes) processes were observed to achieve 50% improvement in biogas generation and upto 80% reduction in sludge production, respectively. MicroSludge® is a physical-chemical method involves caustic soda followed by high pressure homogenization treatment for sludge solubilisation were found to reduce the sludge retention time in anaerobic digester to 9 days mainly due to rapid hydrolysis of pretreated sludge. CROWN® ultrasonic sludge disintegration process based on cavitation phenomenon, where shock waves, by the effect of high local temperature and pressure, causes the sludge disintegration, was found to be effective in increasing the biogas production by 34% and reducing the sludge volume by 24% (Kalogo & Monteith, 2008).

14.6.2 Syngas

Syngas could be used as a clean alternative to fossil fuels in power generation or for the production of liquid fuels like methanol, dimethyl ether and synthetic diesel. The pyrolysis of sewage sludge, which involves heating the sludge in the absence of air or in an oxygen-deficient atmosphere, leading to the production of bio-gaseous fuels. Sludge pyrolysis has several advantages such as; energy recovery (up to 80%) with low emissions of green house gases (NOx and SOx), no toxic organic compounds (e.g. dioxins) formation during pyrolysis and remarkable reduction in sludge volume (upto 93%). Under the appropriate operating conditions (1000°C), the pyrolysis and gasification of the sewage sludge take place and produce a gas with high CO and H_2 contents. Activated carbon was used as microwave receptor in order to accelerate the rate of syngas production. The findings revealed that the use of activated carbon enhanced the concentration of $H_2 + CO$ by 60% in the pyrolysis gas (calorific value of 12.930 MJ/nm^3) (Zuo et al., 2011). To maximise the gas yield and to assess its quality as a source of hydrogen or syngas ($H_2 + CO$), Lv et al. (2007) investigated the pyrolysis of sewage sludge using MW (1000 W and 10 min, 1040°C) as the sources of heat and graphite and char as MW absorbers. Both gases were found to be produce in a higher proportion with maximum values of 38% for H_2 and 66% for $H_2 + CO$. Moreover, this gas showed lower CO_2 and CH_4 concentrations of 50% and 70%, respectively.

14.6.3 Hydrogen

As a sustainable energy source with minimum or zero use of hydrocarbons and high-energy yield (122 kJ/g), hydrogen is a promising alternative to fossil fuels (Rifkin, 2002). Bio-hydrogen production by photosynthetic and dark fermentation methods is considered more energy efficient and environmentally benign process than chemical processes (Han & Shin, 2004). Several researchers studied the hydrogen production from anaerobic fermentation of sludge under different sets of treatment conditions (Table 14.1). Thermal, UV radiation and ultrasonic pretreatment were studied to enhance sludge solubilisation and subsequent hydrogen production during anaerobic digestion of waste sludge. Significantly higher hydrogen production was achieved during anaerobic digestion of pretreated sludge in comparison with non-pretreated sludge.

Table 14.1 Effect of different pretreatment conditions on the bio-hydrogen production.

Sludge type	Anaerobic digestion	Treatment conditions	Hydrogen production	References
Primary sludge	Mesophilic, CSTR, 12 h HRT	$-70°C$, 1 h with enzymatic addition (5% by volume)	• 27 L H_2/kg VS • 82% higher H_2 yield than non-pretreated sludge	Massanet-Nicolau et al. (2010)
Poultry slaughterhouse sludge (TS 5%)	Batch fermentation	– Microwave heating, 850W, 3 min	• 12.77 mL H_2/g TCOD (pretreated) • 0.18 mL H_2/g TCOD (raw)	Thungklin et al. (2010)
Waste activated sludge	Batch fermentation, mesophilic, 35°C	UV irradiation, 25W, 15 min	• 138.8 mL H_2/g TS (pretreated) • 82% higher H_2 yield than non-pretreated sludge	Wang et al. (2010)
Anaerobic digester sludge	–	• Sonication energy-79000 kJ/kg TS • temperature control conditions ($<30°C$),	• 120% increase in H_2 yield	Elbeshbishy et al. (2010)
Anaerobic digester sludge	–	• 130 W/L sonication density, 10 s time	• Increase of 1.30 and 1.48 fold in H_2 yield rate	Guo et al. (2010)

14.6.4 Bio-oil

Bio-oils, components of *n*-alkanes and 1-alkenes, aromatic compounds, which are refined to high-quality hydrocarbon fuels (Dominguez *et al.*, 2003). Sludge pyrolysis at 425–575°C allows the major production of bio-oil, which amounts to 30–40 wt.% of sewage sludge (Cao & Pawlowski, 2012). Earlier studies has been attempted the thermal conversion of sludge to liquid and solid fuels and reported the oil yields ranged from 13% to 46% (Tyagi *et al.*, 2009). Tian *et al.* (2011) reported that thermal (microwaved) pyrolysis (400 W, 6 min) resulted in a maximum oil production of 49.8 wt.% having high calorific value (35.0 MJ/kg), preferable chemical composition (29.5 wt.% monoaromatics) and low density (929 kg/m^3). Dominguez *et al.* (2003) reported that the pyrolysis of sewage sludge can be achieved by thermal (microwaved) treatment using graphite as a microwave absorber at 1000°C, which leads to the production of pyrolysis oils having high calorific. The use of graphite and char as microwave absorbers allow temperatures as high as 1000°C to be reached within a few minutes. The oil produced from microwave pyrolysis is more aliphatic and oxygenated than are oils that are produced by conventional heating at the same temperature (1000°C).

EnerSludge^{TM} and SlurryCarb^{TM} are two commercial methods based on pyrolysis of sludge for bio-oil recovery. In EnerSludge^{TM} process, almost 45% of the energy in the sludge is transformed to oil. However, the process was found cost-intensive due to

Table 14.2 Breakup of estimated cost of biodiesel production (Dufreche et al., 2007).

	Cost per gallon (US$)
Centrifuge O&M	0.43
Drying O&M	1.29
Extraction O&M	0.34
Biodiesel processing O&M	0.60
Labor	0.10
Insurance	0.03
Tax	0.02
Depreciation	0.12
Capital P&I service	0.18
Total cost	3.11

*Assuming 7.0% overall trans-esterification yield. O&M: operation and maintenance. P&I: protection and indemnity.

continuous requirement of liquefied petroleum gas (LPG) during operation (GVRD, 2005). SlurryCarb™ process converts the sludge into a fuel called E-fuel (char), which can be used as fuel in cement kilns (EnerTech, 2006). The SlurryCarb™ was studied well in US and Japan at pilot and demonstration scale (Kalogo & Monteith, 2008).

14.6.5 Bio-diesel

Bio-diesel production from sewage sludge can be a profitable option, since sewage sludge is rich in lipids (triglycerides, diglycerides, monoglycerides, phospholipids, and free fatty acids) the main feedstock for biodiesel production (Kargbo, 2010).

Biodiesel, esters of simple alkyl fatty acids, is produced by the trans-esterification reaction using triglycerides (which are made up of an ester derived from glycerol and three fatty acids and are the main form of fat) and methanol (MeOH) in the presence of homogeneous base catalysts (Zullaikah et al., 2005; Ghadge & Raheman, 2006). Fatty acid methyl ester (FAME) is the term for biodiesel made when methanol is used as the alcohol in the trans-esterification process. Methanol is extensively used as the alcohol for producing biodiesel because it is the least expensive alcohol. The yield of trans-esterification depends on various factors including the type of catalyst (base, acid, enzyme or heterogeneous), alcohol/vegetable oil molar ratio, temperature, and duration of reaction, water content and free fatty acid content (Siddiquee & Rohani, 2011).

Dufreche et al. (2007) reported that combining lipid extraction processes in 50% of all existing municipal wastewater treatment plants in the US and trans-esterification of the extracted lipids could produce approximately 1.8 billion gallons of biodiesel, which is about 0.5% of the yearly national petroleum diesel demand. The approximate cost of biodiesel production from dry sludge is $3.11 per gallon in comparison with recent petro-diesel costs of $3.00 per gallon (Kargbo, 2010). The breakup of estimated cost of biodiesel production is tabulated in Table 14.2.

The main disadvantage in sludge to biodiesel production is the capital cost and complexity of plant operation. However, the key advantages are; recovery of readily usable and storable liquid fuels, generation of greenhouse credits, destruction of organo-chlorine compounds, complete control of heavy metals, small footstep for plant (Spinosa, 2004).

14.6.6 Bio-methanol

Due to higher energy content per volume than fossil fuels, bio-methanol can play a key role as a synthetic fuel in near future. Methanol production from sewage sludge can be a good option, since sewage sludge is rich in carbon-the main constituent of methanol. However, the hydrogen is the second constituent of methanol, which can be obtained from water present in sludge during the gasification process. Ptasinski *et al.* (2002) reported that overall transformation level of carbon present in the sludge into methanol equivalent to 57%. The dry solids content of sludge (optimum: 80 wt.%) and gasifier temperature (optimum: 1000°C) are two key controlling factors in methanol production from sludge.

14.7 ELECTRICITY

The electricity production from microbial decomposition of sludges is another beneficial route of sludge to energy recovery. Microbial fuel cells (MFCs) are used for electricity generation, while achieving the biodegradation of organics or wastes at the same time (Brar *et al.*, 2009).

Microbes in the anodic chamber of a microbial fuel cell oxidise the organic substrates and generate electrons and protons in the process. Unlike in a direct combustion process, the electrons are absorbed by the anode and are transported to the cathode through an external circuit. After crossing an ion exchange membrane or a salt bridge, the protons enter the cathodic chamber where they combine with oxygen to form water. Microbes in the anodic chamber extract electrons and protons in the dissimilative process of oxidizing organic substrates. Overall process is the breakdown of organic matter to water and carbon dioxide with a simultaneous electricity generation as by-product. Microbial fuel cell reactor can produce electricity from the electron flow from the anode to cathode in the external circuit (Rabaey & Verstraete, 2005; Du *et al.*, 2007). Dentel *et al.* (2004) conducted a study by using a single-chamber reactor with graphite foil electrodes in an aerobic (top) and anaerobic sludge zone (bottom). The electrical current of about 60 μA maximum could be achieved with a possibility of several hundreds of millivolts. In another study conducted by Jiang *et al.* (2009) by using a two-chambered microbial fuel cell reactor filled with ultrasonic-pretreated sludge, the electrical current was generated constantly during 10 days operation. Electricity production from sludge seems to be an attractive options, however, the success of the process will depends upon not only the microbial electricity production process itself but also the effect of this process on the sludge composition and volume of residual sludge (Rulkens, 2008).

Two kinds of commercially available fuel cells are: low temperature and high temperature fuel cell with overall efficiency varies between 47–87%. The fuel cell capacities of installations in the US ranges from 200 kW to 1 MW. Despite a very high capital cost, low operating cost ($0.01/kWh) and low emission rates of NOx and SOx are the key advantages of fuel cells (Kalogo & Monteith, 2008).

14.8 PROTEIN AND ENZYMES

Protein (61%) and carbohydrate (11%) are the main component of the sewage sludge, thus it can be used as a good source for protein recovery (Chen *et al.*, 2007). Protein accounts for about 50% of the dry weight of bacterial cells (Shier & Purwono, 1994). Moreover, protein is one of the most important constituents in animal feed, delivering nitrogen and energy. Chishti *et al.* (1992) recovered 91% protein by chemical treatment (NaOH and NaCl) followed by precipitation (ammonium sulphate 40%) of sludge. Under alkaline conditions, the hydrophobic interactions linking proteins to the extracellular polymeric substances matrix was disintegrated. Hwang *et al.* (2008) recovered 80.5% protein from precipitation and drying of sono-alkaline pretreated (1.65×10^{10} kJ/kg VSS; pH 12, 2 h) waste activated sludge, which showed the similar nutrient composition as of commercially available protein. However, production of animal feed from waste sludge is not so popular due to potential metal toxicity and pathogenicity, and availability of cheap nutrients options like agro-industry leftover (Montgomery, 2004).

Enzymes assist as biological catalysts; specifically they accelerate chemical reactions without undergoing any net chemical change during the reaction (Tyagi *et al.*, 2009). The recovered enzymes can be useful to enhance the sludge degradation and subsequent biogas generation during the anaerobic digestion (Nabarlatz *et al.*, 2010). There is a potential to recover the various enzymes such as Protease, Glycosidase, Dehydrogenase Catalase, peroxidase, α-amylase, α-glucosidase from waste sludge (Tyagi *et al.*, 2009). Nabarlatz *et al.* (2010) achieved the higher recovery the protease and lipase (hydrolytic enzymes) from waste activated sludge by ultrasonic (3.9 W/cm^2, 10–20 min) assisted extraction method. Enzymes are commercially used in the food, pharmaceutical, fine chemical industries, detergent and diagnostics industries. However, cost-effective production of enzymes is very crucial from industrial point of view. Upto 40% of the production cost for industrial enzymes is accounted by the cost of the culture medium. Here, high cost benefits could be achieved by replacing the commercial medium with sludge (Tyagi *et al.*, 2009).

14.9 BENEFICIAL REUSE OF SLUDGE ACROSS THE GLOBE

Sludge is recycled to recover the variety of value added products in several countries. Japanese are the pioneer in the production of construction material from sewage sludge ash. Other chief reuse of sludge is electricity generation by pyrolysis of sludge. Sweden has been aimed to recycling 75% of phosphorus from waste and sludge by 2010. Use of sludge derived biogas as biofuel in transportation sector. Approximate 30 buses in Stockholm are running on biogas. In the United States, electricity production through biogas generated from anaerobic digestion of waste sludge at wastewater treatment plant is a commonly practiced. Sludge incineration is another major disposal method.

Incinerated sludge ash used as water absorbent surface amendments in sports fields and horse arena. Ash from thermal oxidation installation used for "brick manufacturing" and as a source of "phosphorus recovery". In China, annual methane generation from feedstock including sludge is 720 million cubic meters. Moreover, efforts have been made for manufacturing the bricks and other building materials. Germany is advancing in developing the novel technologies for phosphorus recovery. They have developed four pilot or bench scale technologies since 2002. UK government proposed a program of energy recovery including the generation of 20% electricity from renewable sources by 2020. In 2005, 11% and 4.2% of all UK renewable energy was recovered by combustion and biogas generation. The Netherlands is one of the first country employed phosphorus recovery at full scale. They aimed to replacing 20% of its current phosphorus rock consumption by phosphorus recovery. Around 32% sewage sludge produced is currently used in cement industry and power stations. In the city of Napier, New Zealand, the government is employing the sludge composting by using the primary sludge and wood chips with the aim of using the sludge compost as soil conditioner and as a source of fuel for energy generation. In Malaysia, the effort has been made to use the municipal sludge as raw material for the production of clay-sludge bricks (Kalogo & Monteith, 2008).

The Hyperion wastewater treatment plant (LA, USA) of 450 MGD (million gallon per day) (dry weather flow, DWF) capacity is a good model of recovering renewable resources i.e. water, biogas and fertilizer from waste. The plant has been expanded and improved numerous times over the last 100+ years. In 1925, it was a simple screening plant. However, the screening plant was not effective in preventing beach closures and highly polluted wastewater was still being discharged into near-shore waters. In 1950, they incorporated a secondary biological treatment unit and sludge digestion system. The effluent is discharged to ocean and a part of treated water is used for landscape irrigation and industrial application. The biogas is captured and piped less than 1 mile to one steam generating station (22.5 MW/day). The energy is returned to Hyperion where it supplies 80% of Hyperion's power supply. An average of 650 tons/day of Class A biosolids are produced, which are used as fertilizer to grow corn, wheat and alfalfa for animal feed.

14.10 TECHNO-ECONOMIC AND SOCIAL FEASIBILITY
(Kalogo & Monteith, 2008)

Energy and resource recovery from sludge is technically feasible through several thermo-chemical (incineration, pyrolysis, gasification), bio-chemical (anaerobic digestion) and mechanical-chemical (ultrasonication) technologies. Energy analysis data revealed that the energy generation was several times higher than the energy requirement. For example, the energy output of Cambi® (biogas recovery) and EnerSludge™ (bio-oil recovery) technologies was 16 times higher energy than the energy input. The resources like phosphorus can be recovered as calcium phosphate or Struvite using extraction-precipitation methods with a recovery efficiency of 60–70% or higher (Levlin et al., 2004; Stendahl & Jäfverström, 2004).

Economic feasibility of any technology can be calculated in terms of capital cost and, operation and maintenance cost. In case of sludge-based bio-refineries, the raw material is available in abundance with almost free of cost, the main cost require for the chemicals used in the process, sophisticated equipments and instruments require and their operation and maintenance. Other important cost governing factors are type of technology, capacity of installation and cost of workforce and land, efficiency of the process and quality of product. Several energy and resources recovery technologies are failed in field mainly due to high capital and operation and maintenance cost. The key issue with resource recovery is associated to manufacturing cost of the products versus the market value. In Sweden, the cost of phosphorus recovery from sludge was observed two times higher than the commercially available phosphorus (Hultman *et al.*, 2003). In Japan, the production of construction material from incinerated sludge ash through thermal solidification process were found technically feasible, however, the production cost is higher than the market value (Okuno *et al.*, 2004).

Social feasibility is another governing factor for the development and wider application of any technology. Though sludge incineration is widely practiced in the US, it is facing pressure from public domain. On other hand, sludge incineration is broadly applied technology in European countries (Germany, Switzerland, The Netherlands) and well accepted by public. The technologies uses hazardous chemicals, are more susceptible to face public protest mainly due to the potential risks to human health and environment. Technologies like gasification includes several processing stages are considered as complex, energy intensive, space exhaustive, and high in capital and operation and maintenance cost. Processes, which result in odor (anaerobic digestion) and noise (mechanical treatment like ultrasonication) generation also face public opposition. Moreover, matured technologies are more likely to perceived positively by the public than those under developing stage or lab scale curiosity.

14.11 SUMMARY
(Tyagi & Lo, 2013; Kalogo & Monteith, 2008)

Building the sludge derived resource recovery system will help to:

- Produce environmentally benign products
- Reduce dependency on non-renewable resources thus enable the conservation of natural resources
- Reduce environmental pollution (GHG emission, soil & ground water contamination by pathogens, heavy metals, emerging contaminants)
- Decrease human health risks
- Offers the routes for sustainable management of waste sludge i.e.

    ○ Environmental friendly
    ○ Economically feasible
    ○ Socially acceptable

However, future challenges are:

- Scale up of sludge to resource recovery technique (several technologies are still lab scale curiosity)

- Cost-effective production of value added products (Several process failed due to high operation and maintenance cost and manufacturing cost)
- Supply chain management
- Environmental Compatibility

Key controlling factors in technology development and its success will be:

- Technical and Economical Feasibility
- Environmental Sustainability
- Marketing Facets
- Public acceptance

QUESTIONS

1 Describe the valuable properties of sludge.
2 What type of resources can be recovered from sludge and what are the uses of them?
3 Explain the scenario of biogas recovery and reuse from waste sludge.
4 Describe the commercialised technologies of resource recovery from sludge.
5 Provide a global scenario i.e. nationwide of sludge reuse and resource recovery.
6 What types of nutrients can be recovered from the sludge? Describe the phosphorus recovery methods in detail.
7 Describe the heavy metals recovery methods from sludge. How much percent recovery is possible by different methods?
8 What types of biofuels can be produced from waste sludge?
9 How biodiesel can be produced from waste sludge? Provide the process details and efficiency.
10 What types of construction material can be prepared from sludge? How sludge can be transformed in construction material?
11 Describe the major hindrances in the scale up of sludge to resource recovery methods.

Problems

1   Establish the annual and daily sludge generation for a wastewater treatment plant with the following operational details:

Flow: $360 \, m^3/h$
Sludge suspended solids: 250 mg/L
Removal efficiency: 50%
Volatile solids: 75%
Specific gravity to volatile solids: 0.970
Fixed solids: 25%
Specific gravity to fixed solids: 2.5
Sludge concentration $= 4.0\%$

2   The water content of sludge is reduced from 99% to 93%. What will be the percent volume reduction by assuming that solids have 80% organic fraction (specific gravity $= 1.01$) and 30% inorganic fraction (specific gravity $= 2.02$)? What will be the specific gravity of 99 and 93% sludge?

3   Determine daily and annual sludge generation for the following removal efficiencies: 40, 50, 60 and 70%.

Flow: $360 \, m^3/h$
Sludge suspended solids: 250 mg/L
Removal efficiency: 50%
Volatile solids: 75%
Specific gravity to volatile solids: 0.970
Fixed solids: 25%
Specific gravity to fixed solids: 2.5
Sludge concentration $= 5.0\%$

Plot annual sludge generation as a function of efficiency.

4   Determine the head loss in a 300 mm diameter pipe 12,000 m long transporting sludge at a flow rate of $150 \, m^3/h$.

Yield stress s_y: $1.5 \, N/m^2$
Rigidity coefficient η: $0.04 \, kg/m \cdot s$
Specific gravity: 1.04

5   What will be the surface area requirement of gravity thickeners having a diameter of 25 m to thicken the sludge from 12 g/L to 4% solids? The sludge flow is 4000 m³/d. The sludge-settling rate is 0.4 m/h.

6   Design a long rectangular sludge-settling basin.

Flow: 15000 m³/d (two tanks of 7500 m³/d)
Overflow rate: 20 m/d
Depth: 3.5 m
Determine the size of settling tank? Calculate surface area, dimension assuming a width ration of 3/1, HRT, horizontal velocity, weir overflow rate.

7   Design a circular sludge-settling basin

Determine the diameter require for circular settling basins?
Surface area, $A_s = 375 \, m^2$

8   Daily inflow: 50,000 m³

Influent TSS: 350 mg/L
Solids removal in primary sedimentation tank: 65%
Solids concentration in primary sludge: 6%

The waste activated sludge flow is 500 m³/d having a solids concentration of 0.8%. If waste activated sludge thickening also takes place in PST, the settled solids concentration will be 3.8%.

How much reduction in sludge volume will be achieved which has to be pump to anaerobic digester on daily basis under the above-said conditions if compare with the direct discharge of primary sludge and waste activated sludge to anaerobic digester? Assume a 100% settling of waste activated sludge in PST.

9   A water treatment plant has a flow rate of 0.6 m³/sec. The settling basin at the plant has an effective settling volume that is 20 m long, 3 m tall and 6 m wide. Will particles that have a settling velocity of 0.004 m/sec be completely removed? If not, what percent of the particles will be removed?

10   How big (area, A) would the basin need to be to remove 100% of the particles that have a settling velocity of 0.004 m/sec?

Flow: 0.6 m³/sec
Width: 6 m

11   Flow rate: 10,000 m³/d

TSS removal by primary sedimentation: 250 mg/L
Volatile fraction in settled solids: 80%
Water content in sludge: 95%
Specific gravity: 1.20 (organic solids) and 2.40 (inorganic solids)
Solids retention time: 15 days
What will be the:

    (a)   Volume of anaerobic digester
    (b)   Minimum capacity of anaerobic digester using recommended loading parameters of kg VS/m³ · d

12 Using the below given data, find out the surface area of gravity thickener for 800 m³/d of sludge flow. The achievable sludge concentration should be 4.5%.

Suspended solids (mg/L)	Settling velocity (m/h)
4000	2.4
6000	1.5
8000	1.0
14000	0.4
29000	0.1
41000	0.05

13 Design a lime stabilisation unit. The sludge flow rate is 150 m³/d with a solids concentration of 6.5% and specific gravity of 1.04. What will be the volume of tank and lime feed rate?

14 The organic loading rate of an anaerobic digester is 420 kg COD/d. By using the waste-utilization efficiency of 80%, what will be the volume of biogas produced upon a SRT of 35 days? Y: 0.15 and k_d: 0.03/d.

15 Design an anaerobic digester. Define the volume of digester, the flow rate of methane and volume of external tank pressurised to 365 kPa. Use the following data for the above said purpose:

Digester temperature: 35°C
VSS: 0.75 TSS
VSS reduction efficiency: 45%
Influent BOD: 3000 mg/L
Fraction of BOD influent consisting primary solids: 0.4
Sludge concentration is 65% of inflow thickened sludge concentration
Flow rate: 75 m³/d
Solids concentration: 4.5%   Specific gravity: 1.03
Design influent bCOD: 3000 mg/L
Design effluent bCOD: 500 mg/L
Design safety factor: 5
No recycling
Gas production from digester at 101.5 kPa atmospheric pressure

16 The sludge flow rate is 75 m³/d. The solids concentration is 4.5% and specific gravity of sludge is 1.03. Design an aerobic digester by using the following data:

Temperature: 30°C
VSS: 0.75 TSS
VSS reduction efficiency: 45%
Influent BOD: 3000 mg/L
Fraction of BOD influent consisting primary solids: 0.4
Sludge concentration is 65% of inflow thickened sludge concentration
Type of mixing: Diffused air

17 How much kg of air will be requiring per kg of biosolids for its complete oxidation? The elemental analysis of a dried biosolids are as follows:

Element	%
C	55.4
O	35.6
H	3.3
N	5.7

18 An anaerobic digester produces $50\,m^3/d$ of sludge with total solids concentration of 4.5%. What quantity of sludge will be disposed yearly if sludge drying beds yields a solids concentration of 45%?

19 What will be the fuel value of sludge from primary settling tank under following conditions (a) no chemical added (b) coagulating solids amount 15% by weight of dry solids. The volatile solids concentration is 80%.

20 The aerobic digester produces a sludge having total solids concentration of 4.5%. A filter press installation at the plant can produce a solids concentration of 20%. If the plant is producing $45\,m^3/d$ of sludge at present, how much sludge production (by volume) can be reduced after installing a filter press?

21 An anaerobic digester produces $2\,m^3/d$ of sludge with total solids concentration of 4.4%. What solids concentration must be achieved by the drying facility to reduce the sludge volume to $0.2\,m^3/d$.

22 Design a gravity thickener for mixed primary and waste activated sludge.

Sludge type	Specific gravity	% solids	Flow, m^3/d
Average design			
Primary sludge	1.04	4.0	500
Waste activated sludge	1.01	0.5	2500
Peak design			
Primary sludge	1.04	4.1	550
Waste activated sludge	1.01	0.55	2800

23 Determine the following:

a. Volume of digester
b. Volumetric loading
c. Percent stabilisation
d. Volume of gas generated per capita

Wastewater flow rate: $40,000\,m^3/d$
dry VS removed: $0.18\,kg/m^3$
bCOD removed: $0.16\,kg/m^3$
SRT: 15 days at 35°C
Waste utilization efficiency E: 0.80
Y: 0.07 kg VSS/kg bCOD k_d: 0.04/d
Methane percentage: 70%

24 Calculated the volatile solids reduction in an anaerobic digester under following condition:

a. Fixed solids weight in digested sludge and untreated sludge are equals
b. Only volatile solids fraction lost during anaerobic digestion

	Volatile solids, %	Fixed solids, %
Untreated sludge	80	20
Digested sludge	60	40

25 A low rate digester is to be designed for waste sludge generated from activated sludge process treating sewage generated from 25000 persons. The raw sludge has 0.11 kg dry solids/capita. day (VS = 70% of ds). The dry solids (ds) is 5% of the sludge and specific gravity is 1.01. During digestion 65% of VS are destroyed and fixed solids remained unchanged. The digested sludge has 7% ds and a wet specific gravity is 1.03. Operating temperature of digester is 35°C and sludge storage time is 45 days. Determine the digester volume required. Assume digestion time of 23 days.

Concluding remarks

Increasing urbanisation and industrialisation have resulted in a dramatic increase in the volume of wastewater produced around the world. Tightening environmental standards have meant that much of this wastewater has to be treated before it can be safely discharged. The wastewater treatment step concentrates the various pollutants in the wastewater into the sludge, normally containing between 1 and 2% by weight dry solids. Because of the dramatic increase in volume of raw water and wastewater being treated, the resultant large volumes of sludge need to be treated and disposed off in an environmentally safe manner (Priestley, 1992).

NATURE AND GENERATION OF SEWAGE SLUDGE
(Brindell & Stephenson, 1996; Clark, 1997; Ottewell, 1990; Priestley, 1992; Scragg, 1999)

Sewage is a complex mixture of waterborne wastes of human, domestic and industrial origin. A list of troublesome components in sewage includes organic matter, emulsified oil and grease, bacteria and virus, nutrients such as nitrate and phosphate, and also heavy metals and organochlorines. A variety of processes have been applied to the treatment of sewage but their overall results are more or less the same. In essence, all they do is to separate the sewage into two streams; a clarified water containing around 20–30 mg/L of suspended solids, and a sludge stream of 1–2% solids dry weight which usually contains 80 to 90% of the contaminants originally present in the raw sewage. Although both streams need to be discharged to the environment, the sludge stream, which is of much smaller volume, has a much greater potential to cause environmental damage.

The nature of the sludge depends, to some extent, on the type of process used to treat the sewage. Sewage treatment processes can be categorised basically into two generic types, one based on biological, and the other on physicochemical techniques. Aerobic biological treatment processes (e.g. activated sludge or trickling filters) largely convert organic matter present in the raw sewage into biomass, with some CO_2 also being produced, largely by endogenous respiration. This biomass is removed in settling tanks to form a sludge with a solids content normally around 1 to 2% dry weight. Sludges from biological treatment processes consist largely of microbial cells and are difficult to dewater. It is both because of the intracellular water content and the fact that microbial cells often exude extracellular polymers, which impart a jelly like consistency to the sludge.

Physicochemical processes for treating sewage can range from simple screening and settling operations, through chemical coagulation processes to sophisticated membrane techniques such as microfiltration. Screening techniques can have screen hole sizes down to 0.1 mm and take out the coarser material in the sewage. Such material can be easily dewatered to a high solids content by simple pressing techniques and is either taken to landfill or incinerated. Chemical coagulation processes are best at removing particles in the colloidal size range (0.01 pm–1 pm). For membrane techniques, the separation depends entirely on the hole size in the membrane e.g. crossflow microfiltration will remove particles down to approximately $0.1\,\mu m$ in size.

SEWAGE SLUDGE TREATMENT AND DISPOSAL
(Fleming, 1986; Karpati, 1989; Kruger et al., 1991; Lee et al., 1997; Marshall, 1988; Ndon & Dageu, 1997; Ra et al., 1998; Rhee et al., 1997)

Just as every water or wastewater treatment plant is different, the sludge produced at each plant is different. Some sludges are anaerobically digested while some are not and some contain higher levels of nutrients than others. Thus, different sludges require different methods of treatment and final disposal. Perhaps that is why a coordinated approach to the treatment and disposal of all forms of sludge has not attracted a great deal of interest. So far, most attention has been centered on the disposal of sewage sludge to land or sea, or by incineration and has resulted in considerable research activity in these areas as covered in this text.

Although individual classes of solids may contain, besides water, other substances of some economic value, they seldom do so in sufficient amount to make isolation or reclamation worthwhile. Examples of paying operations are few, viz. (i) the recovery of lime in water softening; (ii) conversion of wastes from the food and beverage industries into animal feed; and (iii) reclamation of fat and fibers from industrial wastes. Besides this, however different in concept, is the utilization of energy by: (i) burning combustible gases generated by digesting solids, or (ii) incinerating dried highly organic solids.

Generally, digested sludges are utilised in agricultural uses. However, there has been a growing awareness of potentially negative effects of agricultural use of sludge on the environment. The danger of heavy metals and organic micropollutants accumulating in soil has been recognised and the risk of infection has been demonstrated.

As a consequence of this awareness, demand of effective sludge hygienization before its disposal has been raised in case of land application. Conventional methods have failed to fulfil this demand. Nevertheless, newly developed dual digestion process known as thermophilic-aerobic and mesophilic-anaerobic sludge-digestion has proved successful in this regard. During thermophilic stage of this dual digestion system, pasteurization of sludge occurs; while during its mesophilic-anaerobic stage, complete stabilisation takes place.

As far as optimization of biological sludge-treatment systems is concerned, anoxic digestion through endogenous nitrate respiration (ENR; which resembles aerobic digestion except for the use of nitrate than oxygen as the terminal electron acceptor) is expected to prove as a promising alternative over conventional systems when it is associated with an anoxic-aerobic cycle to generate the nitrate in-situ. Anoxic-aerobic sludge digestion significantly reduces power cost over aerobic digestion, because only anoxic mode is incorporated with mixing. From the point of view of by-product recovery and process efficiency, however, the thermophilic-aerobic plus mesophilic-anaerobic digestion system appears to be the best possible alternative, at present, among newly developed and modified biological methods of sludge treatment.

When viewed in fundamental terms, the final resting place for sewage sludge can be any of the three destinations i.e. land, air or water. Land disposal of sludge is a simple physical operation with the main variations depending upon the rates of application and techniques involved. Land spreading, soil injection and landfilling are the three main options, with environmental and safety considerations dictating application rates and the degree of pretreatment. Disposal to air invariably involves some form of oxidation to convert organic matter to CO_2 and water. An indirect route is the use of anaerobic bacteria to convert organic matter to methane, which is then burnt to extract its energy value. The engineering technologies used to achieve complete oxidation of sludge include incineration, smelting, and wet air oxidation. Furthermore, pyrolysis techniques, such as sludge to oil, can produce a gaseous product, which can be burnt to produce energy. All these thermal technologies, however, also produce a solid ash, which has to be disposed off. Disposal to water basically means its dumping into the sea with the main decision being the choice of the site. However, ocean disposal of sludge has been banned in several countries after realizing the adverse impact on ocean environment.

FUTURE TRENDS
(Priestley, 1992; Scragg, 1999; Tyagi & Lo, 2011; Tyagi & Lo, 2013)

Worldwide progress in the treatment and disposal of sewage sludge indicates that in future a variety of approaches will be required to handle the problem according to local, economic, social and environmental considerations. However, it is visibly clear that increased levels of wastewater treatment is considerably increasing the volume of sludges produced. At the same time, available land disposal sites are decreasing rapidly, while ocean dumping is coming under stringent environmental pressures to be banned. These factors are set to ensure the rapid adoption of more intensive sludge processing technologies and techniques, viz. extensive dewatering, incineration, wet air oxidation and even sludge smelting.

Sludge pretreatment by thermal, chemical and mechanical means were also realised to be effective in sludge solubilisation and improvement in biogas generation, pathogens removal, improvement in sludge dewaterability, sludge volume reduction and safe disposal of sludge. However, the high capital, and operation and maintenance cost and energy intensive nature are the major concern. The combined pretreatment (thermal-chemical, thermal-mechanical, mechanical-chemical) shows some promising results, however, the most of them are still lab scale curiosity and need extensive research and development in order to scale up the techniques.

Recovery of value added products i.e. biogas, biofuels, construction materials, metals, nutrients (nitrogen and phosphorus) recovery from sludge also provide an option for sustainable management of sludge. It will support to harvest environmental friendly products, lower the dependency on non-renewable reserves thus helps to conserve the natural reserves, reduce the environmental pollution (water and soil contamination, GHG emission) and human health risk. However, despite a rapid technological development in resource recovery technologies in last few years, there is still a long way to achieve the target of efficient sludge derived resource recovery system, since most of the technologies are still in its infant stage.

References

Abbassi, B., Dusllstein, S. & Rabiger, N. (2000) Minimization of excess sludge production by increase of oxygen concentration in activated sludge flocs: Experimental and theoretical approach. *Water Research*, 34 (1), 139–146.

Act Clean (2013) *ANANOX: Process for Biological Denitrification of Wastewaters Using Anaerobic Pretreatment.* Available from: http://www.act-clean.eu/index.php?node_id= 100.456&lang_id=1 [Accessed 30th June 2013].

Ahel, M., Giger, W. & Koch, M. (1994a) Behaviour of alkylphenol polyethoxylate surfactants in the aquatic environment-I. Occurrence and transformation in sewage treatment. *Water Research*, 28 (5), 1131–1142.

Ahel, M., Giger, W. & Schaffner, C. (1994b) Behaviour of alkylphenol polyethoxylate surfactants in the aquatic environment-II. Occurrence and transformation in rivers. *Water Research*, 28 (5), 1143–1152.

Ahel, M., Hrsak, D. & Giger, W. (1994c) Aerobic transformation of short-chain alkylphenol polyethoxylates by mixed bacterial cultures. *Archives of Environmental Contamination Toxicology*, 26, 540–548.

Ahmad, S., Haque, I., Kaul, S. & Siddiqui, R.H. (1998) Oily sludge farming. *Indian Journal of Environmental Health*, 40 (1), 27–36.

Albertson, O.E. (1991) Bulking sludge control: Progress, practice and problems. *Water Science and Technology*, 23 (4–6), 835–846.

Anderson, B.C. & Mavinic, O.S. (1984) Aerobic sludge digestion with pH control preliminary investigation. *Journal of Water Pollution Control Federation*, 56 (7), 889–897.

Andreottola, G. & Foladori, P. (2006) A review and assessment of emerging technologies for the minimization of excess sludge production in wastewater treatment plants. *Journal of Environmental Science and Health, Part A*, 4, 1853–1872.

Andrews, J.F. (1975) Anaerobic digestion process. *Water Sewage Works*, 122, 62.

APHA (2005) *Standard Methods for the Examination of Water and Wastewater.* 21st edition. Washington, DC, American Public Health Association (APHA).

Appels, L., Baeyens, J., Degreve, J. & Dewil, R. (2008) Principles and potential of the anaerobic digestion of waste-activated sludge. *Progress in Energy Combustion Science*, 34, 755–781.

Apul, O.G. & Sanin, F.D. (2010) Ultrasonic pretreatment and subsequent anaerobic digestion under different operational conditions. *Bioresource Technology*, 101, 8984–8992.

Aqua Reci (2013) *Super Critical Wet Oxidation Process.* Available from: http://www.stowa-selectedtechnologies.nl/Sheets/Sheets/SCWO.Process.html [Accessed 30th June, 2013].

Asian Development Bank (2012) *Promoting Beneficial Sewage Sludge Utilization in the People's Republic of China.* Mandalungyong City, Asian Development Bank (ADB).

Atherton, P.C., Steen, R., Stetson, G., McGovern, T. & Smith, D. (2005) Innovative biosolids dewatering system proved a successful part of the upgrade to the Old Town, Maine water pollution control facility. In: *Proceedings of the 2005 WEFTEC: The Water Quality Event, Washington, DC*. pp. 6650–6665.

Atlas, R.M. (1991) Microbiol hydrocarbon degradation bioremediation of oil spills. *Journal of Chemical Technology and Biotechnology*, 52, 149–156.

Bahadori, A. (2013) *Solid Waste Treatment and Disposal*. In: Waste Management in the Chemical and Petroleum Industries. Chichester, John Wiley & Sons Ltd.

Baier, U. & Schmidheiny, P. (1997) Enhanced anaerobic degradation of mechanically disintegrated sludge. *Water Science and Technology*, 36 (11), 137–143.

Baier, U. & Zwiefelhofer, H.P. (1991) Sludge stabilization, effects of aerobic thermophilic pretreatment. *Water Science and Technology*, 3, 56–61.

Balmer, P. & Frost, R.C. (1990) Managing change in an environmentally conscious society: A case study Gothenburg (Sweden). *Water Science and Technology*, 22 (12), 45–56.

Bandosz, T.J. & Block, K. (2006) Effect of pyrolysis temperature and time on catalytic performance of sewage sludge/industrial sludge-based composite adsorbents. *Applied Catalysis B: Environmental*, 67 (1–2), 77–85.

Banik, S., Bandyopadhyay, S. & Ganguly, S. (2003) Bioeffects of microwave. *Bioresource Technology*, 87 (2), 155–159.

Bargman, R.D., Garber, W.F. & Nagano J. (1958) Sludge filtration and use of synthetic organic coagulants at hyperion. *Journal of Water Pollution Control Federation*, 30, 1079–1100.

Barjenbruch, M. & Kopplow, O. (2003) Enzymatic, mechanical and thermal pretreatment of surplus sludge. *Advanced Environmental Research*, 7 (3), 715–720.

Bell, P. (1991) Status of eutrophication in the Great Barrier Reef lagoon. *Marine Pollution Bulletin*, 23, 89–93.

Bell, P.E., James, B.R. & Chaney R.L. (1991) Heavy metal extractability in long term sewage sludge and metal amendment soil. *Journal of Environmental Quality*, 20, 481–486.

Benefield, L.O. & Randall, C.W. (1978) Design relationship for aerobic digestion. *Journal of Water Pollution Control Federation*, 50, 518–523.

Berg, U. & Schaum, C. (2005) *Recovery of Phosphorus from Sewage Sludge and Sludge Ashes—Applications in Germany and Northern Europe*. Dokuz Eylul Universitesi.

Beurskens, J.E.M., Stams, A.J.M., Zehnder, A.J.B. & Bachmann, A. (1991) Relative biochemical reactivity of three hexachlorocyclohexane isomers. *Ecotoxicology and Environmental Safety*, 21, 128–136.

Bien, J.B., Malina, G., Bien, J.D. & Wolny, L. (2004) Enhancing anaerobic fermentation of sewage sludge for increasing biogas generation. *Journal of Environmental Science and Health Part*, A, 39 (4), 939–949.

Bishop, B. (2004) Use of ceramic membranes in airlift membrane bioreactors. In: *Proceedings of 8th International Conference on Inorganic Membranes (ICIM8), 18–22 July 2004, Cincinnati, USA*.

Bishop, P.L. & Farmer, M. (1978) Fate of nutrients during aerobic digestion. *Journal of Environmental Engineering*, 104, 967.

Booker, N.A., Keir, D., Priestley, A.J., Ritchie, C.B., Sudannana, D.L. & Woods, M.A. (1991) Sewage clarification with magnetite particles. *Water Science and Technology*, 23, 1703–1712.

Boon, A.G. & Burgess, D.R. (1974) Treatment of crude sewage in two high-rate activated sludge plants operated in series. *Journal of Water Pollution Control Federation*, 74, 382.

Bosma, T.N.P., van der Meer, J.R., Schraa, G., Tros, M.E. & Zehnder, A.J.B. (1988) Reductive dechlorination of all trichloro and dichloroisomers. *FEMS Microbiology Ecology*, 53, 223.

Boss, E.E. & Shepherd, S.L. (1999) *Process for Treating a Waste Sludge of Biological Solids*. Available from: http://www.google.ca/patents/US5868942.

Bouchard, J., Naguyen, T.S., Chomet, E. & Overend, R.P. (1990) Analytical methodology for biomass pretreatment. Part 1: Solid residues. *Biomass*, 23, 243–261.

Bouchard, J., Nguyen, T.S., Chornet, E. & Overrend, R.P. (1991) Analytical methodology for biomass pretreatment. Part 2: Characterization of the filtrates and cumulative distribution as a function of treatment severity. *Bioresource Technology*, 36, 121–131.

Bougrier, C., Albasi, C., Delgenes, J.P. & Carrere, H. (2006) Effect of ultrasonic, thermal and ozone pretreatments on WAS solubilization and anaerobic biodegradability. *Chemical Engineering and Processing: Process Intensification*, 45 (8), 711–718.

Bougrier, C., Carrere, H. & Delgenes, J.P. (2005) Solubilisation of waste activated sludge by ultrasonic treatment. *Chemical Engineering Journal*, 106 (2), 163–169.

Bougrier, C., Delgenes, J.P. & Carrere, H. (2008) Effects of thermal treatments on five different WAS samples solubilisation, physical properties and anaerobic digestion. *Chemical Engineering Journal*, 139 (2), 236–244.

Bowen, P.T., Magar, V.S., Lagarenne, W.R., Muise, A.M. & De Bernardi J.R. (1990) Sludge treatment, utilisation and disposal. *Journal of Water Pollution Control Federation*, 62 (2), 425–433.

Braguglia, C.M., Mininni, G. & Gianico, A. (2008) Is sonication effective to improve biogas production and solids reduction in excess sludge digestion? *Water Science and Technology*, 57 (4), 479–483.

Braguglia, C.M., Gianico, A. & Mininni, G. (2012) Comparison between ozone and ultrasound disintegration on sludge anaerobic digestion. *Journal of Environmental Management*, 95, 139–143.

Brar, S.K., Verma, M., Tyagi, R.D. & Surampalli, R.Y. (2009) Value addition of wastewater sludge: Future course in sludge reutilization. *Practice Periodical of Hazardous, Toxic, and Radioactive Waste Management*, 13 (1), 59–74.

Brechtel, H. & Eipper, H. (1990) Improved efficiency of sewage sludge incineration by preceding sludge drying, IAWPRC. *Water Science and Technology*, 222 (12), 269–276.

Brindell, K. & Stephenson, T. (1996) The application of membrane biological reactors for the treatment of wastewaters. *Biotechnology and Bioengineering*, 49, 601–610.

Brown, G.J., Chow, L.K., Landine, R.C. & Cocei, A.A. (1980) Lime use in anaerobic filters. *Journal of Environmental Engineering*, 106 (4), 837–839.

Buser, H. & Muller, M.D. (1995) Isomer and enantioselective degradation of hexachlorocyclohexane isomers in sewage sludge under anaerobic conditions. *Environmental Science & Technology*, 29, 664–672.

Cabelli, V.J. (1983) *Health Effects of Criteria for Marine Recreational Waters*. Pub. No. EPA-600/1-80-031, Washington, DC, US Environmental Protection Agency.

Caccavo Jr., F., Frolund, B., Van Ommen, Koleke, F. & Nielsen, P.H. (1996) Deflocculation of activated sludge by the dissimilatory Fe(III)-reducing bacterium *Shewanella alga* BrY. *Applied Environmental Microbiology*, 62, 1487–1490.

Calabrese, A., Gould, E. & Thurberg, E.P. (1982) Effects of toxic metals in marine animals of the New York bight, In: Mayer, G.F. (ed.) *Ecological Stress and the New York Bight*. Columbia, SC, Estuarine Research Foundation.

Cambi Recycling Energy (2013) *Unleash the Power of Anaerobic Digestion*. Available from: http://www.cambi.no/wip4/detail.epl?cat=10636 [Accessed 1st July 2013].

Campbell, H.W. (1988) A status report on environment Canada's oil from sludge technology, *CEC/EWPCA International Conference on "Sewage Sludge Treatment and Use"*, 19–23 September 1988, Amsterdam.

Campbell, H.W. & Crescuolo, P.J. (1982) The use of rheology for sludge characterization. *Water Science and Technology*, 14 (6/7), 475–489.

Campbell, H.W. & Crescuolo, P.J. (1989) Control of polymer addition for sludge conditioning: A demonstration study. *Water Science and Technology*, 21 (10/11), 1309–1317.

Canales, A., Pareilleux, A., Rols, J., Goma, G. & Huyard, A. (1994) Decreased sludge production strategy for domestic wastewater treatment. *Water Science and Technology*, 30 (8), 97–106.

Cao, Y. & Pawłowski, A. (2012) Sewage sludge-to-energy approaches based on anaerobic digestion and pyrolysis: Brief overview and energy efficiency assessment. *Renewable & Sustainable Energy Reviews*, 16, 1657–1665.

Carballa, M., Omil, F. & Lema, J.M. (2009) Influence of different pretreatments on anaerobically digested sludge characteristics: Suitability for final disposal. *Water Air and Soil Pollution*, 99, 311–321.

Carrere, H., Dumas, C., Battimelli, A., Batstone, D.J., Delgenes, J.P., Steyer, J.P. & Ferrer, I. (2010) Pretreatment methods to improve sludge anaerobic degradability: A review. *Journal of Hazardous Material*, 183, 1–15.

CCME, Canadian Council of Ministers of the Environment (1991) *Interim Canadian Environmental Quality Criteria for Contaminated Sites*. Ottawa, EPC-CS34, Environment Canada.

Cetin, F.D. & Surucu, G. (1990) Effects of temperature and pH on the settleability of activated sludge flocs. *Water Science and Technology*, 22 (9), 249–254.

Chang, C., Tyagi, V.K. & Lo, S.L. (2011) Effects of microwave and alkali induced pretreatment on sludge solubilization and subsequent aerobic digestion. *Bioresource Technology*, 102 (17), 7633–7640.

Chang, J., Chudoba, P. & Capdeville, B. (1993) Determination of the maintenance requirement of activated sludge. *Water Science and Technology*, 28, 139–142.

Chen, G.H., Mo, H.K., Saby, S., Yip, W.K. & Liu, Y. (2000) Minimization of activated sludge production by chemically stimulated energy spilling. *Water Science and Technology*, 42 (12), 189–200.

Chen, G.H., An, K.J., Saby, S., Brois, E. & Djafer, M. (2003) Possible cause of excess reduction in an oxic-settling-anaerobic activated sludge process (OSA process). *Water Research*, 37 (16), 3855–3866.

Chen, Y.G., Jiang, S., Yuan, H.Y., Zhou, Q. & Gu, G.W. (2007) Hydrolysis and acidification of waste activated sludge at different pHs. *Water Research*, 41, 683–689.

Cheng, C.Y., Updergraff, D.M. & Ross L.W. (1970) Sludge dewatering by high rate freezing at small temperature differences. *Environmental Science & Technology*, 4 (12), 1145–1147.

Chishti, S.S., Hasnain, S.N. & Khan, M.A. (1992) Studies on the recovery of sludge protein. *Water Research*, 26 (2), 241–248.

Choi, H.B., Hwan, K.Y. & Shin, E.B. (1997) Effect on anerobic digestion of sewage sludge pretreatment. *Water Science and Technology*, 35 (10), 207–211.

Chou, T.L. (1958) Resistance of sewage sludge to flow in pipes. *Journal of Sanitary Engineering, ASCE*, 84 (SA1), 1557.

Chu, C.P., Chang, B.V., Liao, G.S., Jean, D.S. & Lee, D.J. (2001) Observations on changes in ultrasonically treated waste activated sludge. *Water Research*, 35, 1038–1046.

Chu, L.B., Yan, S.T., Xing, X.H., Sun, X.L. & Jurcik, B. (2009) Progress and perspectives of sludge ozonation as a powerful pretreatment method for minimization of excess sludge production. *Water Research*, 43 (7), 1811–1822.

Chu, L.B., Yan, S.T., Xing, X.H., Yu, A.F., Sun, X.L. & Jurcik, B. (2008) Enhanced sludge solubilization by microbubble ozonation. *Chemosphere*, 72 (2), 205–212.

Chudoba, J., Grau, P. & Ottova, V. (1973) Control of activated sludge filamentous bulking II: Selection of micro-organisms by means of a selector. *Water Research*, 7, 1389–1406.

Chudoba, P., Chudoba, J. & Capdeville, B. (1992) The aspect of energetic uncoupling of microbial growth in the activated sludge process: OSA system. *Water Science and Technology*, 26 (9–11), 2477–2480.

Chung, Y.C. & Neethling, J.B. (1990) Viability of anaerobic digester sludge. *Journal of Environmental Engineering, ASCE*, 116 (2), 330–343.

Clark, R.B. (1997) *Marine Pollution*. 4th edition. Oxford, Oxford University Press.

Coker, C.S., Walden, R.I. & Shea, T.G. (1991) Dewatering municipal wastewater sludge for incineration. *Water Environment and Technology*, 3, 65–67.

Crawford, P.M. (1990) Optimizing polymer consumption in sludge dewatering applications. *Water Science and Technology*, 22 (7/8), 262–267.

Cummings, R.J. & Jewell, W.J. (1977) Thermophilic aerobic digestion of dairy wastes. In: *Proceedings of 9th Cornell University Waste Management Conference, Syracuse, New York, April 28, 1977*.

Cunningham, G.K. & Duwer, R. (1989) Wet oxidation-a new approach to wastewater treatment. In: *Proceedings of 13th Federal Conference on Australian Water and Wastewater Association, Canberra*. pp. 197–203.

Danesh, P., Hong, S.M., Moon, K.W. & Park, J.K. (2008) Phosphorus and heavy metal extraction from wastewater treatment plant sludges using microwaves for generation on exceptional quality bio-solids. *Water Environment Research*, 80 (9), 784–795.

Datar, M.T. & Bhargava, D.S. (1984) Thermophilic aerobic digestion of activated sludge. *Journal of Institution of Public Health Engineers*, TS III, 22–27.

Datar, M.T. & Bhargava, D.S. (1988) Effect of temperature on BOD and COD reductions during aerobic digestion of activated sludge. In: *Proceedings of the Paper Meeting of the Environmental Engineering Division, Institution of Engineers, Mysore, India, 26–27 April 1986*, pp. 1–6 (EN 147).

de Bekker, P.H.A.M.J. & van den Berg, J.J. (1988) Wet oxidation as the alternative for sewage sludge treatment. In: Dirkzwager, A.H. and Hermite, P.L. (eds.) *Conference on Sewage Sludge Treatment and Use*. Elsevier Applied Science, England.

Delgado, J., Aznar, M.P. & Corella, J. (1997) Biomass gasification with steam in fluidized bed: Effectiveness of CaO MgO, and CaO-MgO for hot raw gas cleaning. *Industrial and Engineering Chemistry Research*, 36, 1535–1543.

Demirbas, A. (2009) Biofuels securing the planet's future energy needs. *Energy Conversion and Management*, 50, 2239–2249.

Dentel, S.K., Strogen, B. & Chiu, P. (2004) Direct generation of electricity from sludges and other liquid wastes. *Water Science and Technology*, 50 (9), 161–168.

Deublein, D. & Steinhauser, A. (2008) *Biogas from Waste and Renewable Resources*. Weinheim, Wiley-VCH.

Dewil, R., Appels, L., Baeyens, J. & Degreve, J. (2007) Peroxidation enhances the biogas production in the anaerobic digestion of biosolids. *Journal of Hazardous Material*, 146, 577–581.

Dewling, R.T., Maganelli, R.M. & Baer Jr., G.T. (1980) Fate and Behaviour of selected heavy metals in incinerated sludge. *Journal of Water Pollution Control Federation*, 52, 2552–2557.

Dhuldhoya, D., Lemen, J., Martin, B. & Myers, J. (1996) Cost-effective treatment of organic sludges in a high rate bioreactor. *Environmental Progress and Sustainable Energy*, 15, 135–140.

Doe, P.W., Benn, D. & Bays, L.R. (1965) The disposal of wastewater sludge by freezing. *Journal of the Institute of Water Engineering*, 19 (4), 251–287.

Dogan, I. & Sanin, F.D. (2009) Alkaline solubilization and MW irradiation as a combined sludge disintegration and minimization method. *Water Research*, 43, 2139–2148.

Dohanyos, M., Zabranska, J. & Jenicek, P. (1997) Enhancement of sludge anaerobic digestion by using of a special thickening centrifuge. *Water Science and Technology*, 36 (11), 145–153.

Dohanyos, M., Zabranska, J., Kutil, J. & Jenicek, P. (2004) Improvement of anaerobic digestion of sludge. *Water Science and Technology*, 49 (10), 89–96.

Doi, Y. & Fukuda, K. (eds.) (1994) *Biodegradable Plastic and Polymers*. London, Elsevier Publishing.

Dominguez, A., Menendez, J.A., Inguanzo, M., Bernard, P.L. & Pis, J.J. (2003) Gas chromatographic-mass spectrometric study of the oil fractions produced by microwave-assisted pyrolysis of different sewage sludges. *Journal of Chromatography A*, 1012, 193–206.

Dowdy, R.H., Latterell, J.J., Hinesly, T.D., Gurssman, R.B. & Sullivan, D.L. (1991) Trace metal movement in an aeric ochraqualf following 14 years of annual sludge applications. *Journal of Environmental Quality*, 20, 119–123.

Drews, A. (2013) *A Schematic of Membrane Bioreactor*. Available from: http://en.wikipedia.org/wiki/File:MBRvsASP_Schematic.jpg [Accessed 30th June 2013].

Drier, O.E. & Obma, C.A. (1963) *Aerobic Digestion of Solids*. Aurora, IL, Walker Process Equipment Co., Bulletin No. 26-5-18194.

Du, Z., Li, H. & Gu, T. (2007) A state of the art review on microbial fuel cells: A promising technology for wastewater treatment and bioenergy. *Biotechnology Advances*, 25, 464–482.

Duarte, A.C. & Anderson, G.K. (1982) Inhibition modelling in anaerobic digestion. *Water Science and Technology*, 14, 749–763.

Dufreche, S., Hernandez, R., French, T., Sparks, D., Zappi, M. & Alley, E. (2007). Extraction of lipids from municipal wastewater plant microorganisms for production of biodiesel. *Journal of the American Oil Chemists' Society*, 84 (2), 181–187.

Ejlertsson, J. & Svensson, B.H. (1996) Degradation of bis (2-ethylhexyl) phthalate constituents under methanogenic conditions. *Biodegradation*, 7, 501–506.

Ejlertsson, J., Nilsson, M.L., Kylin, H., Bergman, A., Karlson, L., Oquist, M. & Svensson, B.O.H. (1999) Anaerobic degradation of nonylphenol mono- and diethoxylates in digestor sludge, land filled municipal solid waste, and land filled sludge. *Environmental Science & Technology*, 33 (2), 301–306.

Ekama, G.A. & Marais, G.V.R. (1986) Sludge settleability and secondary settling tanks design procedures. *Journal of Water Pollution Control Federation*, 85 (1), 100–113.

Ekelund, R., Bergman, A., Granmo, A. & Berggren, M. (1990) Bioaccumulation of 4-nonylphenol in marine animals – A reevaluation. *Environmental Pollution, Series A*, 64, 107–120.

Elliott, A. & Mahmood, T. (2007) Pretreatment technologies for advancing anaerobic digestion of pulp and paper biotreatment residues. *Water Research*, 41, 4273–4286.

Emerging Technologies for Biosolids Management (2006) Office of Wastewater Management U.S. Environmental Protection Agency Washington, DC. Available from: http://www.sswm.info/sites/default/files/reference_attachments/EPA%202006%20Emerging%20Technologies%20for%20Biosolids%20Management.pdf.

EnerTech (2006) Environmental Inc. *Company Information Packet*. Available from: www.enertech.com/downloads/InfoPacket.pdf [Accessed May 2007].

Ensminger, D.E. (1986) Acoustic dewatering. In: Muralidhara, H.S. (ed.) *Advances in Solid Liquid Separation*. Columbia, OH, Batelle Press. p. 321.

Eriksson, L. & Alm, B. (1991) Study of flocculation mechanisms by observing effects of a complexing agent on activated sludge properties. *Water Science and Technology*, 24 (7), 21–28.

Eriksson, L., Steen, I. & Tendaj, M. (1992) Evaluation of sludge properties in an activated sludge plant. *Water Science and Technology*, 25 (6), 251–265.

Eskicioglu, C., Kennedy, K.J. & Droste, R.L. (2006) Characterization of soluble organic matter of WAS before and after thermal pretreatment. *Water Research*, 40, 3725–3736.

Eskicioglu, C., Kennedy, K.J. & Droste, R.L. (2008) Initial examination of MW pretreatment on primary, secondary and mixed sludges before and after anaerobic digestion. *Water Science and Technology*, 57 (3), 311–317.

Eskicioglu, C., Kennedy, K.J. & Droste, R.L. (2009) Enhanced disinfection and methane production from sewage sludge by MW irradiation. *Desalination*, 248 (1–3), 279–285.

Eskicioglu, C., Terzian, N., Kennedy, K.J., Droste, R.L. & Hamoda, M. (2007) Athermal MW effects for enhancing digestibility of WAS. *Water Research*, 41, 2457–2466.

Evans, A. (2006) Biosolid reduction and the Deskin quick dry filter bed. In: *Australia Water Industry Operators Association Annual Conference Proceedings*, 220. Available from: www.wioa.org.au/conf_papers/02/paper10.htm.

Fair, G.M. & Moore, E.W. (1937) Relative time required for 90% digestion of plain-sedimentation, primary sludge at different temperatures. *Sewage Works Journal*, 9, 3.

Fair, G.M., Geyer, J.C. & Okun, D.A. (1968) *Water and Wastewater Engineering*, Vol. 2, John Wiley and Sons, Inc.

Fattal, B., Peleg-Olevsky, E., Yoshpe-Purer, Y. & Shuval, H.I. (1986) The association between morbidity among bathers and microbial quality of seawater. *Water Science and Technology*, 18 (11), 59–69.

Finstein, M.S., Miller, F.C., Hogan, J.A. & Strom, P.F. (1987) Analysis of EPA guidance on composting sludge. *BioCycle*, 28 (4), 56–61.

Fleming G. (1986) Sludge: A waste or a resource. In: *Proceedings of the Scottish Centre's Annual Symposium (Sludge Disposal into the 1990s) at Hamilton* on 10 December, 1986.

Florencio, L., Nozhevnikova, A., Van Langerak, A., Stams, A.J.M., Field, J.A. & Lettinga G. (1993) Acidophilic degradation of methanol by a methanogenic enrichment culture. *FEMS Microbiology Letters*, 109, 1–6.

Fonts, I., Azuara, M., Gea, G. & Murillo, M.B. (2009) Study of the pyrolysis liquids obtained from different sewage sludge. *Journal of Analytical and Applied Pyrolysis*, 85, 184–191.

Frolund, B., Palmgren, R., Keiding, K. & Nielsen P.H. (1996) Extraction of extracellular polymers from activated sludge using a cation exchange resin. *Water Research*, 30, 1749–1758.

Gale, R.S. & Baskerville, R.C. (1970) Studies in the vacuum filtration of sewage sludges. *Journal of Water Pollution Control Federation*, 69, 514–532.

Ganaye, V., Fass, S., Urbain, V., Manem, J. & Black J.C. (1996) Biodegradation of volatile fatty acids by three species of nitrate-reducing bacteria. *Environmental Technology*, 17, 1145–1149.

Ganczarczyk, J., Hamoda, M.F. & Wong, H.L. (1980) Performance of aerobic digestion at different sludge solid levels and operation patterns. *Water Research*, 14 (6), 627–633.

Garuti, G., Giordano, A. & Pirozzi, F. (2001) Full-scale ANANOX system performance. *Water SA*, 27 (2) 189–197.

Ghadge, S.V. & Raheman, H. (2006) Process optimization for biodiesel production from mahua (*Madhuca indica*) oil using response surface methodology. *Bioresource Technology*, 97 (3), 379–384.

Gildemeister, H.H. (1988) Sludge dewatering technology in perspective. In: Dirkzwager, A.H. and L'Hermite, P. (eds.) *Proceedings of Conference on Sewage Sludge Treatment and Use*. Elsevier Applied Science, England.

Girovich, M.J. (March 1990) Simultaneous sludge drying and palletizing. *Water Engineering & Management*.

Gore and Storrie Ltd. (1977) *Energy and Economic Considerations of Multiple-Hearth and Fluidised-Bed Incinerators for Sewage Sludge Disposal*. Toronto, ON, Study of Ontario Ministry of Environment Energy Management Program.

Gujer, W. & Kappeller, J. (1992) Modelling population dynamics in activated sludge systems. *Water Science and Technology*, 25 (6), 93–103.

Gujer, W. & Zehnder, A.J.B. (1983) Conversion process in anaerobic digestion. *Water Science and Technology*, 17, 127–167.

Gulas, V., Bond, M. & Benefield, L. (1979) Use of exocellular polymers for thickening and dewatering activated sludge. *Journal of Water Pollution Control Federation*, 51, 798–807.

Gunnerson, C.G. (1963) Mineralisation of organic matter in Santa Monica Bay. In: Oppenheirner, C.H. (ed.) *Marine Microbiology*. Springfield, IL, C.C. Thomas Publishers.

Guyer, J.P. (2011) *Introduction to Sludge Handling, Treatment and Disposal*. Stony Point, NY, Continuing Education and Development, Inc. pp. 1–43.

GVRD (2005) *Review of Alternatives Technologies for Biosolids Management*. Report of the Greater Vancouver Regional District, September 2005.

Hall, J.E. (1988) Methods of applying sewage sludge to land: A review of recent developments. In: Dirkzwager, A.H. and Hermite, P.L. (eds.) *Conference on Sewage Sludge Treatment and Use*. Elsevier Applied Science, England.

Han, S.K. & Shin, H.S. (2004) Bio-hydrogen production by anaerobic fermentation of food waste. *International Journal of Hydrogen Energy*, 29, 569–577.

Han, Y. & Dague, R.R. (1997) Laboratory studies on the temperature-phased anaerobic digestion of domestic primary sludge. *Water Environment Research*, 69, 1139–1143.

Han, Y., Sung, S. & Dague, R.R. (1997) Temperature-phased anaerobic digestion of wastewater sludges. *Water Science and Technology* 36 (6–7), 367–374.

Hao, O.J. & Kim M.H. (1990) Continuous pre-anoxic and aerobic digestion of waste activated sludge. *Journal of Environmental Engineering, ASCE*, 116 (5), 863–879.

Harremoes, P., Bundgaard, E. & Henze, M. (1991) Developments in wastewater treatment for nutrient removal. *Journal of Eueopean Water Pollution Control Federation*, 1 (1), 19–23.

Harrison, S.T.L. (1991) Bacterial cell disruption: A key unit operation in the recovery of intracellular products. *Biotechnology Advances*, 9, 217–240.

Hartman, R.B., Smith, D.G., Bennett, E.R. & Linstedt, K.D. (1979) Sludge stabilization through aerobic digestion. *Journal of Water Pollution Control Federation*, 51, 2353–2365.

Hashimoto, M. & Hiraoka, M. (1990) Characteristics of sewage sludge affecting dewatering by Belt Press Filter. *Water Science and Technology*, 22 (12), 143–152.

Hashimoto, S., Nishimura, K., Iwabi, H. & Shinabe K. (1991) Pilot plant test of electron-beam disinfected sludge composting. *Water Science and Technology*, 23, 1991–1999.

Henze, M., Harrenmoes, P., Janesen, J.C. & Arvin, E. (1995) *Wastewater Treatment Biological and Chemical Processes*. Berlin, Springer-Verlag.

Heo, N., Park, S. & Kang, H. (2003) Solubilization of WAS by alkaline pretreatment and biochemical methane potential (BMP) test for anaerobic co-digestion of municipal organic waste. *Water Science and Technology*, 48 (8), 211–219.

Herandez, L.M., Fernandez, M.A. & Gonzalez, M.J. (1991) Lindane pollution near an industrial source in northeast Spain. *Bulletin of Environmental Contamination and Toxicology*, 46, 9–13.

Herguido, J., Corella, J. & Gonzalez-Saiz, J. (1992) Steam gasification of lignocellulosic residues in a fluidized bed at a small pilot scale. Effect of the type of feedstock. *Industrial and Engineering Chemical Research*, 31, 1274–1282.

Heron, G., Crouzet, C., Bourg, A.C.M. & Christensen T.H. (1995) Speciation of Fe(II) and Fe(III) in contaminated aquifer sediments using chemical extraction techniques. *Environmental Science & Technology*, 8, 1698–1705.

Hills, D.J. & Dykstra, R.S. (1980) Anaerobic digestion of cannery tomato solid wastes. *Journal of Environmental Engineering*, 106, 257–266.

Hogan, F., Mormede, S., Clark, P. & Crane, M. (2004) Ultrasonic sludge treatment for enhanced anaerobic digestion. *Water Science and Technology*, 50 (9), 25–32.

Holliger, C., Schraa, G., Stams, A.J.M. & Zehnder, A.J.B. (1993) A highly purified enrichment culture couples the reductive dechlorination of tetrachloroethene to growth. *Applied Environmental Microbiology*, 59, 2991–2997.

Holt, M.S., Mitchell, G.C. & Watkinson R.J. (1992) *Detergents*. In: de Oude, N.T. (ed.) Springer, Verlag, Berlin. pp. 89–144.

Hong, S.M., Park, J.K. & Lee, Y.O. (2004) Mechanisms of MW irradiation involved in the destruction of fecal coliforms from biosolids. *Water Research*, 38 (6), 1615–1625.

Hosh, S.G. & Pohland, E.C. (1974) Kinetics of substrate assimilations and product formation in anaerobic digestion. *Journal of Water Pollution Control Federation*, 46 (4), 748–759.

Hsieh, C.H., Lo, S.L., Hu, C.Y., Shih, K., Kuan, W.H. & Chen, C.L. (2008) Thermal detoxification of hazardous metal sludge by applied electromagnetic energy. *Chemosphere*, 71 (9), 1693–1700.

Hultman, B., Levlin, E., Plaza, E. & Stark, K. (2003). *Phosphorus Recovery from Sludge in Sweden – Possibilities to Meet Proposed Goals in an Efficient, Sustainable and Economical Way*. No 10: 19–28. Available from: www.lwr.kth.se/forskningsprojekt/Polishproject/JPS10s19.pdf [Accessed August 2007].

Huyard, A., Ferran, B. & Audic, J.M. (2000). The two phase anaerobic digestion process: Sludge stabilization and pathogens reduction. *Water Science and Technology* 42, 41–47.

IEI (1999) *Technology Update*. 3, IEI News, September 1999.

IKA® (2013) *High Pressure Homogenizer*. Available from: http://www.ikaprocess.com/Products/High-press-homogenizer-cph-43/HPH-csb- HPH/ [Accessed 1st July 2013].

Jack, T.R., Francis, M.M. & Stehmeier, L.G. (1994) Disposal of slop oil and sludges by biodegradation. *Research in Microbiology*, 145 (1), 49–53.

Jean, D.S., Chang, B.V., Liao, G.S., Tsou, G.W. & Lee, D.J. (2000) Reduction of microbial density level in sewage sludge through pH adjustment and ultrasonic treatment. *Water Science and Technology*, 42 (9), 97–102.

Jenkins, D. (1992) Towards a comprehensive model of activated sludge foaming and bulking. *Water Science and Technology*, 25 (6), 215–230.

Jewell, W.J. & Kabrick, R.M. (1980) Auto heated aerobic thermophilic sludge-digestion with aeration. *Journal of Water Pollution Control Federation*, 52, 512–523.

Jiang, Y.M., Chen, Y.G. & Zheng, X. (2009) Efficient polyhydroxyalkanoates production from a waste-activated sludge alkaline fermentation liquid by activated sludge submitted to the aerobic feeding and discharge process. *Environmental Science & Technology*, 43, 7734–7741.

Jin, Y., Li, H., Mahar, R.B., Wang, Z. & Nie, Y. (2009) Combined alkaline and ultrasonic pre-treatment of sludge before aerobic digestion. *Journal of Environmental Science*, 21, 279–284.

Johri, A.K., Dua, M., Tuteja, D., Saxena, R., Saxena, D.M. & Lal, R. (1996) Genetic manipulations of microorganisms for the degradation of hexachlorocyclohexane. *FEMS Microbiology Reviews*, 19, 69–84.

Jones, E.W. & Westmoreland, D.J. (1998) Degradation of nonylphenol ethoxylates during the composting of sludge from wool scour effluents. *Environmental Science & Technology*, 32, 2623–2627.

Kalogo, Y. & Monteith, H. (2008) State of science report: Energy and resource recovery from sludge. *Global Water Research Coalition*. Alexandria, VA, Water Environment Research Foundation. p. 238.

Kamiya, T. & Hirotsuki, J. (1998) New combined system of biological process and intermittent ozonation for advanced wastewater treatment. *Water Science and Technology*, 38 (8–9), 145–153.

Kampe, W. (1988) Organic substances in soils and plants after intensive applications of sewage sludge. In: Dirkswager, A.H. and Hermite, P.L. (eds.) *Conference on Sewage Sludge Treatment and Use*. Elsevier Applied Science, England.

Kappeler, J. & Gujer, W. (1992) Bulking in activated sludge: A qualitative simulation model for *Sphaerotilus natans*, Type 021N and Type 0961. *Water Science and Technology*, 26 (3–4), 473–482.

Kargbo, D.M. (2010) Biodiesel production from municipal sewage sludges. *Energy Fuels*, 24, 2791–2794.

Karpati, A. (February 20–22, 1989) *Wastewater Pretreatment and Sludge Utilization in the Dairy Industry*. Viena, Envirotech.

Karr, P.R. & Keinath, T. (1978) Influence of the particle size on sludge dewaterability. *Journal of Water Pollution Control Federation*, 50 (11), 1911–1930.

Katsiris, N. & Kouzeli-Katsiri, A. (1987) Bound water content of biological sludges in relation to filtration and dewatering. *Water Resources*, 21, 1319–1327.

Katz, W.J. & Mason, G.G. (1970) Freezing methods used to condition activated sludge. *Water Sewage Works*, 117 (4), 110–114.

Keey, R.B. (1972) *Drying Principles and Practices*. London, Pergamon Press.

Keith, J.O., Woods Jr., L.A. & Hunt, E.G. (1970) Reproductive failures in brown pelicans on the Pacific coast. In: *Proceedings of 35th North American Wildlife and Natural Resources Conference*. Washington, DC, Wildlife Management Institute. pp. 56–63.

Kepp, U., Machenbach, I., Weisz, N. & Solheim, O.E. (2000) Enhanced stabilization of sewage sludge through thermal hydrolysis–three years of experience with full-scale plant. *Water Science and Technology*, 42 (9), 89–96.

Khursheed, A. & Kazmi, A.A. (2011) Retrospective of ecological approaches to excess sludge reduction. *Water Research*, 45, 4287–4310.

Kim, J., Kaurich, T.A., Sylvester, P. & Martin, A.G. (2006) Enhanced selective leaching of chromium from radioactive sludges. *Separation Science & Technology*, 41, 179–196.

Kim, J., Park, C., Kim, T., Lee, M., Kim, S., Kim, S. & Lee, J. (2003) Effects of various pretreatments for enhanced anaerobic digestion with WAS. *Journal of Bioscience Bioengineering*, 95 (3), 271–275.

Kim, M.H. (1989) *Anoxic Sludge Digestion of Waste Activated Sludge*. ME Thesis. College Park, MD, University of Maryland.

Kim, M.H. & Hao, O.J. (1990) Comparison of activated sludge stabilization under aerobic or anoxic conditions. *Research Journal of the Water Pollution Control Federation*, 62 (2), 160–168.

Kim, Y. & Parker, W. (2008) A technical and economic evaluation of the pyrolysis of sewage sludge for the production of bio-oil. *Bioresource Technology*, 99, 1409–1416.

Kleinig, A.R. & Middelberg, A.P.J. (1990) On the mechanism of microbial cell disruption in high-pressure homogenization. *Chemical Engineering Science* 53, 891–898.

Knocke, W.R., Glosh, M.M. & Novak, J.T. (1980) Vacuum filtration of metal hydroxide sludges. *Journal of Environmental Engineering*, 106 (2), 363.

Koers, D.A. & Mavinic, D.S. (1977) Aerobic digestion of waste activated sludge at low temperatures. *Journal of the Water Pollution Control Federation*, 49 (3), 460.

Kohno, T., Yoshika, K. & Satoh S. (1991) The role of intracellular organic storage materials in the selection of microorganisms in activated sludge. *Water Science and Technology*, 23 (4–6), 889–898.

Kondoh, S. & Hiraoka, M. (1990) Commercialization of pressurized electro-osmotic dehydrator (PED). *Water Science and Technology*, 22 (12), 259–268.

Kopp, J., Muller, J., Dichtl, N. & Schwedes, J. (1997) Anaerobic digestion and dewatering characteristics of mechanically disintegrated sludge. *Water Science and Technology*, 36 (11), 129–136.

Korsaric, N., Blaszczyk, R. & Orphan, L. (1990) Factors influencing formation and maintenance of granules in anaerobic sludge blanket reactors (UASBR). *Water Science and Technology*, 9 (22), 275–282.

Kravetz, L., Salanitro, J.P., Dorn, P.B. & Guin, K.F. (1991) Influence of hydrophobe type and extent of branching on environmental response factors of nonionic surfactants. *Journal of American Oil Chemical Society*, 68 (8), 610–618.

Lane, G.M. (1990) *Process for Treating Oil Sludge.* Available from: http://www.google.co.in/patents/EP0348707A1?cl=en.

Leavitt, M.E. & Brown, K.L. (1994) Bio-stimulation versus bioaugmentation-three case studies. *Hydrocarbon Bioremediation*, Boca Raton, FL, Lewis Publication, 72–79.

LeBlanc, R.J., Allain, C.J., Laughton, P.J. & Henry, J.G. (2004) Integrated, long term, sustainable, cost effective sludge management at a large Canadian wastewater treatment facility. *Water Science and Technology*, 49 (10), 155–162.

Lee, J. (1997) Biological conversion of lignocellulosic biomass to ethanol. *Journal of Biotechnology*, 56, 1–24.

Lee, J.W., Cha, H.Y., Park, K.Y., Song. K.G. & Ahn, K.H. (2005) Operational strategies for an activated sludge process in conjunction with ozone oxidation for zero excess sludge production during winter season. *Water Research*, 39, 1199–1204.

Lee, N.M. & Welander, T. (1996) Reducing sludge production in aerobic wastewater treatment through manipulation of the ecosystem. *Water Research*, 30 (8), 1781–1790.

Lee, Y.D., Shin, E.B., Choi, Y.S., Yoon, H.S., Lee, H.S., Chung, L.J. & Na, J.S. (1997) Biological removal of nitrogen and phosphorus from wastewater by a single sludge reactor. *Environmental Technology*, 18, 975–986.

Levlin, E., Löwé,n M. & Stark, K. (2004). *Phosphorus Recovery from Sludge Incineration Ash and Supercritical Water Oxidation Residues with Use of Acids and Bases.* No 11. pp. 19–28. Available from: www.lwr.kth.se/forskningsprojekt/Polishproject/JPS11p19.pdf [Accessed August 2007].

Lewis, M.A. (1991) Chronic and sublethal toxicities of surfactants to fresh water and marine animals: A review and risk assessment. *Water Research*, 25, 101–113.

Li, C., Xie, F., Ma, Y., Cai, T., Li, H., Huang, Z. & Yuan, G. (2010) Multiple heavy metals extraction and recovery from hazardous electroplating sludge waste via ultrasonically enhanced two-stage acid leaching. *Journal of Hazardous Material*, 178, 823–833.

Li, Y.Y. & Noike, T. (1992) Upgrading of anaerobic digestion of waste activated sludge by thermal pretreatment. *Water Science and Technology*, 26 (3–4), 857–866.

Liao, P.B. (1974) Fluidised-bed sludge incinerator design. *Journal of Water Pollution Control Federation*, 46 (8), 1895–1913.

Liao, P.H., Wong, W.T. & Lo, K.V. (2005a) Release of phosphorus from sewage sludge using microwave technology. *Journal of Environmental Engineering and Science*, 4, 77–81.

Liao, P.H., Wong, W.T. & Lo, K.V. (2005b) Advanced oxidation process using hydrogen peroxide/microwave system for solubilization of phosphate. *Journal of Environmental Science and Health, Part A*, 40, 1753–1761.

Lin, J.G., Chang, C.N. & Chang, S.C. (1997) Enhancement of anaerobic digestion of WAS by alkaline solubilization. *Bioresource Technology*, 62, 85–90.

Lipke, S. (1990) High solids centrifuges turn out to be surprise dewatering choice. *Water Engineering & Management*, 137 (6), 22–24.

Liu, J.C., Lee, C.H., Lai, J.Y., Wang, K.C., Hsu, Y.C. & Chang, B.V. (2001) Extracellular polymers of ozonized waste activated sludge. *Water Science and Technology*, 44 (10), 137–142.

Liu, X., Liu, H., Chen, J., Du, G. & Chen, J. (2008) Enhancement of solubilization and acidification of waste activated sludge by pretreatment. *Waste Management*, 28, 2614–2622.

Liu, Y. (2003) Chemically reduced excess sludge production in the activated sludge system. *Chemosphere*, 50, 1–7.

Liu, Y. & Tay, J.H. (2001) Strategy for minimization of excess sludge production from the activated sludge process. *Biotechnology Advances*, 19 (2), 97–107.

Lo, K.V., Liao, P.H. & Yin, G.Q. (2008) Sewage sludge treatment using MW-enhanced advanced oxidation processes with and without ferrous sulfate addition. *Journal of Chemical Technology and Biotechnology*, 83, 1370–1374.

Lockhart, N.C. (1986) Electro-dewatering of fine suspensions. In: Muralidhara, H.S. (ed.) *Advances in Solid Liquid Separation*. Columbia, OH, Batelle Press. p. 241.

Logsdon, G. & Edgerley Jr., E., (1971) Sludge dewatering by freezing. *Journal of American Water Works Association*, 63 (11), 734–740.

Lotito, V., Mininni, G. & Spinosa, L. (1990) Models of sewage sludge, conditioning. *Water Science and Technology*, 22 (12), 163–172.

Loupy, A. (2002) *Microwaves in Organic Synthesis*. France, Wiley-VCH.

Low, E.W. & Chase, H.A. (1999) Reducing production of excess biomass during wastewater treatment. *Water Research*, 33 (5), 1119–1132.

Low, E.W., Chase, H.A., Milner, M.G. & Curtis, T.P. (2000) Uncoupling of metabolism to reduce biomass production in the activated sludge process. *Water Research*, 34 (12), 3204–3212.

Lue-Hing, C., Zeng, D.R. & Kuchenither, R. (eds.) (1998) Water Quality Management Library, Vol. 4, *Municipal Sewage Sludge Management: A Reference Test on Processing, Utilization and Disposal*. Lancaster, PA, Technomic Publishing Co.

Lv, P., Yuan, Z., Wu, C., Ma, L., Chen, Y. & Tsubaki, N. (2007) Bio-syngas production from biomass catalytic gasification. *Energy Conservation and Management*, 48, 1132–1139.

Lynd, L.R., Cushman, J.H., Nichols, R.J. & Wyman, C.E. (1991) Fuel ethanol from cellulosic biomass. *Science*, 251, 1318–1323.

Magbanua, B. & Bowers, A.R. (1998) Effect of recycle and axial mixing on microbial selection in activated sludge. *Journal of Environmental Engineering*, 124 (10), 970–978.

Marklund, S. (1990) Dewatering of sludge by natural methods. *Water Science and Technology*, 22 (3/4), 239–246.

Marshall, E. (1988) The sludge factor. *Science*, 242, 307.

Martin, M.H. & Bhattarai, R.P. (1991) More mileage from gravity sludge thickeners. *Water Environment and Technology*, 3 (7), 57–60.

Mastin, B.J. & Lebster, G.E. (2006) Dewatering with Geotube® containers: A good fit for a Midwest wastewater facility? In: *Proceedings of the WEF/AWWA Joint Residuals and Biosolids Management Conference, Cincinnati, Ohio*.

Mathiesen, M.M. (1990) Sustained shockwave plasma (SSP) destruction of sewage sludge – A rapid oxidation process. *Water Science and Technology*, 22 (12), 339–344.

Matsumura, Y., Xu, X. & Antal, M.J. (1997) Gasification characteristics of an activated carbon in supercritical water. *Carbon*, 35, 819–824.

Matsuzawa, Y. & Mino, T. (1991) Role of glycogen as an intracellular carbon reserve of activated sludge in the competitive growth of filamentous and non-filamentous bacteria. *Water Science and Technology*, 23 (4–6), 899–905.

Mavinic, D.S. & Koers, D.A. (1979) Performance and kinetics of low temperature, aerobic sludge digestion. *Journal of the Water Pollution Control Federation*, 51, 2088.

Mavinic, O.S. & Koers, O.A. (1982) Fate of nitrogen in aerobic sludge digestion. *Journal of Water Pollution Control Federation*, 54 (4), 352–360.

McBride, M.B., Richards, B.K., Steenhuis, T.S., Peverly, J.H., Russell, J.J. & Suave, S. (1997) Mobility and solubility of toxic metals and nutrients in soil fifteen years after sludge application. *Soil Science*, 162, 487–500.

McCarty, P.L., Bck, L. & Amant, P. St. (1969) Biological denitrification of waste-waters by addition of organic materials. In: *Proceedings of the 24th Purdue Industrial Waste Conference, Lafayette*.

McClintock, S.A., Sherrad, J.H., Novak, J.T. & Randall, C.W. (1988) Nitrate versus oxygen respiration in the activated sludge process. *Journal of Water Pollution Control Federation*, 60 (3), 342–350.

McWhirter, J.R. (1978) The use of high-purity oxygen in the activated sludge process In: *Oxygen and Activated Sludge Process*. Vol. 1. Boca Raton, FL, CRC Press.

MCZM (Massachusetts Coastal Zone Management Office) (1982) *PCB Pollution in the New Bedford, Massachusetts Area, Boston.*

Messenger, J.R., de Villiers, H.A. & Ekama, G.A. (1990) Oxygen utilization rate as a control parameter for the aerobic stage in dual digestion. *Water Science and Technology*, 22 (12), 217–227.

Metcalf & Eddy (2003) *Wastewater Engineering: Treatment Disposal Reuse.* 4th edition. New York, NY, McGraw-Hill, Inc.

Middeidorp, P.J.M., Zehnder, A.J.B. & Schraa, G. (1996) Biotransformation of alpha-, beta-, gamma- and delta-hexachlorocyclohexane under methanogenic conditions. *Environmental Science & Technology*, 30, 2345–2349.

Milne, B.J., Baheri, H.R. & Hill, G.A. (1998) Composting of a heavy oil refinery sludge. *Environmental Progress*, 17 (1), 24–27.

Mitsdorffer, R., Demharter, W. & Bischofsberger, W. (1990) Stabilization and disinfection of sewage sludge by two-stage anaerobic thermophilic/mesophilic digestion. *Water Science and Technology*, 22 (7/8), 269–270.

Montgomery, R. (2004) Development of bio-based products. *Bioresource Technology*, 91, 1–29.

Moseley, J.L., Patterson, L.N. & Sieger, R.B. (April 1990) *Sludge Disposal.* Dallas style, Civil Engineering.

MSST (1987) *Manual on Sewerage and Sewage Treatment.* 1st edition. New Delhi, CPHEE-Organization, Ministry of Urban Development, Government of India.

Müller, J., Günther, L., Dockhorn, T., Dichtl, N., Phan, L.-C., Urban, I., Weichgrebe, D., Rosenwinkel, K.-H. & Bayerle, N. (2007). Nutrient recycling from sewage sludge using the seaborne process. In: *Proceeding of the IWA Conference on Biosolids, Moving Forward Wastewater Biosolids Sustainability: Technical, Managerial, and Public Synergy, June 24–27, 2007, Moncton, New Brunswick, Canada.* pp. 629–633.

Muller, J., Lehne, G., Schwedes, J., Battenberg, S., Naveke, R., Kopp, J., Scheminski, A., Krull, R. & Hempell, D.C. (1998) Disintegration of sewage sludges and influence on anaerobic digestion. *Water Science and Technology*, 38, 425–433.

Murakami, T., Ishida, T., Sasabe, K., Sasaki, K. & Harada, S. (1991) Characteristics of melting process for sewage sludge. *Water Science and Technology*, 23, 2019–2028.

Muralidhara, H.S., Senapati, N. & Beard, R.B. (1986) A novel electroacoustic separation process for fine particle separations. In: Muralidhara, H.S. (ed.) *Advances in Solid Liquid Separation.* Columbia, OH, Batelle Press. pp. 335–374.

Murray, K.C., Tong, A. & Bruce, A.M. (1990) Thermophilic aerobic digestion-A reliable and effective process for sludge treatment at small works. *Water Science and Technology*, 22 (3/4), 225–232.

MWST (March 1991) *Manual on Water Supply and Treatment.* 3rd edition. New Delhi, Ministry of Urban Development, Government of India.

Nabarlatz, D., Vondrysova, J., Janicek, P., Stüber, F., Font, J. & Fortuny, A. (2010) Hydrolytic enzymes in activated sludge: Extraction of protease and lipase by stirring and ultrasonication. *Ultrasonics Sonochemistry*, 17, 923–931.

NACWA (National Association of Clean Water Agencies) (2010) *Renewable Energy Resources: Banking on Biosolids.* Available from: www.nacwa.org.

Nagasaki, K., Akakura, N., Adachi, T. & Akiyama T. (1999) Use of waste-water sludge as a raw material for production of L-lactic acid. *Environmental Science & Technology*, 33, 198–200.

Nagasawa, S., Kukuchi, R., Nagata, Y., Takagi, M. & Matsuo, M. (1993) Aerobic mineraliza-tion of γ-HCH by *Pseudomonas paucimobilis* UT 26. *Chemosphere*, 26, 1719–1728.

Nah, I., Kan, Y., Hwang, K. & Song, W. (2000) Mechanical pretreatment of WAS for anaerobic digestion process. *Water Research*, 34 (8), 2362–2368.

Ndon, U.J. & Dague, R.R. (1997) Ambient temperature treatment of low strength wastewater using anaerobic sequencing batch reactor. *Biotechnology Letters*, 19, 319–323.

Nealson, K.H. & Saffarini, D. (1994) Iron and manganese in anaerobic respiration: Environmental significance, physiology, and regulation. *Annual Review of Microbiology*, 48, 311–343.

Negulescu, M. (1985) *Municipal Wastewater Treatment.* New York, NY, Elsevier Science Publishers.

Neyens, E. & Baeyens, J. (2003) A review of thermal sludge pretreatment processes to improve dewaterability. *Journal of Hazardous Material*, 98 (1–3), 51–67.

Neyens, E., Baeyens, J. & Creemers, C. (2003) Alkaline thermal sludge hydrolysis. *Journal of Hazardous Material*, B97, 295–314.

Neyens, E., Baeyens, J., Dewil, R. & De Heyder, B. (2004) Advanced sludge treatment affects extracellular polymeric substances to improve activated sludge dewatering. *Journal of Hazardous Materials*, 106 (2–3), 83–92.

Nielsen, J.L. & Nielsen, P.H. (1998) Microbial nitrate-dependent oxidation of ferrous Iron in activated sludge. *Environmental Science & Technology*, 32, 3556–3561.

Nielsen, P.H. (1996) The significance of microbial Fe(III) reduction in the activated sludge process. *Water Science and Technology*, 34 (5–6), 129–136.

Nielsen, P.H., Frolund, B., Spring, S. & Caccavo, E. (1997) Microbial Fe (III) reduction in activated sludge. *Systematic and Applied Microbiology*, 20 (4), 645–651.

Nielsen, S. (2003) Sludge treatment in wetland systems. In: Dias, V. & Vymazal, J. (eds.) *Proceedings of the Conference on the Use of Aquatic Macrophytes for Wastewater Treatment in Constructed Wetlands, Lisbon, Portugal.*

Nielsen, S. (2005a) Sludge reed bed facilities: Operation and problems. *Water Science and Technology* 51 (9), 99–107.

Nielsen, S. (2005b) Mineralization of hazardous organic compounds in a sludge reed bed and sludge storage. *Water Science and Technology* 51 (9), 109–117.

Nihlgard, B. (1985) The ammonium hypothesis-An additional explanation to the forest dieback in Europe. *AMBIO*, 14 (1), 2–8.

Nishioka, M., Yanagisawa, K. & Yamasaki, N. (1990) Solidification of sludge ash by hydrothermal hot pressing. *Research Journal Water Pollution Control Federation*, 62 (7), 926–932.

NOAA (1979) *Proceedings of a Workshop on Scientific Problems Relating to Ocean Pollution Environmental Research Laboratories.* Boulder, CO, National Oceanic and Atmospheric Administration.

Nordrum, S.B. (1992) Treatment of production tank bottom sludge by composting. In: *Proceedings of the 67th Annual Technique Conference and Exhibition of the SPE, Cincinnati, Ohio.* pp. 181–191.

Novak, L., Larrea, L., Wanner, J. & Carcia-Heras, J.L. (1993) Non-filamentous activated sludge bulking in a laboratory scale system. *Water Research*, 27, 1339–1346.

Ødegaard, H. (2004) Sludge minimization technologies – An overview. *Water Science and Technology*, 49 (10), 31–40.

Oku, S., Kasai, T., Hiraoka, M. & Takeda, N. (1990) Melting system for sewage sludge. *Water Science and Technology*, 22 (12), 319–321.

Okuno, N., Ishikawa, Y., Shimizu, A. & Yoshida, M. (2004). Utilization of sludge in building material. *Water Science and Technology*, 49 (10), 225–232.

Oman, C. & Hynning, P.A. (1993) Identification of organic compounds in municipal landfill leachates. *Environmental Pollution*, 80 (3), 265–271.

Onyeche, T. (2006) Sewage sludge as source of energy. In: *Proceedings of the IWA Specialized Conference on Sustainable Sludge Management: State-of-the-Art, Challenges and Perspectives. Moscow, Russia, May 29–31 2006.* pp. 235–241.

Ottewell, S. (1990) Sewage sludge incineration. *The Chemical Engineer*, 14, 477.

Park, B., Ahn, J., Kim, J. & Hwang, S. (2004) Use of MW pretreatment for enhanced anaerobiosis of secondary sludge. *Water Science and Technology*, 50 (9), 17–23.

Park, W.J., Ahn, H., Hwang, S. & Lee, C.K. (2010) Effect of output power, target temperature, and solid concentration on the solubilization of waste activated sludge using microwave irradiation. *Bioresource Technology*, 101 (1), S13–S16.

Parker, D.S., Morill, M.S. & Tetreault, M.J. (1992) Wastewater treatment process theory and practice: The emerging convergence. *Water Science and Technology*, 25 (6), 301–315.

Paulsrud, B. (1990) Sludge handling and disposal at small waste-water treatment plants in Norway. *Water Science and Technology*, 22 (3/4), 233–238.

Paulsrud, B. & Eikum, A.S. (1975) Lime stabilization of sewage sludges. *Water Research*, 9 (3), 297–305.

Peimin, Y. (1997a) PhD Dissertation, The United Graduate School of Agricultural Science, Shizuoka University.

Peimin, Y., Nishina, N., Kosakia, Y., Yahiro, K., Park, Y. & Okabe, M.J. (1997b) Enhanced production of L (+)-lactic acid from cornstarch in a culture of *Rhizopus oryzae* using an airlift bioreactor. *Journal of Fermentation and Bioengineering*, 84, 249–253.

Penaud, V., Delgenes, J.P. & Moletta, R. (1999) Thermo-chemical pretreatment of a microbial biomass: Influence of sodium hydroxide addition on solubilization and anaerobic biodegradability. *Enzyme and Microbial Technology*, 25, 258–263.

Perez-Cid, B., Lavilla, I. & Bendicho, C. (1999) Application of microwave extraction for partitioning of heavy metals in sewage sludge. *Analytica Chimica Acta*, 378, 201–210.

Perez-Elvira, S.I., Diez, P.N. & Fdz-Polanco, F. (2006) Sludge minimization technologies. *Reviews in Environmental Science and Bio/Technology*, 5, 375–398.

Perez-Elvira, S.I., Fdz-Polanco, M., Plaza, F.I., Garralon, G. & Fdz-Polanco, F. (2009) Ultrasound pretreatment for anaerobic digestion improvement. *Water Science and Technology*, 60 (6), 1525–1532.

Perez-Elvira, S.I., Fernandez-Polanc, F., Fernandez-Polanco, M., Rodriguez, P. & Rouge, P. (2008) Hydrothermal multivariable approach. Full-scale feasibility study. *Electronic Journal of Biotechnology*, 11, 7–8.

Pergamon PATSEARCHR, Pergamon Orbit Infoline Inc., 8000 Westpark Drive, McLean, Virginia 22102 USA.

Perry, J.H. (1973) *Chemical Engineers, Handbook*. 5th edition. New York, NY, McGraw-Hill.

Persson, N.A. & Welander, T.G. (1994) *Biotreatment of Petroleum Hydrocarbons-Containing Sludge by Land Farming in Hydrocarbon Bioremediation*. In: Hinchee, R.A., Alleman B.C., Hoeppel R.E. & Miller R.N. (eds.) Boco Raton, FL, Lewis Publishers. pp. 335–342.

Philippidis, G.P. (1996) *Handbook on Bioethanol: Production and Utilization*. In: Wyman, C.E. (ed.) Washington, DC, Taylor and Francis. pp. 253–285.

Philp, D.M. (1985) Sludge incineration.The Lower Molonglo water quality control centre. In: *Proceedings 11th Federal Convention, Australian Water and Wastewater Association, Melbourne*. pp. 459–466.

Pierre Le Clech (2013) *Schematic of MBR Process and Membrane Fouling*. Available from: http://en.wikipedia.org/wiki/Membrane_bioreactor [Accessed 3rd July 2013].

Pilli, S., Bhunia, P., Yan, S., LeBlanc, R.J., Tyagi, R.D. & Surampalli, R.Y. (2011). Ultrasonic pretreatment of sludge: A review. *Ultrasonics Sonochemistry*, 18, 1–18.

Pino-Jelcic, S.A., Hong, S.M. & Park, J.K. (2006) Enhanced anaerobic biodegradability and inactivation of fecal coliforms and salmonella spp. in wastewater sludge by using microwaves. *Water Environment Research*, 78 (2), 209–216.

Potter, C.L. & Glasser, J.A. (1995) Design and testing of an experimental In-vessel composting system. In: *Proceedings of the 21st Annual RREL Research Symposium, Cincinnati, Ohio*. pp. 56–60.

Priestley, A.J. (1992) *Sewage Sludge Treatment and Disposal-Environmental Problems and Research Needs from an Australian Perspective*. CSIRO, Division of Chemicals and Polymers.

Prince, M. & Sambasivam, Y. (1993) Bioremediation of petroleum wastes from the refining of lubricant oils. *Environmental Progress*, 12 (1), 5–11.

Ptasinski, K.J., Hamelinck, C. & Kerkhof, P.J.A.M. (2002) Energy analysis of methanol from the sewage sludge process. *Energy Conversion and Management*, 43 (9–12), 1445–1457.

Qiao, W., Wang, W., Wan, X., Xia, A. & Deng, Z. (2010) Improve sludge dewatering performance by hydrothermal treatment. *Journal of Residuals Science and Technology*, 7, 7–11.

Qiao, W., Wang, W., Xun, R., Lu, W. & Yin, K. (2008) Sewage sludge hydrothermal treatment by microwave irradiation combined with alkali addition. *Journal of Material Science*, 43, 2431–2436.

Ra, C.S., Lo, K.V. & Mavinic, D.S. (1998) Real-time control of two-stage sequencing batch reactor system for the treatment of animal wastewater. *Environmental Technology*, 19, 343–356.

Rabaey, K. & Verstraete, W. (2005) Microbial fuel cells: Novel biotechnology for energy generation. *Trends in Biotechnology*, 23, 291–298.

Randall, C.W., Turpin, J.K. & King, P.H. (1971) Activated sludge dewatering: Factors affecting drainability. *Journal of Water Pollution Control Federation*, 43, 102–122.

Rao, M.N. & Datta, A.K. (1987) *Waste Water Treatment*. 2nd edition. New Delhi, Oxford & IBH Publishing Co. Pvt. Ltd.

Rasmussen, H. & Nielsen P.H. (1996) Iron reduction in activated sludge measured with different extraction techniques. *Water Research*, 30, 551–558.

Ratsak, C.H., Kooi, B.W. & van Verseveld, H.W. (1994) Biomass reduction and mineralization increase due to the ciliate *Tetrahymena pyriformis* grazing on the bacterium *Pseudomonas fluorescens*. *Water Science and Technology*, 29 (7), 119–128.

Ratsak, C.H., Kooijman, S.A.L. & Kooi, B.W. (1993) Modeling the growth of an oligochaete on activated sludge. *Water Research*, 27 (5), 739–747.

Ray, B.T., Lin, J.G. & Rajan, R.V. (1990) Low level alkaline solubilization for enhanced anaerobic digestion. *Journal of Water Pollution Control Federation*, 62, 81–87.

Rensink, J.H., Donker, H.J.G.W. & Ijwema, T.S.J. (1982) *The influence of feed pattern on sludge bulking*. In: Chamber, B. & Tomlinson, E.J. (eds.) *Bulking of Activated Sludge: Preventative and Remedial Methods*. Chichester, Ellis Horwood Ltd. pp. 147–163.

Reynolds, D.T., Cannon, M. & Pelton, T. (2001) *Preliminary Investigation of Recuperative Thickening for Anaerobic Digestion*. WEFTEC Paper.

Rhee, S.K., Lee, J.J. & Lee, S.T. (1997) Nitrite accumulation in a sequencing batch reactor during the aerobic phase of biological nitrogen removal. *Biotechnology Letters*, 19, 195–198.

Rich, L.G. (1982) A cost-effective system for the aerobic stabilization and disposal of waste activated sludge. *Water Research*, 16, 535–542.

Richards, B.K., Steenhuis, T.S., Peverly, J.H. & McBride, M.B. (1998) Metal mobility at an old, heavily loaded sludge application site. *Environmental Pollution*, 99, 365–377.

Rifkin, J. (2002) *The Hydrogen Economy: The Creation of the Worldwide Energy Web and the Redistribution of the Power on Earth*. New York, NY, Penguin Putnam.

Riggle, D. (1995) Successful bioremediation with compost. *BioCycle*, 36 (2), 57–59.

Roberts, K. & Olsson, O. (1975) Influence of colloidal particles on dewatering of activated sludge with polyelectrolyte. *Environmental Science & Technology*, 9, 945–948.

Rocher, M., Roux, G., Goma, G., Begue, A.P., Louvel, L. & Rols, J.L. (2001) Excess sludge reduction in activated sludge processes by integrating biomass alkaline heat treatment. *Water Science and Technology*, 44 (2–3), 437–444.

Rosenberger, S., Kraume, M. & Szewzyk, U. (1999). Sludge free management of membrane bioreactors. In: *Proceedings of 2nd Symposium on Membrane Bioreactor for Wastewater Treatment. The School of Water Sciences, Cranfield University, UK*.

Rosenberger, S., Laabs, C., Lesjean, B., Gnirss, R., Amy, G., Jekel, M. & Schrotter, J.C. (2006) Impact of colloidal and soluble organic material on membrane performance in membrane bioreactors for municipal wastewater treatment. *Water Research*, 40, 710–720.

Rulkens, W. (2008) Sewage sludge as a biomass resource for the production of energy: Overview and assessment of the various options. *Energy & Fuels*, 22, 9–15.

Saby, S., Djafer, M. & Chen, G.H. (2002) Feasibility of using a chlorination step to reduce excess sludge in activated sludge process. *Water Research*, 36 (3), 656–666.

Sahu, S., Patnaik, K.K. & Sethunathan, N. (1992) Dehydrochlorination of γ-isomer of hexachlorocyclohexane by a soil bacterium, *Pseudomonas* sp. *Bulletin of Environmental Contamination and Toxicology*, 48, 265–268.

Sahu, S.K., Patnaik, K.K., Bhuyan, S., Sreedharan, B., Kurihari, N., Adhya, T.K. & Sethunathan N.J. (1995) Mineralization of alpha-, gamma-, beta-isomers og hexachlorocyclohexane by a soil bacterium under aerobic conditions. *Journal of Agricultural and Food Chemistry*, 43, 833–837.

Sakai, Y., Aoyagi, T., Shiota, N., Akashi, A. & Hasegawa, S. (2000) Complete decomposition of biological waste sludge by thermophilic aerobic bacteria. *Water Science and Technology*, 42 (9), 81–88.

Salanitro, J.P. & Diaz, L.A. (1995) Anaerobic biodegradability testing of surfactants. *Chemosphere*, 30, 813–830.

Salsabil, M.R., Laurent, J., Casellas, M. & Dagot, C. (2010) Techno-economic evaluation of thermal treatment, ozonation and sonication for the reduction of wastewater biomass volume before aerobic or anaerobic digestion. *Journal of Hazardous Material*, 174 (1–3), 323–333.

Salsabil, M.R., Prorot, A., Casellas, M. & Dagot, C. (2009) Pretreatment of activated sludge: Effect of sonication on aerobic and anaerobic digestibility. *Chemical Engineering Journal*, 148 (2–3), 327–335.

Schanke, C.A. & Wackett, L.P. (1992) Transition-metal coenzymes mimic environmental reductive elimination reactions of polychlorinated ethanes. *Environmental Science & Technology*, 26, 830–833.

Schmidt, S. & Padukone, N.J. (1997) Production of lactic acid from wastepaper as a cellulosic feedstock. *Journal of Industrial Microbiology and Biotechnology*, 18, 10–14.

Schnurer, A., Houwen, E.P. & Svensson, B.H. (1994) Mesophilic syntrophic acetate oxidation during methane formation by a triculture at high ammonium concentration. *Archives of Microbiology*, 162, 70–74.

Scholten, J.C.M. & Stams, A.J.M. (1995) The effect of sulfate and nitrate on methane formation in a freshwater sediment. *Antonie Van Leeuwenhoek*, 68, 309–315.

Scragg, A. (1999) *Environmental Biotechnology*. England, Pearson Education Ltd. pp. 70–77.

Senoo, K. & Wada, H. (1990) y-HCH-decomposing ability of several strains of *Pseudomonas paucimobilis*. *Soil Science and Plant Nutrition*, 36, 677–678.

Sezgin, M., Jeckins, D. & Parker, D.S. (1978) A unified theory of filamentous activated sludge bulking. *Journal Water Pollution Control Federation*, 50, 362–381.

Shen, T.T. (1979) Air pollutants from sewage sludge incinerators. *Journal of Environmental Engineering*, 105 (1), 61–74.

Sherwood, M.J. (1982) Fin erosion, liver condition and trace contaminant exposure in fishes from three coastal regions, In: G.F. Mayer (ed.) *Ecological Stress and the New York Bight*. Columbia, SC, Estuarine Research Foundation.

Shier, W.T. & Purwono, S.K. (1994) Extraction of single-cell protein from activated sewage sludge: Thermal solubilization of protein. *Bioresource Technology*, 49, 157–162.

Siddiquee, M.N. & Rohani, S. (2011) Lipid extraction and biodiesel production from municipal sewage sludges: A review. *Renewable and Sustainable Energy Reviews*, 15 (2), 1067–1072.

Sikora, L.J., Frankos, N.H., Murrary, C.M. & Walker J.M. (1980) Trenching digested sludge. *Journal of Environmental Engineering*, 106 (2), 351–361.

Sinha, R.K. & Heart, S. (2003) *Industrial and Hazardous Wastes: Health Impacts and Management Plan.* India, Pointer Publishers.

Smollen, M. (1990) Evaluation of municipal sludge drying and dewatering with respect to sludge volume reduction. *Water Science and Technology*, 22 (12), 153–161.

Solera, R., Romero, L.I. & Sales, D. (2002) The evolution of biomass in a two-phase anaerobic treatment process during start-up. *Chemical and Biochemical Engineering Quarterly*, 16 (1), 25–29.

Spartan Environmental Technologies (2013) *Ozone Sludge Reduction.* Available from: http://www.spartanwatertreatment.com/ozone-sludge-reduction.html [Accessed 3rd July 2013].

Spinosa, L. (2004) From sludge to resources through biosolids. *Water Science and Technology*, 50 (9), 1–9.

Stathis, T.C. (1980) Fluidized bed for biological wastewater treatment. *Journal of Environmental Engineering*, 106 (1), 227–241.

Steenhuis, T.S., McBride, M.B., Richards, B.K. & Harrison, E. (1999) Trace metal retention in the incorporation zone of land-applied sludge. *Environmental Science & Technology*, 33, 1171–1174.

Steenhuis, T.S., Ritsema, C.J. & Dekker, L.W. (1996) Introduction. *Geoderma*, 70, 83–85.

Stendahl, K. & Jäfverström, S. (2004) Recycling of sludge with the Aqua Reci process. *Water Science and Technology*, 49 (10), 233–240.

Straub, K.L., Benz, M., Schink, B. & Widdel, E. (1996) Anaerobic, nitrate-dependent microbial oxidation of ferrous iron. *Applied and Environmental Microbiology*, 62, 1458–1460.

Strauch, D. (1988) Improvement of the quality of sewage sludge: Microbiological aspects. In: Dirkzwager, A.H. and Hermite, P.L. (eds.) *Conference on Sewage Sludge Treatment and Use.* Elsevier Applied Science, England.

Stuckey, D.C. & McCarty, P.L. (1978) Thermochemical pretreatment of nitrogenous materials to increase methane yield. *Biotechnology and Bioengineering Symposium*, 8, 219–233.

Suprenant, B.A, Lahrs, M.C. & Smith, R.L. (1990) Oil crete. *Civil Engineering*, 60, 61–63.

Swinton, E.A., Eldridge, R.J., Becker, N.S.C. & Smith, A.D. (1988) Extraction of heavy metals from sludges and muds by magnetic ion-exchange. In: Dirkzwager, A.H. and Hermite, P.L. (eds.) *Conference on Sewage Sludge Treatment and Use.* Elsevier Applied Science, England.

Swisher, R.D. (1987) *Surfactant Biodegradation.* New York, NY, Marcel Dekker Inc.

Tanaka, S. & Kamiyama, K. (2002) Thermo-chemical pretreatment in the anaerobic digestion of waste activated sludge. *Water Science and Technology*, 46, 173–179.

Tanaka, S., Kobayashi, T., Kamiyama, K. & Bildan, M. (1997) Effects of thermo-chemical pretreatment on the anaerobic digestion of WAS. *Water Science and Technology*, 35 (8), 209–215.

Tanghe, T., Devriese, G. & Verstraete, W. (1994) Nonylphenol degradation in lab scale activated sludge units is temperature dependent. *Water Research*, 32, 2889–2896.

Tchobanoglous, G., Burton, F.L. & Stensel, H.D. (2003). *Wastewater Engineering: Treatment, Disposal, and Reuse.* 4th edition. New York, NY, McGraw-Hill, Inc. p. 1819.

Tenney, M.W., Echelberger Jr., W.F., Coffey, J.J. & McAloon, T.J. (1970) Chemical conditioning of biological sludges for vacuum filtration. *Journal of Water Pollution Control Federation*, 42, R1–R200.

Tenny, M.W. & Stumm W. (1965) Chemical flocculation of micro-organisms in biological waste treatment. *Journal of Water Pollution Control Federation.* 37, 1370–1388.

Tezel, U., Tandukar, M. & Pavlostathis, S.G. (2011) Anaerobic bio-treatment of municipal sewage sludge. In: Young, M.M. (Editor-in-Chief) & Agathos, S. (eds.) *Comprehensive*

Biotechnology, 2nd edition, Vol. 6, *Environmental Biotechnology and Safety*. Amsterdam, Elsevier.

Theis, T.L., McKieman, M. & Padgets, L.E. (1984) *Analysis and Assessment of Incinerated Municipal Sludge Ashes and Leachates*. Cincinnati, OH, US EPA Report No. 600/52-84-038.

Tian, Y., Zuo, W., Ren, Z. & Chen, D. (2011) Estimation of a novel method to produce bio-oil from sewage sludge by microwave pyrolysis with the consideration of efficiency and safety. *Bioresource Technology*, 102, 2053–2061.

Tiemeyer, E. (2002) *Method of Enhancing Biological Activated Sludge Treatment of Waste Water, and a Fuel Product Resulting Therefrom*. Available from: http://www.google.com.tr/patents/US20020148780.

Tomlinson, E.J. (1982) The emergency of the bulking problem and the current situation in the U.K. In: Chambers, B. & Tomlinson, E.J. (eds.) *Bulking of Activated Sludge: Preventive and Remedial Methods*. Chichester, Ellis Horwood Ltd. pp. 17–23.

Topping, G. (1986) Sewage sludge dumping in Scottish waters: Current practices and future outlook. In: *Proceedings of Scottish Centre's Annual Symposium (Sludge Disposal into the 1990s) at Hamilton on 10 December 1986*.

Tran, E.T. & Tyagi, R.D. (1990) Mesophilic and thermophilic digestion of municipal sludge in a deep-shaft V-shaped bioreactor. *Water Science and Technology*, 22 (12), 205–215.

Tsang, K.R. & Vesilind, P.A. (1990) Moisture distribution in sludges. *Water Science and Technology*, 22 (12), 135–142.

Turovskiy, I.S. & Mathai, P.K. (2006) *Wastewater Sludge Processing*. New Jersey, Wiley Interscience Publication.

Tyagi, R.D., Surampalli, R.Y. & Yan, S. (2009) *Sustainable Sludge Management: Production of Value Added Products*. Reston, VA, American Society of Civil Engineers. p. 72.

Tyagi, V.K. & Lo, S.L. (2011) Application of physico-chemical pretreatment methods to enhance the sludge disintegration and subsequent anaerobic digestion: An up to date review. *Reviews in Environmental Science and Biotechnology*, 10, 215–242.

Tyagi, V.K. & Lo, S.L. (2013) Sludge: A waste or renewable source for energy and resources recovery? *Renewable and Sustainable Energy Reviews*, 25, 708–728.

Uggetti, E., Ferrer, I., Llorens, E. & García, J. (2010) Sludge treatment wetlands: A review on the state of the art. *Bioresource Technology*, 101, 2905–2912.

Uggetti, E., Ferrer, I., Molist, J. & García, J. (2011) Technical, economic and environmental assessment of sludge treatment wetlands. *Water Research*, 45 (2), 573–582.

University of Tennessee (2013) *High Pressure Homogenization*. Available from: http://web.utk.edu/~fede/high%20pressure%20homogenization.html [Accessed 3rd July 2013].

Urbain, V., Block, J.C. & Manem, J. (1993) Bio-flocculation in an activated sludge: An analytical approach. *Water Research*, 27, 829–838.

USEPA, United States Environmental Protection Agency (1979) *Process Design Manual for Sludge Treatment and Disposal*. EPA 625/1-79/011.

USEPA, United States Environmental Protection Agency (1982) *Guide to the Disposal of Chemically Stabilized and Solidified Waste*, SW-872, Washington, DC, Office Solid Waste Emergency Response.

USEPA, United States Environmental Protection Agency (1993) *Standards for the Use or Disposal Sewage Sludge*. Final Rules 40 CFR Part 257.

USEPA, United States Environmental Protection Agency (2000a) *Biosolids Technology Fact Sheet: Alkaline Stabilization of Biosolids*. Washington, DC, Office of Water, EPA 832-F-00-052, September 2000.

USEPA, United States Environmental Protection Agency (2000b) *Biosolids Technology Fact Sheet: Centrifuge Thickening and Dewatering*. Washington, DC, Office of Water, EPA 832-F-00-053, September 2000.

Vail, R.L. (1991) Refiner biodegrades separator type sludge to BDAT standards. *Oil and Gas Journal*, 89, 53–57.

Valo, A., Carrere, H. & Delgene, J. (2004) Thermal, chemical, and thermo-chemical pretreatment of WAS for anaerobic digestion. *Journal of Chemical Technology and Biotechnology*, 79, 1197–1203.

Van Eekert, M.H.A., Schroder, T.J., Stams, A.J.M.M., Schraa, G. & Field, J.A. (1998a) Degradation and fate of carbon tetrachloride in unadapted methanogenic granular sludge. *Applied and Environmental Microbiology*, 64, 2350–2356.

Van Eekert, M.H.A., Van Ras, N.J.P., Mentink, G.H., Rijnaarts, H.H.M., Stams, A.J.M. Field, J.A. & Schraa, G. (1998b) Anaerobic transformation of p-hexachloroyclohexane by methanogenic granular sludge and soil microflora. *Environmental Science & Technology*, 32, 3299–3304.

van Loosdrecht, M.C.M. & Henze, M. (1999) Maintenance, endogenous respiration, lysis, decay and predation. *Water Science and Technology*, 39 (1), 107–117.

Vecchioli, G.I. Del Panno, M.T. & Painceira, M.T. (1990) Use of selected autochthonous soil bacteria to enhance degradation of hydrocarbons in soil. *Environmental Pollution*, 67, 249–258.

Venkateswaran, K., Hoaki, T., Kato, M. & Maruyama, T. (1995) Microbial degradation of resins fractionated from Arabian light crude oil. *Canadian Journal of Microbiology*, 41 (4–5), 418–424.

Venosa, A.D., Haines, J.R., Nisamaneepong, W., Govind, R., Pradhan, S. & Siddique, B. (1992) Efficacy of commercial products in enhancing oil biodegradation in closed laboratory reactors. *Journal of Industrial Microbiology*, 10 (1), 13–23.

Vesilind, P.A. (1988) The capillary suction time as a fundamental measure of sludge dewatering. *Journal of Water Pollution Controel Federation*, 60 (2), 215–220.

Vesilind, P.A. & Martel, C.J. (1990) Freezing of water and waste water sludges. *Journal of Environmental Engineering*. 116 (5), 854–862.

Wahlberg, C., Renberg, L. & Wideqvist, D. (1990) Determination of nonylphenol ethoxylates as theirpentafluoro-benzoates in water, sewage sludge and biota. *Chemosphere*, 20, 179–195.

Wang, Q., Kuninobo, M., Kakimoto, K., Ogawa, H.I. & Kato, Y. (1999) Upgrading of anaerobic digestion of WAS by ultrasonic pretreatment. *Bioresource Technology*, 68, 309–313.

Wanner, J. (1992) Comparison of biocenosis from continuous and sequencing batch reactors. *Water Science and Technology*, 25 (6), 239–249.

Wanner, J. (1994) *Activated Sludge Bulking and Foaming Control*. Lancaster, PA, Technomic Publishing Co. Inc.

Webb, P.C. (1964) Dehydration. In: *Biochemical Engineering*. London, D. Nostrand Company Ltd.

Wedag, A.G. (1990) *HI-COMPACT Method-A Purely Mechanical Process for Maximum Secondary Dewatering of Sludges*. KHD Humboldt Brochure No. 5, 400.

WEF (1998) Design of municipal wastewater treatment plants, 4th edition, *Manual of Practice 8 (ASCE 76)*, Alexandria, VA, Water Environment Federation.

WEF (2007) *Pumping of Wastewater and Sludge*, Chapter 8. Alexandria, VA, Water Environment Federation. pp. 1–88.

Wei, Y., Van Houten, R.T., Borger, A.R., Eikelboom, D.H. & Fan, Y. (2003a) Comparison performances of membrane bioreactor (MBR) and conventional activated sludge (CAS) processes on sludge reduction induced by Oligochaete. *Environmental Science & Technology*, 37 (14), 3171–3180.

Wei, Y., Van Houten. R.T., Borger, A.R., Eikelboom, D.H. & Fan, Y. (2003b) Minimization of excess sludge production for biological wastewater treatment. *Water Research*, 37, 4453–4467.

Wilhelm, A. & Knopp, P. (1979) Wet air oxidation-An alternative to incineration, *Chemical Engineering Progress*, 75, 46–52.

Willett, I.R., Jakobsen, P., Cunningham, R.B. & Gunthorpe, J.R. (1984) *Effects of Lime Treated Sewage Sludge on Soil Properties and Plant Growth*. CSIRO Division of Soils, Report No. 67.

Wilson, S.C. & Jones, K.C. (1993) Bioremediation of soil contaminated with polynuclear aromatic hydrocarbons (PAHs): A review. *Environmental Pollution*, 81, 229–249.

Wong, W.T., Chan, W.I., Liao, P.H., Lo, K.V. & Mavinic, D.S. (2006) Exploring the role of hydrogen peroxide in the microwave advanced oxidation process: Solubilization of ammonia and phosphates. *Journal of Environmental Engineering Science*, 5, 459–465.

Wong, W.T., Lo, K.V. & Liao, P.H. (2007) Factors affecting nutrient solubilization from sewage sludge using microwave-enhanced advanced oxidation process. *Journal of Environmental Science and Health, Part A*, 42, 6, 825–829.

Woodard, S. & Wukasch, R. (1994) A hydrolysis/thickening/filtration process for the treatment of waste activated sludge. *Water Science and Technology*, 30 (3), 29–38.

Worner, H.K. (1990) Revolutionary new use for sewage sludges in smelting. *Australian and New Zealand Association for the Advancement of Science Journal*, 21 (8), 28–29.

Wu, Q., Wang, Y. & Chen, G.Q. (2009) Medical application of microbial biopolyesters polyhydroxyalkanoates. *Artificial Cells, Blood Substitutes, and Biotechnology*, 37 (1), 1–12.

Wunderlich, R., Barry, J., Greenwood, D. & Carry, C. (1985) Startup of a high-purity, oxygen-activated sludge system at the Los Angeles County Sanitation Districts' Joint water pollution control plant. *Journal of Water Pollution control Federation*, 57, 1012–1018.

Wymann, C.E. (1994) Ethanol from lignocellulosic biomass: Technology, economics, and opportunities. *Bioresource Technology*, 50, 3–15.

Wymann, C.E. & Goodmann, B. (1993) Biotechnology for production of fuels, chemicals, and materials. *Applied Biochemistry and Biotechnology*, 39/40, 41–59.

Xavier, S. & Lonsane B.K. (1994), Sugar-cane pressmud as a novel and inexpensive substrate for production of lactic acid in a solid-state fermentation system. *Applied Microbiology and Biotechnology*, 41, 291–295.

Xie, B., Liu, H. & Yan, Y. (2009) Improvement of the activity of anaerobic sludge by low-intensity ultrasound. *Journal of Environmental Management*, 90, 260–264.

Xu, G., Chen, S., Shi, J., Wang, S. & Zhu, G. (2010) Combination treatment of ultrasound and ozone for improving solubilization and anaerobic biodegradability of waste activated sludge. *Journal of Hazardous Material*, 180, 340–346.

Xu, X. & Anal, M.I. (1998) Gasification of sewage sludge and other biomass for hydrogen production in supercritical water. *Environmental Progress*, 17 (4), 215–220.

Xu, X., Matsumara, Y., Stenberg, J. & Antal, M.J. (1996) Carbon catalyzed gasification of organic feed-stocks in supercritical water. *Industrial and Engineering Chemistry Research*, 35, 2522–2530.

Yamamoto, K., Hiasa, M., Mahmood, T. & Matsuo, T. (1989) Direct solid-liquid separation using hollow fibre membrane in an activated sludge aeration tank. *Water Science and Technology*, 21, 43–54.

Yang, X., Wang, X. & Wang, L. (2010) Transferring of components and energy output in industrial sewage sludge disposal by thermal pretreatment and two-phase anaerobic process. *Bioresource Technology*, 101 (8), 2580–2584.

Yasuda, Y. (1991) Sewage sludge utilisation technology in Tokyo. *Water Science and Technology*, 23, 1743–1752.

Yasui, H. & Shibata, M. (1994) An innovative approach to reduce excess sludge production in the activated sludge process. *Water Science and Technology*, 30 (9), 11–20.

Yasui, H., Nakamura, K., Sakuma, S., Iwasaki, M. & Sakai, Y. (1996) A full-scale operation of a novel activated sludge process without excess sludge production. *Water Science and Technology*, 34 (3–4), 395–404.

Ye, F. & Li, Y. (2010) Oxic-settling-anoxic (OSA) process combined with 3,3′,4′,5-tetrachlorosalicylanilide (TCS) to reduce excess sludge production in the activated sludge system. *Biochemical Engineering Journal*, 49 (2), 229–234.

Young, D.R., Heeson, T.C. Esra, G.M. & Howard, E.B. (1979) ODE contaminated fish off Los Angeles are suspected cause in deaths of marine birds. *Bulletin of Environmental Contaminant Toxicology*, 21, 584–590.

Zullaikah, S., Lai, C.-C., Vali, S.R. & Ju, Y.-H. (2005) A two-step acidcatalyzed process for the production of biodiesel from rice bran oil. *Bioresource Technology*, 96 (17), 1889–1896.

Zuo, W., Tian, Y. & Ren, N. (2011) The important role of microwave receptors in bio-fuel production by microwave-induced pyrolysis of sewage sludge. *Waste Management*, 31 (6), 1321–1326.

Subject index